Frontiers in Clinical Drug Research - CNS and Neurological Disorders

(Volume 5)

Edited by

Atta-ur-Rahman, *FRS*
Kings College, University of Cambridge, Cambridge, UK

General:

1. Any dispute or claim arising out of or in connection with this License Agreement or the Work (including non-contractual disputes or claims) will be governed by and construed in accordance with the laws of the U.A.E. as applied in the Emirate of Dubai. Each party agrees that the courts of the Emirate of Dubai shall have exclusive jurisdiction to settle any dispute or claim arising out of or in connection with this License Agreement or the Work (including non-contractual disputes or claims).

2. Your rights under this License Agreement will automatically terminate without notice and without the need for a court order if at any point you breach any terms of this License Agreement. In no event will any delay or failure by Bentham Science Publishers in enforcing your compliance with this License Agreement constitute a waiver of any of its rights.

3. You acknowledge that you have read this License Agreement, and agree to be bound by its terms and conditions. To the extent that any other terms and conditions presented on any website of Bentham Science Publishers conflict with, or are inconsistent with, the terms and conditions set out in this License Agreement, you acknowledge that the terms and conditions set out in this License Agreement shall prevail.

Bentham Science Publishers Ltd.
Executive Suite Y - 2
PO Box 7917, Saif Zone
Sharjah, U.A.E.
Email: subscriptions@benthamscience.org

BENTHAM SCIENCE

CONTENTS

PREFACE

Frontiers in Clinical Drug Research - CNS and Neurological Disorders presents the recent developments for the treatment of central nervous system (CNS) and nerve disorders. The book is a valuable resource for pharmaceutical scientists, postgraduate students and researchers seeking updated and critical information for devising research plans in the field of neurology. The chapters are written by eminent authorities in the field. The contents of this volume represent exciting recent researches on spinal cord injury, essential tremor, alzheimer's disease, pain management, neurotransmitters and neurodegenerative diseases, NMDA receptors, and plant alkaloids as antidepressants.

I hope that the readers will find these reviews valuable and thought provoking so that they may trigger further research in the quest for the new and novel therapies against CNS disorders.

I am grateful for the efforts made by the editorial personnel, especially Mr. Mahmood Alam (Director Publications), and Mr. Shehzad Naqvi (Senior Manager Publications) at Bentham Science Publishers.

Atta-ur-Rahman, *FRS*
Honorary Life Fellow
Kings College
University of Cambridge
UK

List of Contributors

Alissa Rubinfeld Suburban Pulmonary & Sleep Associates, Elmhurst College, North Riverside, IL, USA
Chicagoland Advanced Pain Specialists, Westmont, IL, USA
The Chicago School of Professional Psychology, Chicago, IL, USA

Antoine Chami Suburban Pulmonary & Sleep Associates, Elmhurst College, North Riverside, IL, USA
Chicagoland Advanced Pain Specialists, Westmont, IL, USA
The Chicago School of Professional Psychology, Chicago, IL, USA

Dimitrios P. Nikolelis Laboratory of Environmental Chemistry, Department of Chemistry, University of Athens, Athens, Greece

Dolores Viña Center for Research in Molecular Medicine and Chronic Diseases (CIMUS), University of Santiago de Compostela, Santiago de Compostela, Spain

Fernanda Rodríguez-Enríquez Center for Research in Molecular Medicine and Chronic Diseases (CIMUS), University of Santiago de Compostela, Santiago de Compostela, Spain

Gennady A. Evtugyn Analytical Chemistry Department, Kazan Federal University, Kazan, Russian Federation
OpenLab "DNA-Sensors", Kazan Federal University, Kazan, Russian Federation

Georgia-Paraskevi Nikoleli Laboratory of Inorganic & Analytical Chemistry, School of Chemical Engineering, Dept. 1, Chemical Sciences, National Technical University of Athens, Athens, Greece

Gino Giannaccini Department of Pharmacy, University of Pisa, Pisa, Italy

Haroon Khan Department of Pharmacy, Abdul Wali Khan University, Mardan, Pakistan

Helena Carla Castro LabiEMol, Laboratório de Antibióticos, Bioquímica, Ensino e Modelagem molecular, Instituto de Biologia, Universidade Federal Fluminense, Niterói, RJ, Brazil

Hiroyuki Yoshihara State University of New York, University Hospital of Brooklyn, Brooklyn, NY, USA

Iria Torres Center for Research in Molecular Medicine and Chronic Diseases (CIMUS), University of Santiago de Compostela, Santiago de Compostela, Spain

Jason H. Oh State University of New York, University Hospital of Brooklyn, Brooklyn, NY, USA

Kathy Sexton-Radek Suburban Pulmonary & Sleep Associates, Elmhurst College, North Riverside, IL, USA
Chicagoland Advanced Pain Specialists, Westmont, IL, USA
The Chicago School of Professional Psychology, Chicago, IL, USA

Laura Betti Department of Pharmacy, University of Pisa, Pisa, Italy

Lionella Palego Department of Clinical and Experimental Medicine, University of Pisa, Pisa, Italy

Marcos Vinicius Santana LabiEMol, Laboratório de Antibióticos, Bioquímica, Ensino e Modelagem molecular, Instituto de Biologia, Universidade Federal Fluminense, Niterói, RJ, Brazil

Martin Kronenbuerger School of Medicine, Department of Neurology, Johns Hopkins University, Baltimore, MD, USA
Department of Neurology, University Medicine Greifswald, Greifswald, Germany

Mohammad K. Siddiqi Interdisciplinary Biotechnology Unit, A.M.U., Aligarh, India

Munazza T. Fatima Department of Biochemistry and Tissue Biology, Institute of Biology, State University of Campinas (UNICAMP), Campinas, SP, Brazil

Parveen Salahuddin DISC, Interdisciplinary Biotechnology Unit, A.M.U., Aligarh, India

Paula Alvarez Abreu LAMCIFAR, Laboratório de Modelagem Molecular e Pesquisa em Ciências Farmacêuticas, Universidade Federal do Rio de Janeiro, Rio de Janeiro, RJ, Brazil

Rizwan H. Khan Interdisciplinary Biotechnology Unit, A.M.U., Aligarh, India

Tibor Hianik OpenLab "DNA-Sensors", Kazan Federal University, Kazan, Russian Federation
Faculty of Mathematics, Physics and Informatics, Comenius University, Bratislava, Slovakia

Yasser E. Shahein Molecular Biology Department, Genetic Engineering and Biotechnology Division, National Research Centre, Cairo, Egypt

Zoltan Mari School of Medicine, Department of Neurology, Johns Hopkins University, Baltimore, MD, USA
Cleveland Clinic Lou Ruvo Center for Brain Health, West Bonneville Ave, Las Vegas, NV, USA

CHAPTER 1

Drug Treatment for Spinal Cord Injury

Jason H. Oh[*] and **Hiroyuki Yoshihara**[*]

State University of New York – University Hospital of Brooklyn, Brooklyn, New York, USA

Abstract: Spinal cord injury (SCI) is a devastating event that often leads to profound disability. Traditionally, the treatment for such injury consisted of steroids, spinal decompression and stabilization surgery, and physical therapy. Despite all these treatments, however, prognoses for meaningful functional recovery remained grim. Recently, laboratory-based advancements in our understanding of central nervous system injuries at the cellular and molecular levels have ushered in new drug treatment strategies for neuroprotection and regeneration following SCI. Emerging strategies include pharmacotherapy to reduce spinal cord ischemia, cellular excitotoxicity, demyelination, and free radical-mediated peroxidation and ensuing cell death. In this chapter, we review traditional avenues of drug therapy following traumatic SCI including methylprednisolone, naloxone, and monosialotetrahexosyl (GM-1) ganglioside. We also discuss pharmacotherapy options currently under investigation for the treatment of SCI, with attention given to those that are actively under human clinical trials: riluzole, minocycline, Rho protein antagonist, magnesium chloride in polyethylene glycol formulation, granulocyte colony stimulating factor (G-CSF), and fibroblast growth factor (FGF), and lithium. Far more work remains to be done to further characterize the efficacy, safety, and practicability of these pharmaceutical therapies.

Keywords: Antioxidation, Blood spinal cord barrier, Corticosteroid, Excitotoxicity, GM-1, Gacyclidine, Ganglioside, Granulocyte colony stimulating factor, Growth factor, Immunomodulation, Lipid peroxidation, Lithium, Magnesium, Methylprednisolone, Minocycline, NASCIS, Naloxone, Neuroprotection, Nimodipine, RISCIS, Rho antagonist, Riluzole, Spinal cord injury, Thyrotropin releasing hormone, Tirilazad.

EPIDEMIOLOGY

According to the United States National Spinal Cord Injury Center, spinal cord injury (SCI) affects approximately 40 per million people per year in the US,

[*] **Corresponding author Jason H. Oh and Hiroyuki Yoshihara:** State University of New York – University Hospital of Brooklyn, Brooklyn, New York, USA; Tel: 718.270.8995; Fax: 718.270.3983; E-mails: jho6md@gmail.com and hiroyoshihara55@yahoo.co.jp

accounting for an incidence of approximately 12,500 new cases of survivable SCI per year [1]. The total number of people currently living with SCI in the US is estimated to be between 240,000 to 337,000. Traumatic SCI was historically the affliction of young people, with an average age of onset of 29 years old in the 1970s; currently, the average age at injury is 42 years old. There is a strong predilection for male gender, with men accounting for 80% of new SCI cases. Motor vehicle accidents account for a plurality of new cases at 38%. Falls from height account for an additional 30%, and violence (particularly gun violence), sporting accidents, and miscellaneous causes account for the remainder. Slightly more than half of all SCI cases are at the level of the cervical spine, with the remainder of cases distributed between the thoracic, thoracolumbar, and lumbar spine (including the cauda equina and the exiting nerve roots in the lumbar spine). The average lifetime costs directly attributable to SCI have been estimated to be as high as $4.7 million for a young person with tetraplegia; this figure does not account for indirect costs such as lost wages and other productivity [1].

PATHOPHYSIOLOGY

SCI is initiated with a primary phase of injury, characterized by mechanical disruption of axons by an abnormal force from the spinal column, followed by a delayed secondary phase of injury, which is mediated by inflammation, vascular injury, and excitotoxicity [2].

In the primary phase of SCI, the most typical mode of injury is a compressive insult to the spinal cord that can cause shearing, laceration, or traction injury to the axons [3]. Complete cord transection is rare; usually, there are at least some intact (but potentially demyelinated) axons left in continuity at the subpial rim [4]. Since animal models have demonstrated that there is some potential for clinical recovery with the preservation of even as few as 5-10% of the original axons, it is hoped that even a profoundly spine-injured patient may be able to recover some function with timely neuroprotective measures [5].

In the secondary phase of SCI, a constellation of factors including ischemia, inflammation, dysvascularity, excitotoxicity, and oxidative stresses can contribute to ongoing neuronal and glial cell death by a combination of necrosis and apoptosis [2]. In addition to direct and indirect promotion of cell death, these processes affecting the microenvironment of the injured spinal cord may also inhibit local cell regeneration and remyelination [6]. The secondary phase of SCI can be further categorized into subphases that are roughly based upon time elapsed from initial injury: these are the immediate, acute, subacute, intermediate, and chronic stages of SCI. The phases on SCI are summarized below in Table **1**.

Table 1. Phases of spinal cord injury.

Phase of Injury	Immediate	Acute	Subacute	Intermediate	Chronic
Time elapsed from injury	<2hr	2 – 48hr	2 – 14d	2wk – 6mo	>6mo
Physiologic events	• Ischemia • Hemorrhage • Edema • Thrombosis • Vasospasm • Acidosis • Necrosis • Early recruitment of inflammatory cytokines (IL-1, IL-6, TNF-alpha]	• Increased permeability of the BSCB • Neutrophil infiltration • Ionic dysregulation • Glutamate-mediated cellular excitotoxicity • Free radical production; oxidative stresses • Axonal demyelination • Apoptosis • Microglial activation	• Macrophage infiltration and maturation; phagocytosis • Reactive astrogliosis • Initiation of glial scar production	• Glial scar maturation • Cyst/syrinx formation	• Scar stabilization • Cyst/syrinx maturation • Wallerian degeneration
Therapeutic aims	• Hemostasis • Surgical decompression • Gross mechanical stabilization • Neuroprotection	• Neuroprotection • Immune modulation	• Neuroprotection • Immune modulation	Glial scar inhibition and degradation	Functional rehabilitation

IL-1: interleukin-1; IL-6: interleukin-6; TNF-alpha: tumor necrosis factor alpha; BSCB: blood spinal cord barrier; MPSS: methylprednisolone sodium succinate; TM: tirilazad mesylate; GM-1: monosialotetrahexosylganglioside; TRH: thyrotropin releasing hormone; G-CSF: granulocyte colony stimulating factor; FGF: fibroblast growth factor.

The immediate phase of SCI occurs within the first two hours of injury and is characterized by neuronal and glial cell death associated with the clinical picture of spinal shock [7, 8]. Cell death is caused by necrosis secondary to mechanical rupture of cell membranes, ischemia, hemorrhage, and subsequent edema [5]. During the immediate phase, the inflammatory cascade is initiated with the upregulation of tumor necrosis factor (TNF)-alpha and interleukin (IL)-beta [9]. Furthermore, the levels of extracellular glutamate rise to potentially excitotoxic levels [10]. In general, medical treatment strategies are not effective in addressing the physiological changes associated with the immediate phase of SCI because of the time required to transport a spine-injured patient to a trauma center and make the relevant diagnoses.

The acute phase of SCI follows the immediate phase and continues to about 48 hours post-injury. The acute phase is characterized by cord ischemia [11], which in turn is correlated with acute changes in the permeability of the blood-spinal cord barrier (BSCB). The source of cord ischemia in this critical period is the subject of ongoing research and is likely multifactorial secondary to systemic hypotension in combination with local microvascular disruption and local interstitial edema [12]. Inflammatory cytokines and chemokines released in the immediate and early acute phases are believed to contribute to lability in the permeability of the BSCB [13]. In rat models, BSCB permeability reaches a maximum at about 24 hours post-injury and gradually declines back to pre-injury levels by about 2 weeks post-injury [14]. The porous BSCB allows for the infiltration of neutrophils into the CNS lesion, where they produce cytokines and chemokines as well as matrix metalloproteinases (MMPs), superoxide dismutase, and myeloperoxidase [15]. The neutrophils are followed by monocytes after about 48 hours post-injury. These colonize the lesion and differentiate into macrophages, which in turn produce glutamate, TNF-alpha, IL-1, IL-6, and prostanoids [16]. These inflammatory mediators have both neuroprotective and neurotoxic, proinflammatory effects.

Free radicals play an important role in the acute phase of SCI. The reactive oxygen and nitrogen species that are produced by immune cells, or as a direct byproduct of ischemia and reperfusion injury, have been noted to peak at approximately 12 hours post-injury and remain persistently elevated for one week in animal models [15]. Hall [17] discussed the manner in which free radicals such as peroxynitrite interact with high iron concentrations in the central nervous system (CNS) to induce intracellular lipid peroxidation (LP). LP, in turn, has several deleterious effects that exacerbate SCI: it disrupts cellular ionic homeostasis by contributing to intracellular sodium and calcium ion accumulation, it interferes with mitochondrial function, it enhances glutamate-mediated excitotoxicity, and it impairs microvascular perfusion by inducing damage to microvascular endothelium, enhancing platelet-leukocyte adhesion, and promoting the formation of microemboli. Free radical species are also known to contribute directly to neuronal apoptosis from experimental animal models [18].

Disruption of ionic homeostasis, especially sodium and calcium, propagates cellular losses through positive feedback-loop production of more free radical species, mitochondrial dysfunction, and calpain activation [19]. Ensuing increases in the levels of extracellular glutamate contributes to rising intracellular sodium and calcium ion concentrations *via* the N-methyl-D-aspartic acid (NMDA) and alpha-amino-3-hydroxy-5-methyl-4-isoxazoleproprionic acid (AMPA) receptors followed by intracellular acidosis, cytotoxic edema, and excitotoxic cell death [20].

The cumulative effect of the above mentioned changes during the acute phase of SCI leads to neuronal cell death primarily through cell necrosis, though apoptosis also plays a role [21]. Glial cell death occurs in parallel, though apoptosis tends to take a more dominant role.

The subacute phase of SCI follows the acute phase, lasting from 2 days post-injury to 2 weeks post-injury. This phase is characterized by an initial wave of astrocyte necrosis followed by hypertrophy and repopulation; these astrocytes tend to conglomerate into a large glial scar that forms a biochemical barrier for neuronal regeneration [2, 22]. In addition to its inhibitory mechanical properties, the glial scar produces inhibitory molecules such as chondroitin sulfate proteoglycans that further slow the process of neuronal regeneration [23].

The intermediate phase of SCI follows the subacute phase and lasts for approximately 6 months post-injury. In this time period, the glial scar continues to evolve and produce reactive gliosis. Some axonal sprouting has been observed during the intermediate phase in rat models [24]. More research still needs to be done to establish whether this axonal sprouting can be correlated with clinically meaningful functional recovery.

Finally, the chronic phase of SCI lasts from 6 months post-injury through the remainder of the patient's life. As the scar becomes fully mature, a cyst can develop within the spinal cord lesion as neuronal cells continue to die by Wallerian degeneration, and the ensuing necrotic and apoptotic debris is resorbed [25]. Like the glial scar, this cyst presents a mechanical barrier to neuronal regeneration [26]. Relatively few medical modalities have been proposed to assist in functional recovery for the chronic phase. Instead, physical rehabilitation measures typically predominate, with an emphasis on preserving and expanding the function of the surviving axons through neuroplasticity.

DRUG THERAPIES

In this section, we discuss individual drug therapies for the treatment of SCI. Drugs with completed human randomized trials (RCTs) are reviewed first, and organized by chronological order from oldest to newest. Drugs without completed human RCTs will be discussed at the end, again in chronological order. Table **2** summarizes those treatments that have completed human RCTs.

Table 2. Completed randomized control trials of pharmacotherapy for spinal cord injury.

Trial (year)	Patients enrolled	Deficit	Time to treatment	Treatment Arms	Conclusion
NASCIS I (1984)	330	Incomplete	48h	MPSS 100mg/d x10d MPSS 1000mg/d x10d	No difference in primary analysis
NASCIS II (1990)	487	Complete and incomplete	12h	MPSS 30mg/kg bolus + 5.4mg/kg/hr x24hr Naloxone 5.4mg/kg bolus + 4.0 mg/kg/hr x24hr Placebo	No difference in primary analysis Secondary analyses showed statistically significant improvement of neurologic recovery when MPSS given <8hr No benefit from naloxone
Maryland GM-1 (1991)	34	Incomplete	72h	GM-1 100mg/d x18-32d Placebo	Improved neurologic recovery with GM-1
Japan MPSS (1994)	158	Not specified	8h	MPSS 30mg/kg bolus + 5.4mg/kg/hr x24hr Placebo	Sensory improvements noted with MPSS-treated arm; no significant difference in motor recovery
TRH (1995)	20	Complete and incomplete	2h	TRH 0.2 mg/kg bolus + 0.2mg/kg/hr x6hr Placebo	No difference in primary analysis Secondary analysis showed statistically significant benefit in incompletely-injured subset
NASCIS III (1997)	499	Incomplete	12h	MPSS 30mg/kg bolus + 5.4mg/kg/hr x24hr MPSS 30mg/kg bolus + 5.4mg/kg/hr x48hr MPSS 30mg/kg bolus + tirilazad mesylate 2.5mg/kg q6hr x48hr	If treated within 3hr, beneficial only to give MPSS x24hr If treated between 3-8hr, beneficial to given MPSS x48hr TM not demonstrably superior to MPSS
Nimo-dipine (1998)	106	Complete and incomplete	6h	MPSS 30mg/kg bolus + 5.4mg/kg/hr x24hr Nimodipine 0.015mg/kg/hr x2hr + 0.03mg/kg/hr x7d MPSS 30mg/kg bolus + nimodipine 0.015mg/kg/hr x2hr + MPSS 5.4mg/kg/hr x24hr + nimodipine 0.03mg/kg/hr x7d Placebo	No difference in primary analysis Significantly more infection-related complications noted in MPSS-treated groups
Gacycli-dine (1999)	280	Complete and incomplete	2h	Gacyclidine 0.005mg/kg x 2 doses Gacyclidine 0.01mg/kg x 2 doses Gacyclidine 0.02mg/kg x 2 doses Placebo	No difference in primary analysis Suggestion of improved motor recovery for cervical incomplete SCI patients

(Table 2) contd.....

Trial (year)	Patients enrolled	Deficit	Time to treatment	Treatment Arms	Conclusion
Sygen GM-1 (2001)	797	Incomplete	72h	MPSS 30mg/kg bolus + GM-1 300mg bolus + MPSS 5.4mg/kg/hr x24hr + GM-1 100mg/d x56d MPSS 30mg/kg bolus + GM-1 600mg bolus + MPSS 5.4mg/kg/hr x24hr + GM-1 200mg/d x56d MPSS 30mg/kg bolus + 5.4mg/kg/hr x24hr + placebo	No difference in primary outcomes Secondary analyses suggested trend for earlier improvement with GM-1
Lithium (2012)	40	Not specified	>12mo	Lithium oral administration titrated to serum concentration 0.6-1.2mmol/L x6wk Placebo	No serious adverse events No difference in functional or neurological classification scale improvement Significant benefit in pain scale improvement
Mino-cycline (2012)	52	Complete and incomplete	12h	Minocycline 200mg q12hr x7d Minocycline 800mg bolus + tapering dose -100mg q12hr until 400mg q12hr x7d Placebo	No serious adverse events; one episode of asymptomatic elevation of liver enzymes in high-dose minocycline group No difference in primary analysis of efficacy Secondary analysis trend toward benefit for incomplete cervical injury

NASCIS: National Acute Spinal Cord Injury Study; SCI: spinal cord injury; MPSS: methylprednisolone sodium succinate; TM: tirilazad mesylate; GM-1: monosialotetrahexosylganglioside; TRH: thyrotropin releasing hormone.

Methylprednisolone

Methylprednisolone is a corticosteroid that is administered intravenously in its sodium succinate form (MPSS) (Fig. **1**). Its proposed mechanism of neuroprotection is immunomodulatory, reducing the rate of infiltration of immune cells through the BSCB by inhibiting the process of lipid peroxidation (Fig. **2**) during the acute phase of injury. It has also been observed to reduce TNF-alpha and nuclear factor kappa B activity [27].

Methylprednisolone is the most widely studied drug therapy for SCI, having been tested in three large, well-designed double-blind, randomized control trials named the National Acute Spinal Cord Injury Studies (I, II, III) [28 - 30].

NASCIS I sought to establish the role of MPSS as a neuroprotective agent by testing its efficacy in low and high dose protocols for incomplete acute SCI patients who were able to initiate treatment within 48 hours of injury. The drug was administered as either a 100mg or 1000mg loading dose followed by 25mg or 250mg boluses, respectively, administered every 6 hours thereafter for 10 days. Due to ethical concerns at the time of study design, there was no placebo control group. Outcome measures included gross motor function as well as sensibility to pinprick and light touch, and patients were followed for one year post-injury. The results of the study were that there was no significant difference appreciated between treatment groups on the basis of neurological improvement – however, there was a notable increase in the incidence of infectious adverse events in the high-dose subset [28]. Understandably, the results of NASCIS I tempered enthusiasm for MPSS administration and the drug saw considerably less widespread clinical use in the months following its publication.

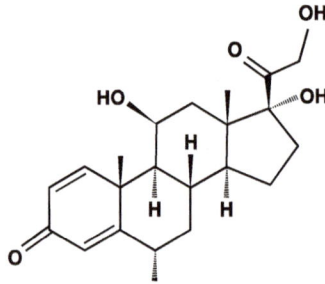

Fig. (1). Molecular structure of methylprednisolone.

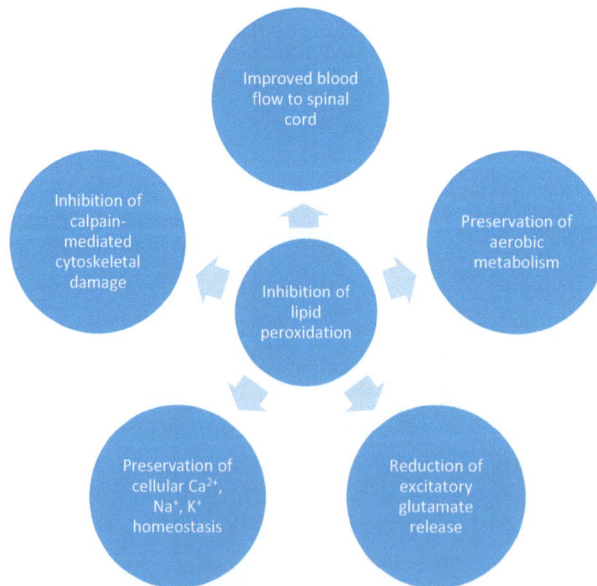

Fig. (2). Protective effects of inhibition of lipid peroxidation.

However, follow-up animal studies performed by Hall *et al.* [31] subsequently found that even the "high-dose" arm of the NASCIS I study had used steroid concentrations that were well below the required threshold for clinically effective neuroprotection. In response, the NASCIS II trial was designed to compare the effectiveness of truly high-dose MPSS (given as a 30mg/kg bolus followed by a 5.4mg/kg/hr infusion for the following 23 hours] as compared against placebo and the opioid receptor antagonist naloxone [32]. Patients received their initial treatment no more than 12 hours post-injury, and each individual's time to initial treatment was recorded to further subcategorize the effectiveness of each subset. Initial review of the NASCIS II data revealed that there was no significant difference between treatment arms in the 6-week, 6-month or 1-year outcomes of all trial participants. However, subsequent subgroup analysis revealed a statistically significant improvement in motor and sensory function up to 1 year post-injury in those patients that received MPSS within 8 hours of injury.

With the publication of the NASCIS II 1992 subgroup analysis, high-dose administration of MPSS became the de facto standard of care for SCI patients within 8 hours of their injury in the US. However, there was a substantial number of clinicians who remained skeptical of MPSS treatment, citing the poorly standardized manner of data collection in the NASCIS II trial (particularly the lack of a functional outcome measure] and a suspicion for selection bias considering the retrospective nature of determining clinical relevance in the sub-8-hour administration arm. A follow-up study was conducted in Japan by Otani *et al.* [33] that confirmed the effectiveness of MPSS if administered within 8 hours of injury; however, this study was also criticized for low sample sizes, a short 6-month follow-up period, and continuing concerns regarding adverse effects of high-dose corticosteroid administration.

Finally, the NASCIS III trial was designed to refute some of the above critiques. The outcome measures were changed to ASIA scores in combination with a functional independence measure (FIM). All patients received a 30mg/kg bolus of MPSS followed by one of three subsequent treatment strategies: 1) MPSS 5.4mg/kg/hr infusion for 24hr; 2) MPSS 5.4mg/kg/hr for 48hr; and 3) tirilazad mesylate (TM, a synthetic glucocorticoid with inhibitory effects on lipid peroxidation with fewer adverse effects than traditional glucocorticoids) at doses of 2.5mg/kg administered every 6hr for 48hr. Analysis of the NASCIS III data revealed that patients who received their initial MPSS bolus within 3 hours of injury benefited from 24hr MPSS administration but did not receive any additional benefit thereafter. Patients who received their initial treatment within 3-8 hours of their injury did show significant improvement with 48hr MPSS administration.

The potential adverse effects of high-dose corticosteroids have been well documented in the medical literature and include infectious sequelae such as wound infections, pneumonia, sepsis, and death, as well as gastrointestinal complications such as peptic ulcer disease and ensuing hemorrhage [34]. In addition, it is unclear whether the anti-inflammatory effects of steroids such as MPSS have deleterious effects on neuronal regeneration and axonal sprouting in the subacute and intermediate phases on injury. It is likely that these delayed harmful effects offset any clinical benefits of MPSS administration, thereby leading to the clinical ineffectiveness of MPSS given more than 8 hours post-injury [3].

The standardized use of MPSS remains a subject of fierce debate among clinicians. Although the NASCIS studies have been the most complete clinical trials to date regarding any drug therapy for SCI, their therapeutic level of evidence is only grade III due to the above mentioned flaws in study design, data presentation, and data analysis. Other studies have since emerged that frankly contest the efficacy of MPSS administration, such as a 2015 study by Evaniew *et al.* [35] that found no significant benefit from NASCIS-II protocol MPSS administration compared to placebo when matching patients based upon anatomical level and severity of injury. Furthermore, critical reviews concerning the clinical application of MPSS administration have found that there is still ongoing confusion regarding the indications for treatment, as well as instances of overtreatment resulting from fear of litigation [36]. Officially, the American Academy of Neurological Surgeons' (AANS) current clinical practice guidelines have concluded that there is insufficient evidence to support the use of MPSS as anything other than a treatment option, with a pointed reminder that the evidence of harmful side effects of high-dose MPSS administration has historically been far more consistent than evidence of clinically significant benefit.

Naloxone

The opioid peptide, dynorphin A, is released endogenously in response to traumatic SCI [3]. In addition to providing pain relief, the dynorphins in high concentrations produce a paradoxical hyperalgesia and allodynia. Furthermore, prolonged elevations in dynorphin-derived peptides have been found to be directly neurotoxic, and associated activation of the kappa-type opioid receptor has been noted to reduce blood flow to the spinal cord [37]. Initial animal models investigating the use of the nonselective opioid receptor antagonist naloxone (Fig. 3) for SCI demonstrated improved spinal cord electrophysiology as well as reduced edema, free radical generation, and allodynia with naloxone administration.

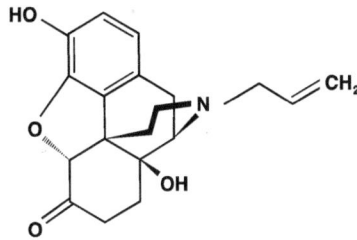

Fig. (3). Molecular structure of naloxone.

Naloxone was subsequently used in NASCIS II as an independent treatment arm, administered as an intravenous 5.4 mg/kg bolus followed by 4 mg/kg/hr infusion for 23 hours. Although there was no statistically significant benefit over placebo in primary outcomes, secondary analysis showed that there was significantly increased recovery below the level of injury in patients who had received naloxone within 8 hours of injury [38]. Despite these promising findings, more human research has yet to be performed to assess naloxone's clinical utilization potential.

Monosialotetrahexosylganglioside

The gangliosides are a group of glycosphingolipids with sialic acid moieties that are found in the membranes of nervous tissue [39]. In vertebrates, gangliosides are found most predominantly in the cellular plasma membrane of central nervous system cells. They are found in varying concentrations depending on the type of nervous tissue (*i.e.*, white matter *vs.* gray matter) and location (*i.e.*, brain cortex *vs.* brain stem *vs.* spinal cord). Although their function is still under investigation, prior basic science studies have demonstrated that exogenous administration of these compounds can promote neurite outgrowth *in vitro*. *In vivo* animal studies have shown that ganglioside administration can potentiate neurotrophic effects. The proposed mechanism for the neuroprotective effects of gangliosides is that they may block the formation of nitric oxide and other reactive oxidative species, thereby reducing toxic effects to the injured spinal cord during the acute phase of injury. Human studies have previously been conducted investigating the efficacy of ganglioside administration for the treatment of stroke and peripheral neuropathy, with promising results [40, 41].

In 1991, Geisler et al conducted the Maryland monosialotetrahexosylganglioside (GM-1) Study to determine GM-1's clinical impact in patients with cervical and thoracic SCI (Fig. **4**) [42]. Patients within 72 hours of injury were recruited for the study and given either 100 mg of GM-1 or placebo *via* daily intravenous administration for 18 to 32 total doses. The subjects were followed for a year with successive Frankel scale and American Spinal Cord Injury Association (ASIA) score assessments. There was a statistically significant increase in clinical improvement in the GM-1 treated patients, with the caveat that significant improvements were only seen in the lower extremities. Based on these findings, it was proposed that GM-1 may improve the survival and function of damaged white matter tracts passing through the level of injury; however, there is no evidence to suggest that GM-1 has any significant impact on the gray matter at the zone of injury. It was noted, with some excitement, that GM-1 demonstrated benefit even when administered 48hr post-injury. Moreover, there were no adverse reactions documented as a result of the drug administration, fueling ongoing interest in further research.

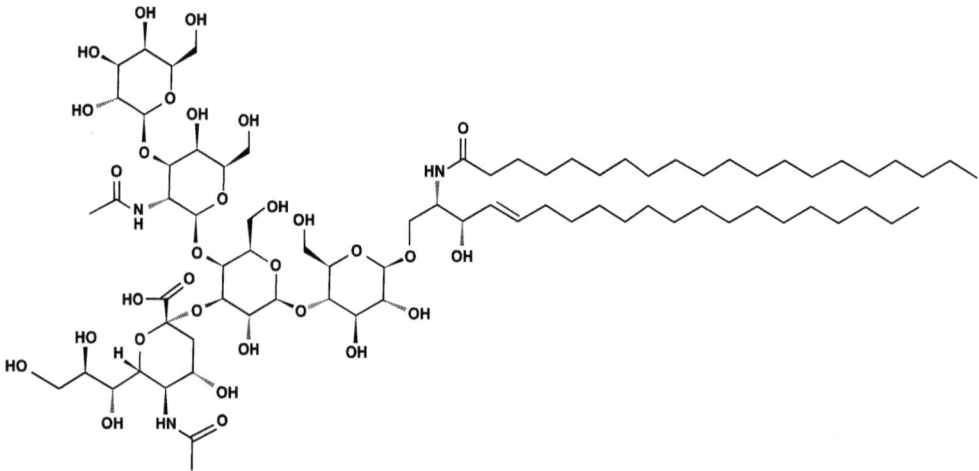

Fig. (4). Molecular structure of monosialotetrahexosyl (GM-1) ganglioside.

The Maryland study was followed by the Sygen GM-1 study: a multi-center, prospective, double-blind randomized and stratified trial that enrolled 797 cervical or thoracic SCI patients in total [43]. All patients were initially managed with MPSS as per the NASCIS II protocol, and then were split into one of three treatment arms: 1) placebo, 2) low dose GM-1 (300mg loading dose followed by 100mg daily for 56 days), or 3) high dose GM-1 (600mg loading dose followed by 200mg daily for 56 days). The outcome measures were the modified Benzel Classification and ASIA motor/sensory scores, with reassessments at 4, 8, 16, 26, and 52 weeks. The primary efficacy assessment of the trial was called "marked recovery" and defined as an improvement in two functional grades by week 26 of the study.

There were several interesting observations noted at the Sygen study's conclusion. First, the high-dose GM-1 arm was discontinued relatively early in the trial because of markedly higher mortality, thus casting doubt on the previously touted "safety" of GM-1 proposed by the Maryland study. Second, the primary study outcome was ultimately negative: there was no statistically significant difference between study arms in the proportion of patients that achieved marked recovery at the six-month time period. However, secondary subgroup analyses found that patients in both the low-dose and high-dose GM-1 arms demonstrated a more accelerated recovery within the first three months of treatment [43]. Third, the patients included in the placebo arm of the Sygen trial failed to display the same levels of neurological improvement as their equally-treated counterparts in the NASCIS II and III trials, thus calling into question the legitimacy of the NASCIS findings.

As with MPSS, the clinical application of GM-1 remains the subject of some debate. Because the only positive effects of GM-1 in the Sygen study could only be appreciated based on secondary analysis, the current AANS guidelines recognize GM-1 as a treatment option when used in combination with NASCIS II/III protocols; however, there is no recommendation in favor of its routine use [44].

Thyrotropin Releasing Hormone

Thyrotropin releasing hormone (TRH) is a tripeptide best known for its role in regulating thyroid hormone homeostasis *via* its effects on the pituitary gland (Fig. **5**). In addition to this important function, TRH may also protect against secondary neurological injury in SCI. Dumont *et al.* [45] hypothesized that TRH can provide neuroprotection by antagonizing the effects of endogenous opioids, platelet activating factor, peptidoleukotrienes, and excitatory amino acids. In a rat model of SCI, Hashimoto and Fukuda [46] found that the daily subcutaneous administration of TRH for 7 days, initiated either 24hr or 7 days following injury, resulted in improved neurologic function.

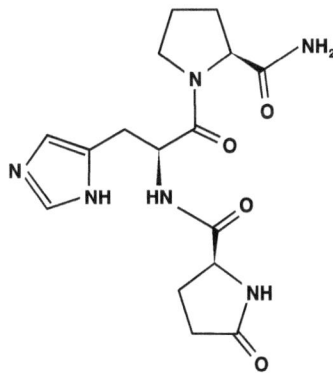

Fig. (5). Molecular structure of thyrotropin releasing hormone.

In 1995, Pitts *et al.* [47] reported the results of a Phase I/II study of TRH in human SCI in which they had recruited 20 patients with complete or incomplete SCI who were able to receive initial treatment within 12 hours of injury. The subjects in the experimental group received an initial bolus of 0.2mg/kg TRH followed by an infusion of 0.2mg/kg/hr for 6 hours. Saline infusion was administered to the control group. The patients were followed for 4 months and evaluated with NASCIS standard motor/sensory scales as well as the Sunnybrook scale. The patients with incomplete SCI who received TRH treatment had statistically significant improvements in all outcome measures as compared to those in the placebo group. No significant difference in outcomes was appreciated among the complete SCI patients. Criticisms against this study included the small sample sizes and the relatively large amount of variability within the placebo-treated incomplete SCI group. No follow-up human trials of TRH have yet been initiated.

Tirilazad

Tirilazad mesylate (TM) is a type of "lazaroid", or 21-aminosteroid (Fig. **6**): a synthetic glucocorticoid analog that has been proposed to provide neuroprotection in cases of SCI without incurring the harms associated with systemic glucocorticoid receptor activation. Hall [48] proposed three mechanisms of neuroprotection: 1) antioxidation; 2) preservation of endogenous vitamin E; and 3) neuronal cell membrane stabilization *via* inhibition of iron-dependent lipid peroxidation.

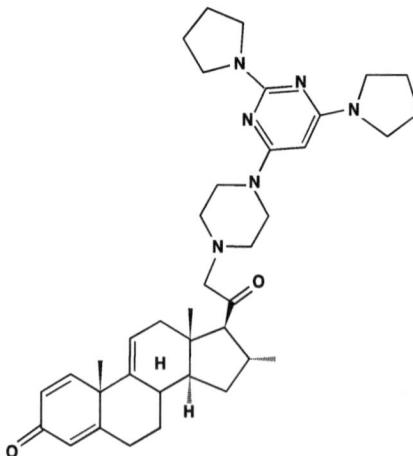

Fig. (6). Molecular structure of tirilazad.

Intermittent IV infusion of tirilazad was one of the treatment arms of NASCIS III following the global loading bolus of 30mg/kg MPSS, and it was found to be as effective as 24hr administration of MPSS [49]. However, since its superiority was never established over a placebo control group, and since it did not demonstrate clinical superiority over MPSS, TM and other lazaroids have not been adopted into clinical practice.

Nimodipine

Continuing on the therapeutic strategy of preventing acute and subacute-phase SCI secondary to calcium ion-mediated excitotoxicity, direct calcium channel blockade has been investigated. In 1989, Fehlings *et al.* [50] proposed that the L-type calcium channel blocking agent nimodipine may reduce excitotoxicity-mediated cell damage as well as vasospasm-induced local cellular ischemia (Fig. 7). In a rat model of SCI, they found that administration of nimodipine with dextran improved spinal cord blood flow and axonal function. A subsequent baboon study by Pointillart and Petitjean [51] confirmed that nimodipine administration improved neurologic outcomes on the basis of somatosensory evoked potentials as compared to placebo.

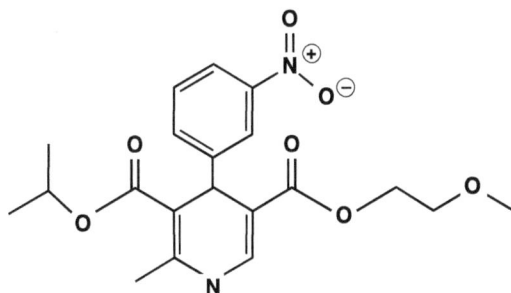

Fig. (7). Molecular structure of nimodipine.

In 1996, Pointillart *et al.* [52] went on to conduct a human study including 106 complete and incomplete SCI patients that were split into four treatment arms: 1) NASCIS II MPSS protocol; 2) nimodipine administered at 0.015mg/kg/hr for 2 hours followed by 0.03mg/kg/hr for 7 days; 3) combination of MPSS and nimodipine as above; and 4) placebo. The study subjects also underwent early decompression and stabilization surgery in addition to their medical treatment. Primary outcomes at one year follow-up were negative – although each treatment group experienced statistically significant improvement from baseline, no benefit was seen from any pharmacological intervention (including "standard" NASCIS II MPSS treatment). These results have been criticized for the relatively low powering of the study, but no further human trials of nimodipine have yet been performed.

Gacyclidine

Following acute spinal cord injury, the sudden rise in intracellular sodium levels *via* voltage-gated sodium channels triggers the profuse extracellular release of glutamate, an excitatory neurotransmitter [10]. Glutamate activates postsynaptic and NMDA receptors, contributing to Na^+ and Ca^{2+} influx into the postsynaptic cells. Axonal edema and neuronal cell death quickly follow. Glutamate antagonism was initially perceived as a difficult therapeutic strategy since systemic administration of conventional anti-glutamatergic agents such as Selfotel has been complicated by severe cognitive adverse effects, including hallucinations, memory loss, and agitation in prior human stroke and brain injury studies [53].

Interest in direct NMDA antagonism for the treatment of SCI was renewed with the development of the noncompetitive NMDA receptor antagonist, gacyclidine (GK-11, Beaufour-Ipsen Pharma), in the 1990s (Fig. **8**). Rodent studies showed that administration of gacyclidine was associated with improved functional, histological, and electrophysiological recovery in rats after spinal cord contusion, and these benefits were realized with significantly fewer adverse effects than with the older, competitive NMDA antagonists [54]. Tadie *et al.* [55] performed a double-blinded Phase II human randomized control trial on over 200 patients in which the study participants received two bolus injections of either 0.005mg/kg, 0.01mg/kg, or 0.02mg/kg of gacyclidine within 2 hours of injury; these were compared against a placebo group. Although overall recovery at one year was not statistically significant in between groups, there were trends noted toward earlier motor recovery as well as toward better overall recovery in the subset of patients that had suffered incomplete cervical injuries. Despite conjecture that statistical significant might have been achieved with a larger study, no additional human trials on gacyclidine have yet been conducted.

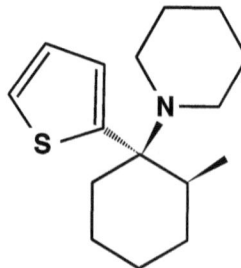

Fig. (8). Molecular structure of gacyclidine.

Lithium

For decades, lithium has been used as a mood-stabilizing agent in the treatment of depression and bipolar disorder. It has also been used against Alzheimer's disease and Huntington's disease with good effect. Lithium has many physiologic effects, some of which have been implicated for its role in neuroprotection. Its inhibition of glycogen synthetase kinase-3-beta (GSK3β), an enzyme that activates various nuclear factors including nuclear factor of activated T-cells (NFAT) and Wnt/β-catenin [56]. Furthermore, lithium upregulates neurotrophic factors such as brain-derived neurotrophic factors (BDNF), nerve growth factor (NGF), and neurotrophin-3 (NT3) [57]. Lithium also inhibits glutamate-mediated excitotoxicity, upregulates antiapoptotic intracellular genetic regulators such as Bcl-2, and downregulates proapoptotic factors such as p53 and Bax. In a 2007 study, Su *et al.* [58] found that treatment with lithium improved survival, proliferation, and differentiation of neural progenitor cells that had been implanted into injured rodent spinal cords. It also reduced the rate of microglia and macrophage activation, suggesting that it has an immunomodulatory role as well.

In 2007, Wong *et al.* performed a Phase I clinical trial of lithium administration on 20 chronic SCI patients [59]. These subjects, all of whom were beyond 12 months of injury, were treated with 6 weeks' duration of oral lithium carbonate titrated to attain a serum lithium level of 0.6-1.2mmol/L. Although there was a relatively high incidence of minor adverse events (predominantly nausea and vomiting), no severe adverse events were noted. This study was succeeded by a Phase II study by Yang *et al.* [60], in which 40 chronic SCI patients received the same 6-week duration of oral lithium therapy and were compared against placebo for a 6 month follow-up period. The primary outcome parameters measured in the study included functional outcome measures, neurological classification scales, and visual analog pain scales. There were no significant differences in functional or neurological classification improvement between the experimental and control groups at 6 months; however, there was a significant improvement in pain reduction in the lithium-treated group. Larger scale trials have yet to be performed to assess lithium's potential for SCI treatment.

Minocycline

Minocycline is a tetracycline antibiotic that has most often been used for the treatment of acne in humans (Fig. **9**) [61]. Yong *et al.* [62] proposed that minocycline may have neuroprotective properties based upon the reduction of excitotoxicity, matrix metalloproteinase inhibition, mitochondrial stabilization, antioxidation, and calcium regulation. Stirling *et al.* [63] added that minocycline may be particularly well-suited for SCI treatment since it can cross the BSCB with a relatively long half-life, and may have anti-inflammatory properties. In addition, Kobayashi *et al.* [64] found that minocycline may inhibit harmful gliosis in M1 microglia. In a rodent model of cervical SCI, minocycline treatment reduced apoptosis of astrocytes and microglia, improved microglial density, and retarded the rate of corticospinal tract dieback; these findings were associated with an overall improvement in functional recovery [65]. There have been several proposed mechanisms for minocycline's inhibition of apoptosis. Teng *et al.* [66] found that minocycline treatment reduces the rate of cytochrome c release from mitochondria; cytochrome c activates pro-apoptotic caspases. Another mechanism proposed by Yune *et al.* [67] focuses on p38 mitogen-activated protein kinase (p38MAPK), an enzyme found in microglia that becomes activated *via* phosphorylation upon spinal cord trauma. The activated p38MAPK forms pro-nerve growth factor (pro-NGF), which in turn promotes oligodendrocyte apoptosis. The researchers found that minocycline administration reduces the phosphorylation of p38MAPK and thereby reduces the downstream rate of oligodendrocyte apoptosis.

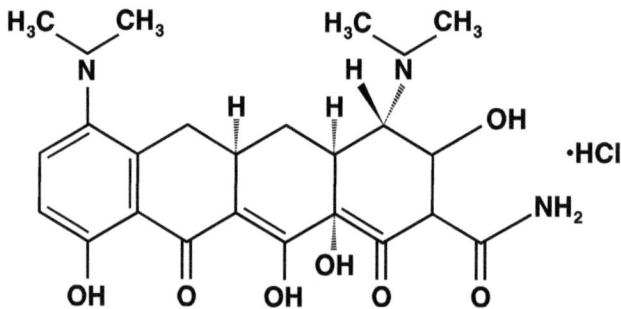

Fig. (9). Molecular structure of minocycline.

Minocycline has already been the subject of some human studies relating to its usefulness in other neurological disorders such as stroke, multiple sclerosis, amyotrophic lateral sclerosis (ALS), and schizophrenia [68 - 71]. These have shown relatively low rates of adverse events, though there have been some instances of drug-induced lupus with prolonged use. Casha *et al.* [72] executed a Phase I/II randomized control trial involving a total of 52 patients with complete or incomplete cervical or thoracic SCI who received either low dose minocycline, high dose minocycline, or placebo. Subjects in the low dose arm of the study were given the previously published "maximum allowable" dose of minocycline, 200mg every 12 hours, for seven days. The high dose subjects were given an initial bolus of 800mg followed by successive doses tapered by 100mg each 12 hours until 400mg; the 400mg dose was carried through until the end of 7 days.

They found that there were no significant adverse events attributable to minocycline at either dosing level, though there was one instance of asymptomatic elevation in liver enzymes in one of the patients in the high-dose arm. Although there was no statistically significant difference in functional recovery after 1 year, secondary analyses suggested a trend toward improved outcomes among the incomplete cervical SCI patients that received minocycline treatment. This trend was not statistically significant, possibly due to inherent lack of power in the initial study design. Phase III research is ongoing.

Magnesium

Magnesium is an NMDA receptor antagonist that can offset the NMDA receptor overstimulation caused by high glutamate levels in acute SCI. By doing so, it can help to prevent the massive intracellular calcium influx that leads to excitotoxic cell damage and death [73]. Consequently, it protects the integrity of the BSCB, reduces lipid peroxidation, and promotes functional recovery. Although the amount of raw magnesium chloride required to attain therapeutic levels in the CNS by systemic administration would be toxic, a polyethylene glycol (PEG)-associated formulation has been developed. In rodent SCI models, the PEG-associated MgCl demonstrated improved neuroprotection and superior motor function recovery as compared to placebo [74].

MgCl-PEG was evaluated in Phase I human trials in 2009 and found to generate no adverse effects [75]. Phase II trials are currently underway.

Fibroblast Growth Factor

Fibroblast growth factors (FGFs) are another agent for SCI treatment. Basic FGF (FGF-2) has previously been shown to increase neural stem cell and progenitor cell proliferation [76] and decrease cell death rates in excitotoxic conditions [77]. FGF-2 also enhances functional recovery in SCI by neuroprotection, stimulation of angiogenesis, and inhibition of cavitation [78]. Furthermore, treatment with FGF-2 appears to reduce BSCB permeability in SCI models, thereby reducing the rates of immune cell infiltration, cytokine production, glial scar formation, and astrogliosis in rodent models [79 - 81]. FGF also appears to have promise for the treatment of chronic SCI, as rodent models of complete spinal cord transection have demonstrated benefit from FGF directly impregnated alongside peripheral nerve bridge grafts [82].

Wu *et al.* [83] performed an uncontrolled human clinical trial with acidic FGF (FGF-1, closely related to FGF-2) in which the compound was surgically administered to SCI lesions at an average of 26 months post-injury, with additional FGF-1 spinal injections given 3 months and 6 months post-operatively.

Over the 24 month follow-up period, study enrollees demonstrated improved ASIA sensorimotor scores and functional independence measures. No significant adverse events were reported. Although the study suffered from lack of control and possible confounding due to concomitant physical rehabilitation protocols, Phase II research on both FGF-1 and FGF-2 are ongoing.

Rho Protein Antagonist: BA-210

Rho is an intracellular GTPase whose activation of the Rho-associated kinases (ROCKs) leads to actomyosin contraction, growth cone collapse, neurite retraction, and eventual neuronal cell death. Rho is activated by various myelin-associated debris including Nogo, myelin-associated glycoprotein (MAG), oligodendrocyte myelin glycoprotein (OMgp), and chondroitin sulfate glycoproteins, which are commonly expressed by the glial scar [84]. Several Rho antagonists have been studied for the potential promotion of axonal regeneration and prevention of cell death in SCI (Fig. **10**). The most promising of these is C3 transferase (C3), an enzyme derived from Clostridium botulinum. C3 has been studied in animal models of optic nerve injury and SCI, and has been shown to increase long-distance axonal regeneration and improve motor function; it also reduces the rate of cellular apoptosis [85, 86].

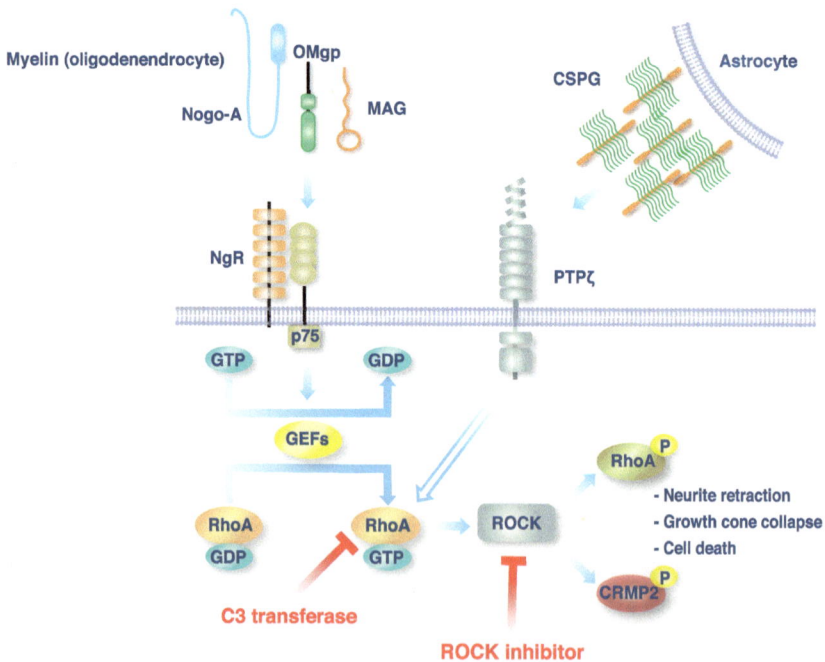

Fig. (10). Role of Rho inhibition in cellular signal transduction for neuroprotection.

One of the clinical challenges encountered with C3 is relatively low cellular permeability. To address this, a more-permeable variant of C3 called BA-210 was developed. Epidural administration routes are used to bypass the BSCB [87]. Fehlings *et al.* [88] executed a multicenter Phase I/II study using BA-210 to treat a total of 48 complete (ASIA A) cervical or thoracic SCI patients. BA-210 was used as an adjunct in combination with decompression surgery performed within 7 days of injury; the compound was impregnated onto a fibrin-mediated delivery system and directly administered onto the dura mater overlying the spinal cord lesion at the time of surgery. The drug was believed to diffuse locally into the spinal cord in a dose-dependent manner, with relatively low rates of release into the systemic circulation. Although this study was relatively small and did not include a control group, the results were promising: over 30% of cervical SCI patients improved to at least an ASIA grade C after 1 year, and there were no major adverse events associated with drug administration. It is important to note that BA-210 is one of the only compounds reviewed in this chapter that relies upon direct surgical administration to achieve therapeutic efficacy. Therefore, its indications and scope of use are currently limited to those spine-injured patients who are indicated for surgical treatment. BA-210 is currently under further investigation with a Phase IIb/III double-blinded, placebo-controlled RCT.

Granulocyte Colony Stimulating Factor

Granulocyte colony stimulating factor (G-CSF) is an endogenously produced glycoprotein named for its roles in the promotion of neutrophil differentiation and the inhibition of neutrophil apoptosis [89]. In addition to these functions, G-CSF also has important neuroregulatory functions. G-CSF is expressed in large quantities on microglia, where it promotes the expression of the phenotype markers Arg 1 and CD206 as well as various neurotrophic factors [90]. Due to its relatively small size (19.6 kDa), G-CSF is able to cross the BSCB under normal physiological conditions [91]. G-CSF also has a neuroprotective role through the inhibition of inflammatory cytokines including TNF-alpha, IL-1-beta, and nuclear factor kappa B. Rodent models have demonstrated that G-CSF can also inhibit neuronal and oligodendrocytic apoptosis, and that it promotes angiogenesis [92].

Takahashi *et al.* [93] performed a Phase I/IIa human clinical trial in which 16 cervical and thoracic SCI subjects within 48hrs of their injury received daily intravenous injections of either 5 or 10 microg/kg/day for 5 days. The drug was well tolerated at both dosing levels, and significant improvements were seen in ASIA scores in the 10 microg/kg/day group, especially for injuries at the cervical levels. This work was built upon by the same group as 28 more patients were recruited to receive the 5-day regimen of 10 microg/kg/day [94]. All the data were retrospectively compared to the outcomes of 34 similar patients who had received standard-of-care MPSS treatment. Based upon this retrospective analysis, G-CSF appeared to promote improved functional recovery with fewer adverse events as compared to MPSS. However, a prospective randomized control trial to directly compare G-CSF against MPSS has yet to be conducted.

Riluzole

Derangement of sodium ion homeostasis in the zone of SCI is an important mediator of secondary cell death. Increasing concentrations of intracellular sodium are associated with reversal of function of sodium/calcium antiporters, leading to pathologic influx of calcium intracellularly. In order to prevent this toxic progression, Rosenberg *et al.* [95] initially investigated the use of tetrodotoxin, a voltage-gated sodium channel blocker, in rat models of SCI. They found that there were significantly lower axonal losses with tetrodotoxin administration.

Since tetrodotoxin is extremely toxic, the benzothiazole anticonvulsant sodium channel blocker, riluzole, is currently under investigation for potential use in human models of SCI (Fig. **11**). Riluzole has already been Food and Drug Administration (FDA)-approved in the US for the treatment of ALS, and previous animal studies have demonstrated its neuroprotective properties *via* its modulation of glutamate release by preventing over activation of sodium channels (Fig. **12**) [96, 97]. Riluzole is also believed to stimulate neurotrophic factor expression [98]. Other rodent studies with riluzole have demonstrated improved functional recovery, reduced CNS cavitation [99], and increased preservation of white matter, mitochondrial function, and motor neurons as measured by somatosensory-evoked potentials [100].

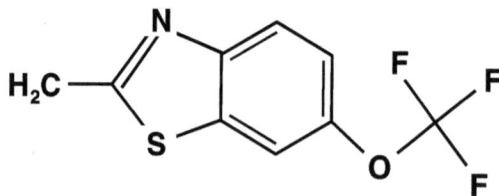

Fig. (11). Molecular structure of riluzole.

The Riluzole in Spinal Cord Injury Study (RISCIS) is an ongoing multicenter clinical trial designed to test the safety and efficacy of riluzole in humans. The Phase I portion of the study involved 36 subjects and found that there were no serious adverse events associated with the administration of 50mg of riluzole administered every 12 hours for 14 days, with first administration within 12 hours of injury in cervical and thoracic SCI patients. Within the first 90 days of the

Phase I trial, there were some improved functional improvements noted especially within the cervical SCI subgroup treated with riluzole [101]. Phases II and III of RISCIS are ongoing.

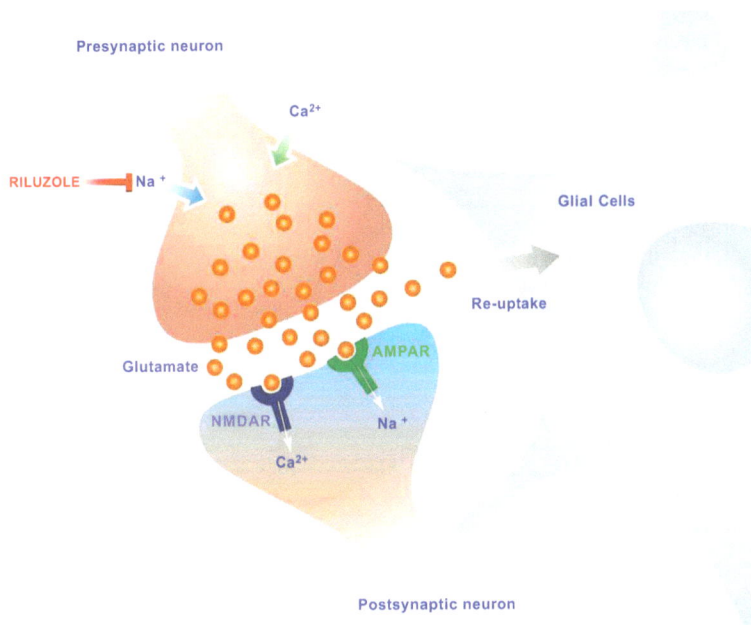

Fig. (12). Role of riluzole in reducing glutamate-mediated cellular excitotoxicity.

Table **3** below reviews the mechanism of action and most advanced level of human trial for all drugs discussed in this section. They are listed in chronological order based upon date of their first trials in human subjects.

Table 3. Summary of pharmacotherapy options for spinal cord injury with completed human trials.

Drug [year]	Mechanism	Phase of action	Level of investigation	Comments
MPSS [1984]	Immune modulation *via* lipid peroxidation inhibition; anti-inflammation; reduce BSCB permeability	Acute	Phase III	Significant adverse effects associated with MPSS over placebo Clinical application still controversial
Naloxone [1990]	Inhibition of dynorphin-mediated neurotoxicity and ischemia; reduced edema; anti-oxidation	Acute	Phase III	No benefit realized compared to placebo at 1 year No ongoing human trials

(Table 3) contd.....

Drug [year]	Mechanism	Phase of action	Level of investigation	Comments
GM-1 [1991]	Antioxidation; promotion of neurite outgrowth; potentiation of neurotrophic factors	Acute, Subacute	Phase III	Secondary analyses suggest some acceleration of recovery with medium-dose GM-1 when given in combination with standard MPSS Possible benefit even after 48hr post-injury
TRH [1995]	Modulation of secondary injury mechanisms including excitotoxic amino acids and endogenous opioids	Acute	Phase II	Secondary analyses showed statistically significant improvement in incompletely-injured patients Very small study size (20 total) No ongoing human trials
Tirilazad [1997]	Immune modulation *via* lipid peroxidation inhibition; anti-inflammation; vitamin E preservation; reduce BSCB permeability	Acute	Phase III	Equivalent clinical outcomes as compared to standard MPSS treatment; no evidence of superiority No ongoing human trials
Nimodipine [1998]	Inhibition of vasospasm *via* direct calcium channel blockade; reduction in excitotoxicity	Acute	Phase III	No significant benefit over placebo at 1 year Study likely underpowered No ongoing human trials
Gacyclidine [1999]	Inhibition of glutamate-mediated excitotoxicity	Acute	Phase II	No significant benefit at 1 year compared to placebo; lack of statistical significance possibly due to underpowered study design No ongoing human trials
Magnesium [2009]	Inhibition of glutamate-mediated excitotoxicity *via* NMDA receptor inhibition; BSCB protection; inhibition of lipid peroxidation	Acute	Phase I	No adverse events Phase II trials ongoing
FGF [2011]	Inhibition of glutamate-mediated excitotoxicity Promotion of axonal regrowth after chronic injury	Acute, Intermediate, Chronic	Phase I/II	Drug delivered *via* direct surgical implantation No placebo group No significant adverse events Suggestion of benefit from therapy Phase II trials underway

(Table 3) contd.....

Drug [year]	Mechanism	Phase of action	Level of investigation	Comments
Rho protein antagonist (BA-210) [2011]	Inhibition of Rho/ROCK-mediated neuronal apoptosis, neurite retraction, and growth cone collapse	Acute, Subacute	Phase I/II	Drug delivered *via* direct surgical implantation No placebo group No significant adverse events Suggestion of benefit from therapy Phase II/III trials ongoing
G-CSF [2012]	Upregulation of neurotrophic factors; anti-inflammation; promotion of angiogenesis; inhibition of neuronal and glial apoptosis	Acute	Phase I/II	No significant adverse events Improved functional outcomes with fewer adverse events when retrospectively compared to similar patients treated with MPSS Prospective RCT still required
Lithium [2012]	Inhibition of GSK3β-mediated inflammation; neurotrophic factor regulation; inhibition of glutamate-mediated excitotoxicity; anti-apoptosis	Chronic	Phase I/II	No major adverse events; many minor adverse events including nausea/vomiting Low powered Phase II study
Minocycline [2012]	Anti-inflammation; inhibition of cytochrome c-mediated glial cell apoptosis	Acute	Phase I/II	No serious adverse events Statistically insignificant trends toward improved sensory and functional recovery from secondary analysis of incomplete cervical injury; study underpowered Phase III trials ongoing
Riluzole [2014]	Protection of Na^+ homeostasis, secondary prevention of Ca^{2+} and glutamate dysregulation	Acute	Phase I	No serious adverse events in Phase I trials Phase II/III trials (RISCIS) ongoing

TNF-alpha: tumor necrosis factor alpha; BSCB: blood spinal cord barrier; MPSS: methylprednisolone sodium succinate; TM: tirilazad mesylate; GM-1: monosialotetrahexosylganglioside; TRH: thyrotropin releasing hormone; G-CSF: granulocyte colony stimulating factor; RCT: randomized control trial; FGF: fibroblast growth factor; ROCK: Rho-associated protein kinase; RISCIS: Riluzole in Spinal Cord Injury Study; PEG: polyethylene glycol; NMDA: N-methyl-D-aspartate; GSK3β: glycogen synthetase kinase-3-beta.

CONCLUSIONS

The treatment of spinal cord injury is a formidable challenge both for the basic science researcher as well as for the practicing clinician. Many pharmacological treatment modalities have been investigated, but none have received consensus approval for use in clinical practice. Much of the difficulty involved in discovering effective new therapies is rooted in our incomplete understanding of central nervous system regeneration at the cellular level. Without this basic understanding, it is extremely difficult to predict how an individual drug therapy will take effect.

Furthermore, high quality research is plagued by technical difficulties in attaining adequate sample sizes, eliminating confounding variables, and establishing adequate experimental controls. Far more work has yet to be done to investigate upcoming treatment strategies and build upon our understanding of existing treatment options. To attain optimal patient outcomes, it will be necessary to incorporate pharmacological treatment options into a multidisciplinary approach which also includes cell-based therapies, surgery, and functional rehabilitation.

CONFLICT OF INEREST

The authors (editor) declares no conflict of interest, financial or otherwise.

ACKNOWLEDGMENTS

None declared.

REFERENCES

[1]   Center NS. Facts and Figures at a Glance. In: Birmingham UoAa, editor. Birmingham, AL2015.

[2]   Rowland JW, Hawryluk GW, Kwon B, Fehlings MG. Current status of acute spinal cord injury pathophysiology and emerging therapies: promise on the horizon. Neurosurg Focus 2008; 25(5): E2.
       [http://dx.doi.org/10.3171/FOC.2008.25.11.E2] [PMID: 18980476]

[3]   Baptiste DC, Fehlings MG. Pharmacological approaches to repair the injured spinal cord. J Neurotrauma 2006; 23(3-4): 318-34.
       [http://dx.doi.org/10.1089/neu.2006.23.318] [PMID: 16629619]

[4]   Radojicic M, Reier PJ, Steward O, Keirstead HS. Septations in chronic spinal cord injury cavities contain axons. Exp Neurol 2005; 196(2): 339-41.
       [http://dx.doi.org/10.1016/j.expneurol.2005.08.009] [PMID: 16153640]

[5]   Kakulas BA. Neuropathology: the foundation for new treatments in spinal cord injury. Spinal Cord 2004; 42(10): 549-63.
       [http://dx.doi.org/10.1038/sj.sc.3101670] [PMID: 15346131]

[6]   Dasari VR, Veeravalli KK, Dinh DH. Mesenchymal stem cells in the treatment of spinal cord injuries: A review. World J Stem Cells 2014; 6(2): 120-33.
       [http://dx.doi.org/10.4252/wjsc.v6.i2.120] [PMID: 24772239]

[7]   Norenberg MD, Smith J, Marcillo A. The pathology of human spinal cord injury: defining the

problems. J Neurotrauma 2004; 21(4): 429-40.
[http://dx.doi.org/10.1089/089771504323004575] [PMID: 15115592]

[8]   Boland RA, Lin CS, Engel S, Kiernan MC. Adaptation of motor function after spinal cord injury: novel insights into spinal shock. Brain 2011; 134(Pt 2): 495-505.
[http://dx.doi.org/10.1093/brain/awq289] [PMID: 20952380]

[9]   David S, Kroner A. Repertoire of microglial and macrophage responses after spinal cord injury. Nat Rev Neurosci 2011; 12(7): 388-99.
[http://dx.doi.org/10.1038/nrn3053] [PMID: 21673720]

[10] Park E, Velumian AA, Fehlings MG. The role of excitotoxicity in secondary mechanisms of spinal cord injury: a review with an emphasis on the implications for white matter degeneration. J Neurotrauma 2004; 21(6): 754-74.
[http://dx.doi.org/10.1089/0897715041269641] [PMID: 15253803]

[11] Tator CH, Koyanagi I. Vascular mechanisms in the pathophysiology of human spinal cord injury. J Neurosurg 1997; 86(3): 483-92.
[http://dx.doi.org/10.3171/jns.1997.86.3.0483] [PMID: 9046306]

[12] Ng MT, Stammers AT, Kwon BK. Vascular disruption and the role of angiogenic proteins after spinal cord injury. Transl Stroke Res 2011; 2(4): 474-91.
[http://dx.doi.org/10.1007/s12975-011-0109-x] [PMID: 22448202]

[13] Pardridge WM. Biologic TNFα-inhibitors that cross the human blood-brain barrier. Bioeng Bugs 2010; 1(4): 231-4.
[http://dx.doi.org/10.4161/bbug.1.4.12105] [PMID: 21327054]

[14] Figley SA, Khosravi R, Legasto JM, Tseng YF, Fehlings MG. Characterization of vascular disruption and blood-spinal cord barrier permeability following traumatic spinal cord injury. J Neurotrauma 2014; 31(6): 541-52.
[http://dx.doi.org/10.1089/neu.2013.3034] [PMID: 24237182]

[15] Donnelly DJ, Popovich PG. Inflammation and its role in neuroprotection, axonal regeneration and functional recovery after spinal cord injury. Exp Neurol 2008; 209(2): 378-88.
[http://dx.doi.org/10.1016/j.expneurol.2007.06.009] [PMID: 17662717]

[16] Schwab JM, Brechtel K, Nguyen TD, Schluesener HJ. Persistent accumulation of cyclooxygenase-1 (COX-1) expressing microglia/macrophages and upregulation by endothelium following spinal cord injury. J Neuroimmunol 2000; 111(1-2): 122-30.
[http://dx.doi.org/10.1016/S0165-5728(00)00372-6] [PMID: 11063829]

[17] Hall ED. Antioxidant therapies for acute spinal cord injury. Neurotherapeutics 2011; 8(2): 152-67.
[http://dx.doi.org/10.1007/s13311-011-0026-4] [PMID: 21424941]

[18] Xiong Y, Rabchevsky AG, Hall ED. Role of peroxynitrite in secondary oxidative damage after spinal cord injury. J Neurochem 2007; 100(3): 639-49.
[http://dx.doi.org/10.1111/j.1471-4159.2006.04312.x] [PMID: 17181549]

[19] Vosler PS, Sun D, Wang S, *et al.* Calcium dysregulation induces apoptosis-inducing factor release: cross-talk between PARP-1- and calpain-signaling pathways. Exp Neurol 2009; 218(2): 213-20.
[http://dx.doi.org/10.1016/j.expneurol.2009.04.032] [PMID: 19427306]

[20] Gerardo-Nava J, Mayorenko II, Grehl T, Steinbusch HW, Weis J, Brook GA. Differential pattern of neuroprotection in lumbar, cervical and thoracic spinal cord segments in an organotypic rat model of glutamate-induced excitotoxicity. J Chem Neuroanat 2013; 53: 11-7.
[http://dx.doi.org/10.1016/j.jchemneu.2013.09.007] [PMID: 24126226]

[21] Yu WR, Liu T, Fehlings TK, Fehlings MG. Involvement of mitochondrial signaling pathways in the mechanism of Fas-mediated apoptosis after spinal cord injury. Eur J Neurosci 2009; 29(1): 114-31.
[http://dx.doi.org/10.1111/j.1460-9568.2008.06555.x] [PMID: 19120440]

[22] Young W. Spinal cord regeneration. Cell Transplant 2014; 23(4-5): 573-611.

[http://dx.doi.org/10.3727/096368914X678427] [PMID: 24816452]

[23]   Fitch MT, Silver J. CNS injury, glial scars, and inflammation: Inhibitory extracellular matrices and regeneration failure. Exp Neurol 2008; 209(2): 294-301.
[http://dx.doi.org/10.1016/j.expneurol.2007.05.014] [PMID: 17617407]

[24]   Hill CE, Beattie MS, Bresnahan JC. Degeneration and sprouting of identified descending supraspinal axons after contusive spinal cord injury in the rat. Exp Neurol 2001; 171(1): 153-69.
[http://dx.doi.org/10.1006/exnr.2001.7734] [PMID: 11520130]

[25]   Li J, Lepski G. Cell transplantation for spinal cord injury: a systematic review. BioMed Research International 2013; 2013: 786475.
[http://dx.doi.org/10.1155/2013/786475]

[26]   Kramer AS, Harvey AR, Plant GW, Hodgetts SI. Systematic review of induced pluripotent stem cell technology as a potential clinical therapy for spinal cord injury. Cell Transplant 2013; 22(4): 571-617.
[http://dx.doi.org/10.3727/096368912X655208] [PMID: 22944020]

[27]   Xu J, Fan G, Chen S, Wu Y, Xu XM, Hsu CY. Methylprednisolone inhibition of TNF-alpha expression and NF-kB activation after spinal cord injury in rats. Brain Res Mol Brain Res 1998; 59(2): 135-42.
[http://dx.doi.org/10.1016/S0169-328X(98)00142-9] [PMID: 9729336]

[28]   Bracken MB, Collins WF, Freeman DF, *et al.* Efficacy of methylprednisolone in acute spinal cord injury. JAMA 1984; 251(1): 45-52.
[http://dx.doi.org/10.1001/jama.1984.03340250025015] [PMID: 6361287]

[29]   Bracken MB, Shepard MJ, Holford TR, *et al.* Methylprednisolone or tirilazad mesylate administration after acute spinal cord injury: 1-year follow up. Results of the third National Acute Spinal Cord Injury randomized controlled trial. J Neurosurg 1998; 89(5): 699-706.
[http://dx.doi.org/10.3171/jns.1998.89.5.0699] [PMID: 9817404]

[30]   Bracken MB. Methylprednisolone and acute spinal cord injury: an update of the randomized evidence. Spine 2001; 26(24) (Suppl.): S47-54.
[http://dx.doi.org/10.1097/00007632-200112151-00010] [PMID: 11805609]

[31]   Hall ED, Wolf DL, Braughler JM. Effects of a single large dose of methylprednisolone sodium succinate on experimental posttraumatic spinal cord ischemia. Dose-response and time-action analysis. J Neurosurg 1984; 61(1): 124-30.
[http://dx.doi.org/10.3171/jns.1984.61.1.0124] [PMID: 6374068]

[32]   Bracken MB, Shepard MJ, Collins WF Jr, *et al.* Methylprednisolone or naloxone treatment after acute spinal cord injury: 1-year follow-up data. Results of the second National Acute Spinal Cord Injury Study. J Neurosurg 1992; 76(1): 23-31.
[http://dx.doi.org/10.3171/jns.1992.76.1.0023] [PMID: 1727165]

[33]   Otani K, Abe H, Kadoya S. Beneficial effect of methylprednisolone sodium succinate in the treatment of acute spinal cord injury. Sekitsui Sekizui 1996; 7(3): 335-43. [in Japanese].

[34]   Matsumoto T, Tamaki T, Kawakami M, Yoshida M, Ando M, Yamada H. Early complications of high-dose methylprednisolone sodium succinate treatment in the follow-up of acute cervical spinal cord injury. Spine 2001; 26(4): 426-30.
[http://dx.doi.org/10.1097/00007632-200102150-00020] [PMID: 11224891]

[35]   Evaniew N, Noonan VK, Fallah N, *et al.* Methylprednisolone for the Treatment of Patients with Acute Spinal Cord Injuries: A Propensity Score-Matched Cohort Study from a Canadian Multi-Center Spinal Cord Injury Registry. J Neurotrauma 2015; 32(21): 1674-83.
[http://dx.doi.org/10.1089/neu.2015.3963] [PMID: 26065706]

[36]   Hugenholtz H, Cass DE, Dvorak MF, Fewer DH, Fox RJ, Izukawa DM, *et al.* High-dose methylprednisolone for acute closed spinal cord injury--only a treatment option. The Canadian Journal of Neurological Sciences Le Journal Canadien des Sciences Neurologiques 2002; 29(3): 227-35.

[37]   Long JB, Martinez-Arizala A, Petras JM, Holaday JW. Endogenous opioids in spinal cord injury: a critical evaluation. Central Nervous System Trauma: Journal of the American Paralysis Association 1986; 3(4): 295-315.
[http://dx.doi.org/10.1089/cns.1986.3.295]

[38]   Bracken MB, Holford TR. Effects of timing of methylprednisolone or naloxone administration on recovery of segmental and long-tract neurological function in NASCIS 2. J Neurosurg 1993; 79(4): 500-7.
[http://dx.doi.org/10.3171/jns.1993.79.4.0500] [PMID: 8410217]

[39]   Vorwerk CK. Ganglioside patterns in human spinal cord. Spinal Cord 2001; 39(12): 628-32.
[http://dx.doi.org/10.1038/sj.sc.3101232] [PMID: 11781858]

[40]   Battistin L, Cesari A, Galligioni F, *et al.* Effects of GM1 ganglioside in cerebrovascular diseases: a double-blind trial in 40 cases. Eur Neurol 1985; 24(5): 343-51.
[http://dx.doi.org/10.1159/000115823] [PMID: 3902480]

[41]   Gorio A, Aporti F, Di Gregorio F, Schiavinato A, Siliprandi R, Vitadello M. Ganglioside treatment of genetic and alloxan-induced diabetic neuropathy. Adv Exp Med Biol 1984; 174: 549-64.
[http://dx.doi.org/10.1007/978-1-4684-1200-0_46] [PMID: 6204518]

[42]   Geisler FH, Dorsey FC, Coleman WP. Recovery of motor function after spinal-cord injury--a randomized, placebo-controlled trial with GM-1 ganglioside. N Engl J Med 1991; 324(26): 1829-38.
[http://dx.doi.org/10.1056/NEJM199106273242601] [PMID: 2041549]

[43]   Geisler FH, Coleman WP, Grieco G, Poonian D. The Sygen multicenter acute spinal cord injury study. Spine 2001; 26(24) (Suppl.): S87-98.
[http://dx.doi.org/10.1097/00007632-200112151-00015] [PMID: 11805614]

[44]   Walters BC, Hadley MN, Hurlbert RJ, *et al.* Guidelines for the management of acute cervical spine and spinal cord injuries: 2013 update. Neurosurgery 2013; 60 (Suppl. 1): 82-91.
[http://dx.doi.org/10.1227/01.neu.0000430319.32247.7f] [PMID: 23839357]

[45]   Dumont RJ, Verma S, Okonkwo DO, *et al.* Acute spinal cord injury, part II: contemporary pharmacotherapy. Clin Neuropharmacol 2001; 24(5): 265-79.
[http://dx.doi.org/10.1097/00002826-200109000-00003] [PMID: 11586111]

[46]   Hashimoto T, Fukuda N. Effect of thyrotropin-releasing hormone on the neurologic impairment in rats with spinal cord injury: treatment starting 24 h and 7 days after injury. Eur J Pharmacol 1991; 203(1): 25-32.
[http://dx.doi.org/10.1016/0014-2999(91)90786-P] [PMID: 1797554]

[47]   Pitts LH, Ross A, Chase GA, Faden AI. Treatment with thyrotropin-releasing hormone (TRH) in patients with traumatic spinal cord injuries. J Neurotrauma 1995; 12(3): 235-43.
[http://dx.doi.org/10.1089/neu.1995.12.235] [PMID: 7473798]

[48]   Hall ED. Lipid antioxidants in acute central nervous system injury. Ann Emerg Med 1993; 22(6): 1022-7.
[http://dx.doi.org/10.1016/S0196-0644(05)82745-3] [PMID: 8503522]

[49]   Bracken MB, Shepard MJ, Holford TR, *et al.* Administration of methylprednisolone for 24 or 48 hours or tirilazad mesylate for 48 hours in the treatment of acute spinal cord injury. Results of the Third National Acute Spinal Cord Injury Randomized Controlled Trial. National Acute Spinal Cord Injury Study. JAMA 1997; 277(20): 1597-604.
[http://dx.doi.org/10.1001/jama.1997.03540440031029] [PMID: 9168289]

[50]   Fehlings MG, Tator CH, Linden RD. The effect of nimodipine and dextran on axonal function and blood flow following experimental spinal cord injury. J Neurosurg 1989; 71(3): 403-16.
[http://dx.doi.org/10.3171/jns.1989.71.3.0403] [PMID: 2475595]

[51]   Pointillard V, Petitjean ME. Medical treatment of spinal cord injury during the acute phase. Effect of a calcium inhibitor. Agressologie 1993; 34(Spec No 2): 93-5.

[PMID: 7802154]

[52]   Pointillart V, Petitjean ME, Wiart L, *et al.* Pharmacological therapy of spinal cord injury during the acute phase. Spinal Cord 2000; 38(2): 71-6.
[http://dx.doi.org/10.1038/sj.sc.3100962] [PMID: 10762178]

[53]   Morris GF, Bullock R, Marshall SB, Marmarou A, Maas A, Marshall LF. Failure of the competitive N-methyl-D-aspartate antagonist Selfotel (CGS 19755) in the treatment of severe head injury: results of two phase III clinical trials. J Neurosurg 1999; 91(5): 737-43.
[http://dx.doi.org/10.3171/jns.1999.91.5.0737] [PMID: 10541229]

[54]   Gaviria M, Privat A, d'Arbigny P, Kamenka J, Haton H, Ohanna F. Neuroprotective effects of a novel NMDA antagonist, Gacyclidine, after experimental contusive spinal cord injury in adult rats. Brain Res 2000; 874(2): 200-9.
[http://dx.doi.org/10.1016/S0006-8993(00)02581-6] [PMID: 10960605]

[55]   Tadie M, Gaviria M, Mathe J-F, Menthonnex P, Loubert G, Lagarrigue J, *et al.* Early care and treatment with a neuroprotective drug, gacyclidine, in patients with acute spinal cord injury. Rachis 2003; 15(6): 363-76.

[56]   Young W. Review of lithium effects on brain and blood. Cell Transplant 2009; 18(9): 951-75.
[http://dx.doi.org/10.3727/096368909X471251] [PMID: 19523343]

[57]   Su H, Zhang W, Guo J, Guo A, Yuan Q, Wu W. Lithium enhances the neuronal differentiation of neural progenitor cells *in vitro* and after transplantation into the avulsed ventral horn of adult rats through the secretion of brain-derived neurotrophic factor. J Neurochem 2009; 108(6): 1385-98.
[http://dx.doi.org/10.1111/j.1471-4159.2009.05902.x] [PMID: 19183259]

[58]   Su H, Chu TH, Wu W. Lithium enhances proliferation and neuronal differentiation of neural progenitor cells *in vitro* and after transplantation into the adult rat spinal cord. Exp Neurol 2007; 206(2): 296-307.
[http://dx.doi.org/10.1016/j.expneurol.2007.05.018] [PMID: 17599835]

[59]   Wong YW, Tam S, So KF, *et al.* A three-month, open-label, single-arm trial evaluating the safety and pharmacokinetics of oral lithium in patients with chronic spinal cord injury. Spinal Cord 2011; 49(1): 94-8.
[http://dx.doi.org/10.1038/sc.2010.69] [PMID: 20531359]

[60]   Yang ML, Li JJ, So KF, *et al.* Efficacy and safety of lithium carbonate treatment of chronic spinal cord injuries: a double-blind, randomized, placebo-controlled clinical trial. Spinal Cord 2012; 50(2): 141-6.
[http://dx.doi.org/10.1038/sc.2011.126] [PMID: 22105463]

[61]   Cunliffe WJ. Minocycline for acne. Doctors should not change the way they prescribe for acne. BMJ 1996; 312(7038): 1101.
[http://dx.doi.org/10.1136/bmj.312.7038.1101a] [PMID: 8616440]

[62]   Yong VW, Wells J, Giuliani F, Casha S, Power C, Metz LM. The promise of minocycline in neurology. Lancet Neurol 2004; 3(12): 744-51.
[http://dx.doi.org/10.1016/S1474-4422(04)00937-8] [PMID: 15556807]

[63]   Stirling DP, Koochesfahani KM, Steeves JD, Tetzlaff W. Minocycline as a neuroprotective agent. Neuroscientist 2005; 11(4): 308-22.
[http://dx.doi.org/10.1177/1073858405275175] [PMID: 16061518]

[64]   Kobayashi K, Imagama S, Ohgomori T, *et al.* Minocycline selectively inhibits M1 polarization of microglia. Cell Death Dis 2013; 4: e525.
[http://dx.doi.org/10.1038/cddis.2013.54] [PMID: 23470532]

[65]   Stirling DP, Khodarahmi K, Liu J, *et al.* Minocycline treatment reduces delayed oligodendrocyte death, attenuates axonal dieback, and improves functional outcome after spinal cord injury. J Neurosci 2004; 24(9): 2182-90.
[http://dx.doi.org/10.1523/JNEUROSCI.5275-03.2004] [PMID: 14999069]

[66] Teng YD, Choi H, Onario RC, *et al.* Minocycline inhibits contusion-triggered mitochondrial cytochrome c release and mitigates functional deficits after spinal cord injury. Proc Natl Acad Sci USA 2004; 101(9): 3071-6.
[http://dx.doi.org/10.1073/pnas.0306239101] [PMID: 14981254]

[67] Yune TY, Lee JY, Jung GY, *et al.* Minocycline alleviates death of oligodendrocytes by inhibiting pro-nerve growth factor production in microglia after spinal cord injury. J Neurosci 2007; 27(29): 7751-61.
[http://dx.doi.org/10.1523/JNEUROSCI.1661-07.2007] [PMID: 17634369]

[68] Chaudhry IB, Hallak J, Husain N, *et al.* Minocycline benefits negative symptoms in early schizophrenia: a randomised double-blind placebo-controlled clinical trial in patients on standard treatment. J Psychopharmacol (Oxford) 2012; 26(9): 1185-93.
[http://dx.doi.org/10.1177/0269881112444941] [PMID: 22526685]

[69] Gordon PH, Moore DH, Miller RG, *et al.* Efficacy of minocycline in patients with amyotrophic lateral sclerosis: a phase III randomised trial. Lancet Neurol 2007; 6(12): 1045-53.
[http://dx.doi.org/10.1016/S1474-4422(07)70270-3] [PMID: 17980667]

[70] Lampl Y, Boaz M, Gilad R, *et al.* Minocycline treatment in acute stroke: an open-label, evaluator-blinded study. Neurology 2007; 69(14): 1404-10.
[http://dx.doi.org/10.1212/01.wnl.0000277487.04281.db] [PMID: 17909152]

[71] Zhang Y, Metz LM, Yong VW, Bell RB, Yeung M, Patry DG, *et al.* Pilot study of minocycline in relapsing-remitting multiple sclerosis 2008.
[http://dx.doi.org/10.1017/S0317167100008611]

[72] Casha S, Zygun D, McGowan MD, Bains I, Yong VW, Hurlbert RJ. Results of a phase II placebo-controlled randomized trial of minocycline in acute spinal cord injury. Brain 2012; 135(Pt 4): 1224-36.
[http://dx.doi.org/10.1093/brain/aws072] [PMID: 22505632]

[73] Süzer T, Coskun E, Islekel H, Tahta K. Neuroprotective effect of magnesium on lipid peroxidation and axonal function after experimental spinal cord injury. Spinal Cord 1999; 37(7): 480-4.
[http://dx.doi.org/10.1038/sj.sc.3100874] [PMID: 10438114]

[74] Kwon BK, Roy J, Lee JH, *et al.* Magnesium chloride in a polyethylene glycol formulation as a neuroprotective therapy for acute spinal cord injury: preclinical refinement and optimization. J Neurotrauma 2009; 26(8): 1379-93.
[http://dx.doi.org/10.1089/neu.2009.0884] [PMID: 19317592]

[75] Kwon BK, Sekhon LH, Fehlings MG. Emerging repair, regeneration, and translational research advances for spinal cord injury. Spine 2010; 35(21) (Suppl.): S263-70.
[http://dx.doi.org/10.1097/BRS.0b013e3181f3286d] [PMID: 20881470]

[76] Shihabuddin LS, Ray J, Gage FH. FGF-2 is sufficient to isolate progenitors found in the adult mammalian spinal cord. Exp Neurol 1997; 148(2): 577-86.
[http://dx.doi.org/10.1006/exnr.1997.6697] [PMID: 9417834]

[77] Jin K, LaFevre-Bernt M, Sun Y, *et al.* FGF-2 promotes neurogenesis and neuroprotection and prolongs survival in a transgenic mouse model of Huntington's disease. Proc Natl Acad Sci USA 2005; 102(50): 18189-94.
[http://dx.doi.org/10.1073/pnas.0506375102] [PMID: 16326808]

[78] Kang CE, Baumann MD, Tator CH, Shoichet MS. Localized and sustained delivery of fibroblast growth factor-2 from a nanoparticle-hydrogel composite for treatment of spinal cord injury. Cells Tissues Organs (Print) 2013; 197(1): 55-63.
[http://dx.doi.org/10.1159/000339589] [PMID: 22796886]

[79] Teng YD, Mocchetti I, Taveira-DaSilva AM, Gillis RA, Wrathall JR. Basic fibroblast growth factor increases long-term survival of spinal motor neurons and improves respiratory function after experimental spinal cord injury. J Neurosci 1999; 19(16): 7037-47.
[PMID: 10436058]

[80]   Kang CE, Clarkson R, Tator CH, Yeung IW, Shoichet MS. Spinal cord blood flow and blood vessel permeability measured by dynamic computed tomography imaging in rats after localized delivery of fibroblast growth factor. J Neurotrauma 2010; 27(11): 2041-53.
[http://dx.doi.org/10.1089/neu.2010.1345] [PMID: 20799884]

[81]   Goldshmit Y, Frisca F, Pinto AR, *et al.* Fgf2 improves functional recovery-decreasing gliosis and increasing radial glia and neural progenitor cells after spinal cord injury. Brain Behav 2014; 4(2): 187-200.
[http://dx.doi.org/10.1002/brb3.172] [PMID: 24683512]

[82]   Cheng H, Cao Y, Olson L. Spinal cord repair in adult paraplegic rats: partial restoration of hind limb function. Science 1996; 273(5274): 510-3.
[http://dx.doi.org/10.1126/science.273.5274.510] [PMID: 8662542]

[83]   Wu JC, Huang WC, Chen YC, *et al.* Acidic fibroblast growth factor for repair of human spinal cord injury: a clinical trial. J Neurosurg Spine 2011; 15(3): 216-27.
[http://dx.doi.org/10.3171/2011.4.SPINE10404] [PMID: 21663406]

[84]   He Z, Koprivica V. The Nogo signaling pathway for regeneration block. Annu Rev Neurosci 2004; 27: 341-68.
[http://dx.doi.org/10.1146/annurev.neuro.27.070203.144340] [PMID: 15217336]

[85]   Dergham P, Ellezam B, Essagian C, Avedissian H, Lubell WD, McKerracher L. Rho signaling pathway targeted to promote spinal cord repair. J Neurosci 2002; 22(15): 6570-7.
[PMID: 12151536]

[86]   Dubreuil CI, Winton MJ, McKerracher L. Rho activation patterns after spinal cord injury and the role of activated Rho in apoptosis in the central nervous system. J Cell Biol 2003; 162(2): 233-43.
[http://dx.doi.org/10.1083/jcb.200301080] [PMID: 12860969]

[87]   Lord-Fontaine S, Yang F, Diep Q, *et al.* Local inhibition of Rho signaling by cell-permeable recombinant protein BA-210 prevents secondary damage and promotes functional recovery following acute spinal cord injury. J Neurotrauma 2008; 25(11): 1309-22.
[http://dx.doi.org/10.1089/neu.2008.0613] [PMID: 19061375]

[88]   Fehlings MG, Theodore N, Harrop J, *et al.* A phase I/IIa clinical trial of a recombinant Rho protein antagonist in acute spinal cord injury. J Neurotrauma 2011; 28(5): 787-96.
[http://dx.doi.org/10.1089/neu.2011.1765] [PMID: 21381984]

[89]   Welte K, Platzer E, Lu L, *et al.* Purification and biochemical characterization of human pluripotent hematopoietic colony-stimulating factor. Proc Natl Acad Sci USA 1985; 82(5): 1526-30.
[http://dx.doi.org/10.1073/pnas.82.5.1526] [PMID: 3871951]

[90]   Guo Y, Zhang H, Yang J, *et al.* Granulocyte colony-stimulating factor improves alternative activation of microglia under microenvironment of spinal cord injury. Neuroscience 2013; 238: 1-10.
[http://dx.doi.org/10.1016/j.neuroscience.2013.01.047] [PMID: 23419550]

[91]   Pitzer C, Krüger C, Plaas C, *et al.* Granulocyte-colony stimulating factor improves outcome in a mouse model of amyotrophic lateral sclerosis. Brain 2008; 131(Pt 12): 3335-47.
[http://dx.doi.org/10.1093/brain/awn243] [PMID: 18835867]

[92]   Kawabe J, Koda M, Hashimoto M, *et al.* Neuroprotective effects of granulocyte colony-stimulating factor and relationship to promotion of angiogenesis after spinal cord injury in rats: laboratory investigation. J Neurosurg Spine 2011; 15(4): 414-21.
[http://dx.doi.org/10.3171/2011.5.SPINE10421] [PMID: 21721873]

[93]   Takahashi H, Yamazaki M, Okawa A, Sakuma T, Kato K, Hashimoto M, *et al.* Neuroprotective therapy using granulocyte colony-stimulating factor for acute spinal cord injury: a phase I/IIa clinical trial. European spine journal : official publication of the European Spine Society, the European Spinal Deformity Society, and the European Section of the Cervical Spine Research Society 2012; 21(12): 2580-7.

[94]   Kamiya K, Koda M, Furuya T, Kato K, Takahashi H, Sakuma T, *et al.* Neuroprotective therapy with granulocyte colony-stimulating factor in acute spinal cord injury: a comparison with high-dose methylprednisolone as a historical control 2015.
[http://dx.doi.org/10.1007/s00586-014-3373-0]

[95]   Rosenberg LJ, Teng YD, Wrathall JR. Effects of the sodium channel blocker tetrodotoxin on acute white matter pathology after experimental contusive spinal cord injury. J Neurosci 1999; 19(14): 6122-33.
[PMID: 10407048]

[96]   Bellingham MC. A review of the neural mechanisms of action and clinical efficiency of riluzole in treating amyotrophic lateral sclerosis: what have we learned in the last decade? CNS Neurosci Ther 2011; 17(1): 4-31.
[http://dx.doi.org/10.1111/j.1755-5949.2009.00116.x] [PMID: 20236142]

[97]   Schwartz G, Fehlings MG. Evaluation of the neuroprotective effects of sodium channel blockers after spinal cord injury: improved behavioral and neuroanatomical recovery with riluzole. J Neurosurg 2001; 94(2) (Suppl.): 245-56.
[PMID: 11302627]

[98]   Palace J. Neuroprotection and repair. J Neurol Sci 2008; 265(1-2): 21-5.
[http://dx.doi.org/10.1016/j.jns.2007.08.039] [PMID: 17964604]

[99]   Wu Y, Satkunendrarajah K, Teng Y, Chow DS, Buttigieg J, Fehlings MG. Delayed post-injury administration of riluzole is neuroprotective in a preclinical rodent model of cervical spinal cord injury. J Neurotrauma 2013; 30(6): 441-52.
[http://dx.doi.org/10.1089/neu.2012.2622] [PMID: 23517137]

[100]   Tator CH, Hashimoto R, Raich A, *et al.* Translational potential of preclinical trials of neuroprotection through pharmacotherapy for spinal cord injury. J Neurosurg Spine 2012; 17(1) (Suppl.): 157-229.
[http://dx.doi.org/10.3171/2012.5.AOSPINE12116] [PMID: 22985382]

[101]   Grossman RG, Fehlings MG, Frankowski RF, *et al.* A prospective, multicenter, phase I matched-comparison group trial of safety, pharmacokinetics, and preliminary efficacy of riluzole in patients with traumatic spinal cord injury. J Neurotrauma 2014; 31(3): 239-55.
[http://dx.doi.org/10.1089/neu.2013.2969] [PMID: 23859435]

CHAPTER 2

Essential Tremor: The Current State and Challenges

Martin Kronenbuerger[1,2,*] and **Zoltan Mari**[1,3]

[1] School of Medicine, Department of Neurology, Johns Hopkins University, 600 N. Wolfe Street, Meyer 6-181B, Baltimore, MD 21287, USA

[2] Department of Neurology, University Medicine Greifswald, Fleischmannstr. 8, 17475 Greifswald, Germany

[3] Cleveland Clinic Lou Ruvo Center for Brain Health, 888 West Bonneville Ave, Las Vegas, NV 89106, USA

Abstract: Essential tremor (ET) is one of the most common movement disorders. The classification of ET and its uniform diagnosis have become subjects of intense debate. ET does not have a diagnostic marker. The diagnosis of ET depends on history, clinical examination, and exclusion of mimicking illnesses. There is strong evidence that ET is a hereditary disorder, but the gene defect is elusive. Making the correct diagnosis greatly impacts treatment. ET affects the hands or the hands in combination with head tremor. Isolated tremor elsewhere, such as of the head, tongue, lips, voice or face may be seen less commonly. Such heterogeneity sometimes pointed to discussions whether ET may be maintained as a unified disease entity. The tremor in ET is present when the muscles are rhythmically and abnormally activated, resulting in postural and/or kinetic tremor. Upon careful clinical examination, many patients with ET are found to have signs of cerebellar impairment, such as impairment of tandem stance and gait, as well as goal directed tremor. But action tremor is the predominant finding on clinical examination at any stage of ET. Other neurological deficits such as rigidity, bradykinesia or dystonia should prompt alternative diagnoses. The most important differential diagnoses of ET are enhanced physiologic tremor, idiopathic Parkinson's disease, dystonic tremor, Wilson disease, orthostatic tremor, and psychogenic tremor. Many ET patients do not require treatment as the tremor does not impair daily functioning. If treatment is required, beta-blockers and/or anticonvulsive drugs are the first treatment options. Deep Brain Stimulation is a treatment for ET patients who are impaired in their daily life because of tremor and who did no benefit from drug treatment.

Keywords: Action tremor, Alcohol Responsiveness, Anticonvulsives, Betablocker, Botulinum Toxin, Brainstem, cerebellar syndrome, Cerebellar

* **Corresponding author Martin Kronenbuerger:** School of Medicine, Johns Hopkins University, 600 N. Wolfe Street, Meyer 6-181B, Baltimore, MD 21287, USA; Tel: 410-502-0133; Fax: 410-502-6737; E-mail: martin_kronenbuerger@hotmail.com

Atta-ur-Rahman (Ed.)

tremor, Cerebellum, Deep Brain Stimulation, Definition, Epidemiology, Etiology, Gene defect, Hand tremor, Head tremor, Hereditary tremor, Kinetic tremor, Neuropathology, Parkinson's disease, Pathophysiology, Physiologic tremor, Postural tremor, Primidone, Propranolol, Psychogenic tremor, Risk Factors, Thalamus, Tremor Classification, Tremor at rest.

DEFINITON OF ESSENTIAL TREMOR

The clinical symptoms and signs of tremor define the Essential tremor (ET) syndrome, rather than an etiologic diagnosis. The most important sign of ET is action tremor, which is a progressive, mainly symmetric, rhythmic, oscillatory, and involuntary movement disorder of the hands and forearms (69% of ET patients) [1]. Throughout the course of ET, tremor remains the predominant symptom and source of impairment [2]. The tremor is usually absent at rest, but present during posture and intentional movements. Additionally, the tremor of ET can also involve voice (62% of ET patients), the head (48% of ET patients), and jaw [1, 3, 4]. In contrast, it is very unusual to find tremor of the legs or isolated head tremor in ET [5]. There are patients who have isolated tremor in one body part other than arms or head (*e.g.* voice tremor or facial tremor) or task-specific tremor [5]. Whether or not these patients have a variant of ET is currently uncertain.

There is a debate as to what ET is, given conflicting clues from epidemiological, pathological, genetic and clinic studies [6]. ET is not a single entity and may be a syndrome of tremulousness with different, specific causes which lead to a distinct bilateral postural tremor. Thus, at the present time, phenomenology appears to be more important than etiology to diagnose ET. As our etiological understanding of ET is expected to grow in the future, reclassification of ET with greater role of etiology is expected.

CLASSIFICATION OF TREMOR

Clinical Definitions of the Different Types of Tremor

Rest or resting tremor is observed when the respective part of the body is resting and fully supported against gravity (*e.g.*, the patient rest his/her hands in the lap) [5, 7]. In this type of tremor, the tremor amplitude increases during distractions (*e.g.*, counting backward from 10 to 1), with other movements such as walking or movements of non-affected body parts and diminishes with target-directed movements of the affected body part (*e.g.*, finger-to-nose test) [5]. While the rate of tremor remains relatively constant, the amplitude of rest tremor may vary depending on the patient's activity [5]. Please note the fact that the examined body part (*e.g.* hand) is supported against gravity or appears to be still does NOT necessarily mean that the muscles in that body part are actually fully relaxed in the sense of full electromyography (EMG) silence. A body part that may appear to be resting could still be quite tense with actively contracted muscles. Tremor in this state could then be erroneously interpreted as resting tremor. To exclude this possibility, surface EMG confirmation of completely relaxed muscle may be needed. In Parkinson's disease (PD) resting tremor typically starts in one hand and later spreads to other limbs [5]. However, tremor in PD could begin bilaterally and as action rather than the typical unilateral rest tremor. While PD is the illness that is most commonly associated with rest tremor, rest tremor may also be found in Wilson's disease, non-Wilson's hepatocerebral degeneration, drug induced Parkinson syndrome, or midbrain injury due to stroke, trauma or demyelinating disease [8].

In contrast to rest tremor, *action tremor* is produced by voluntary limb movements leading to muscle contraction and includes tremors such as postural, isometric and kinetic tremors [5]. *Postural tremor* is observed when the respective body part maintains a position against gravity (*e.g.*, holding the upper limbs in front of the chest), but is not actively moving. *Isometric tremor* are observed when muscles contract against stationary objects (*e.g.*, grasping the examiner's fingers) [5]. *Kinetic tremor* is seen with voluntary movement and is subdivided as either simple kinetic tremor or intention tremor [5]. *Simple kinetic tremor* is associated with movements of extremities that are not goal directed (*e.g.*, pronation-supination or flexion-extension wrist movements) [5]. *Intention tremor or target-oriented tremor* is typically seen during visually guided movement toward a target (*e.g.*, finger-to-nose or finger-to-finger testing), with a vast amplitude fluctuation on approaching the target [5]. That is, the tremor amplitude typically increases as a function of proximity to the target. Intention tremor is therefore also labeled as goal directed kinetic tremor, while simple kinetic tremor as non-goal directed tremor [5]. *Global tremor* occurs when the tremor is seen with both conditions, rest and action [5]. Global tremor may be develop in the advanced stages of either (originally) resting tremor disorders such as PD, or action tremor disorders such as ET [5]. For instance, as ET progresses, it may increasingly produce more severe tremor affecting every condition including rest and action [5, 7]. Other examples of global tremor are Holmes' tremor (also known as rubral tremor) or psychogenic tremor [5].

Syndromes and Diseases Presenting with Action Tremor

Action tremor (postural and kinetic tremor) may be the largest group of tremor and therefore physicians in many different specialties and subspecialties will see this type of tremor among their patients. Please see Fig. (1) for an overview.

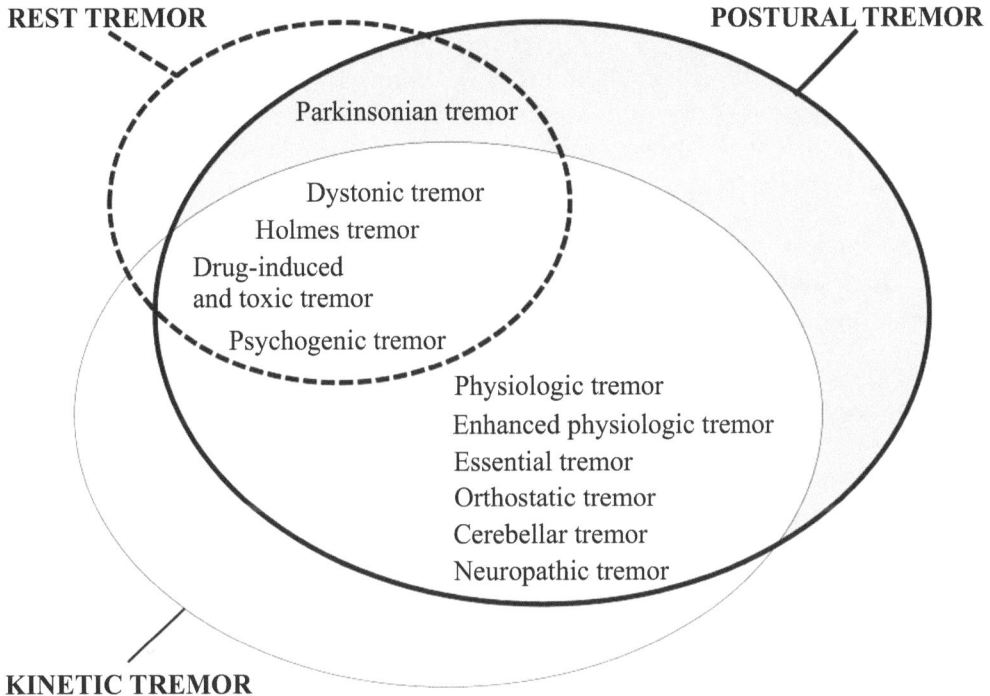

Fig. (1). Overview of the various tremor subtypes [5].

Tremor in ET is a usually bilateral, often symmetric or minimally asymmetric postural and/or kinetic tremor involving hands and forearms that is visible and persistent. It may also include tremor of the head, voice, and/or lips in addition to tremor of the upper limbs [5, 7]. Opinions differ whether isolated tremor of one hand or the head may be consistent with ET.

Physiologic tremor is present in any healthy human subject. It is a small amplitude, high frequency action tremor that is usually not seen under ordinary conditions [5]. *Enhanced or exaggerated physiologic tremor* may be seen equally in the outstretched arms and legs [5]. The underlying cause is believed to be the abnormal activation of the myotatic reflex loop from oversensitive stretch receptors. Increased sensitivity of stretch receptors may be caused by sympathetic activity such as thyrotoxicosis, hypoglycemia, use of certain drugs, or withdrawal from extended use of alcohol or benzodiazepines [8]. Additionally, caffeine, tobacco withdrawal, and exercise induced tremors are usually of this kind [8]. Because of its high prevalence, it is more commonly seen by primary care physicians than Movement Disorder specialists.

Tremor in dystonia is often irregular and jerky rather than regular and oscillatory. Thus, there is an ongoing discussion if this type of movement disorder is a true tremor, as it is not purely rhythmic [5]. Familiar myoclonus-dystonia is a syndrome characterized by myoclonus and dystonia affecting the axial muscles and arms. The myoclonic jerks may appear rhythmic like a tremor and can be confused with action tremor [9].

Tremor in Wilson's disease is often characterized by global (action and resting) tremor. Other neurologic abnormalities in Wilson's disease may be dystonia, dysarthria, and Parkinsonism [5].

The typical *tremor in cerebellar disorders* is intention tremor, often seen together with other cerebellar deficits such as ataxia and dysarthria [5].

Primary writing tremor is a specific type of action tremor, in which the forearm causes a pronation or supination tremor during writing [10]. In a subgroup of patients, the tremor is also present when the patients try to hold their hand in a position that is usually used for writing [10]. Over the course of the disease, the tremor may progress to the other side and there may be a slowing of movements [10]. In contrast, in ET the tremor is present independently of the task [10].

Holmes' tremor (or rubral) is typically seen with focal midbrain injuries such as strokes or traumatic brain injuries [5]. Holmes' tremor is characterized by global (rest and action) tremor [5]. In most cases the tremor in Holmes' tremor has a lower tremor frequency than is typically seen in PD or ET [5].

Orthostatic tremor is a rare condition that usually does not produce visible tremor [5]. However, patients report inability to stand still [5]. The uniquely high frequency (14-16 Hz) tremor which usually affects the legs is best proven by surface EMG recording from the quadriceps femoris muscle and fast Fourier transform analysis showing a characteristic peak around 15 Hz [5].

Neuropathic tremor develops in patients with a peripheral neuropathy [11]. Thus, sensory deficits in a patient with tremor should raise suspicion about the presence of this type of tremor [11]. The tremor presents as a postural and kinetic tremor with mid to low frequency (3-6 Hz) at the limbs [11]. For unclear reasons, only a small percentage of patients with severe neuropathies of the peripheral nervous system develop this type of tremor [11]. In contrast to other forms of tremor, which evolve over several months to years, this neuropathic tremor develops within weeks to a few months [11]. Especially demyelinating neuropathies including Guillain-Barre syndrome, and chronic inflammatory demyelinating polyneuropathy (CIDP) cause such a tremor and therefore the approach to treatment is very different from other tremor syndromes [11]. That is, the underlying disease is the primary target of the treatment [11]. It is not entirely understood why only a small proportion of neuropathic patients have tremor, while the majority will not have tremor. This observation may question neuropathic tremor's specificity and the underlying mechanism of action.

Psychogenic tremor is one of the more commonly seen tremors in movement disorder specialty clinics [12]. It typically presents as action tremor, but also could be resting or global tremor [5]. The main examination features that distinguish this tremor from organic tremors are distractibility, suggestibility, and entrainability [5]. It is very important to note that psychogenic tremor may be simultaneously present with organic tremors [5]. Therefore, it is critical to never dismiss the possibility of coexisting organic disease in patients with clinically evident psychogenic tremor and *vice versa*.

Tremor in Parkinson disease may present with resting and/or action tremor [5], with resting tremor more commonly seen. However, there is also rigidity, bradykinesia, and postural instability. Classically, PD presents with "pill-rolling" rest tremor, yet "re-emergent" postural tremor is also often seen [5]. As opposed to psychogenic tremor, distraction often enhances tremor in PD [5].

Essential Tremor: A Multi-Symptomatic Disorder

While ET typically is characterized by postural tremor, resting tremor has occasionally been reported [13, 14]. For instance, two studies found that about one fifth of ET patients had a tremor at rest, and these patients had a more longstanding and more severe disease than those without rest tremor [13, 14]. Thus, rest tremor may be part of a global tremor in ET as outlined above. It is critical to apply careful observations when characterizing tremor: what may appear "resting" could well be the result of incomplete relaxation of the limb despite its appearance of rest. Therefore, surface EMG confirmation of lack of muscle activity may be necessary to confirm true rest.

While ET is identified by tremor and that is vastly the main source of disability, other motor deficits have also been described in patients with ET. For instance, patients with ET have balance impairment affecting stance and gait [2]. That is, ET patients can have a gait pattern that is similar to patients with cerebellar disease. Most importantly, they make more missteps with tandem gait [2, 15] and ET patients exhibit an increased sway and have more falls when performing challenging stance tests such as tandem stance than matched healthy controls [2, 16].

There are subtle eye movement abnormalities in ET. Compared to healthy controls, ET patients (especially those with intention tremor) have abnormal 'otolith dumping' and impairments of their smooth pursuit as revealed by special technique [17]. However, on routine clinical neuro-ophthalmological examination ET patients do not show specific oculomotor abnormalities [17].

Besides voice tremor that can be found in a substantial proportion of this population, ET patients can have dysarthria. For instance, ET patients had prolonged syllable repetition times in a syllable repetition task than did controls [2, 18], a finding similar to patients with dysarthria due to cerebellar disease such as cerebellar stroke or hereditary ataxia [19, 20].

Moreover, there is evidence for impaired motor learning in ET patients. ET patient showed a lower learning rate than healthy controls as assessed by eye blink conditioning paradigms [21, 22].

In addition to non-tremor motor deficits, there is growing evidence for non-motor deficits in ET patients. There are reports of cognitive difficulties (*i.e.*, verbal, memory, and mental adaptiveness) and personality changes in ET patients [23 - 26]. Most importantly, diminished verbal fluency was found [27 - 29]. Furthermore, there are olfactory impairments [30, 31] and hearing impairments [32 - 34].

These non-motor deficits are subtle and in any stage of ET, tremor is the predominant symptom of ET as outlined above [11]. There is little consensus regarding the clinical significance and the degree of any disability stemming from symptoms and abnormal findings other than tremor.

EPDEMIOLOGY OF ESSENTIAL TREMOR

As there is no established diagnostic marker for it, there is a vast variability of the reported prevalence of ET [35]. However, a recent meta-analysis of 28 population-based prevalence studies revealed an overall prevalence over all age groups assessed of 0.4 to 0.9% [36].

Prevalence of ET increases with age. While ET can rarely be found in children, 4.6% of people older than 65 years were found to have ET and in people older 95 years, 21.7% have ET [36]. Therefore age is the risk-factor for ET most commonly accepted [36].

Many patients with onset of ET in young age have a family history of ET, which shows how important the genetics are [1, 37, 38]. Linkage studies revealed abnormalities in the ETMG (essential tremor monogenetic) locus 2 or 3, the FUS gene (fused in sarcoma), LINGO1 gene (leucine-rich repeat and Ig domain containing 1), or the SLC1A2 gene [39, 40]. Reproduction of these findings is warranted. At present, there is no commercially genetic testing for ET available.

A slightly higher prevalence of ET has been found among men (men: women = 1.08:1) [36]. Consistent with such a slight difference between the sexes, two-thirds of the studies did not reveal a gender difference in ET, while one-third indicated a higher prevalence of ET among men than women [36]. In this latter one-third of studies, the median ration between male and female ET patients was 1.65:1 [36]. Because of the small difference of the prevalence of ET between men and women, gender may not be considered as a proven risk factor for ET [36]. With regards to ethnicity, current available data suggest the highest prevalence of ET is in Caucasians and is lowest in African-Americans, but more data are needed to make a clear assumption because of differing study designs and definitions of ET used in these studies [36, 41, 42]. There are several factors that appear to be associated with tremor severity [43]. These are, tremor of the voice, tremor is more severe on one body side or is present on one body side only, an advanced age, and a longer duration disease [43]. The use of movement disorders drugs was also found to be associated with a higher tremor severity, but this could be the consequence rather than the cause of a more severe tremor [43].

ESSENTIAL TREMOR VARIANTS

The phenotypical heterogeneity of patients diagnosed with ET has long been perplexing clinicians and academicians alike. Most of the patients who are considered to have ET have a postural tremor in the arms and many of them have tremor of the head and voice in addition to tremor of the arms. This phenotype has been called "classic essential tremor" [5]. However, ET is known to be a disorder with a vast variability in temporal course, tremor character and association with other neurological deficits [6]. Additionally, some ET run in families, others are sporadic. There has been a recently intensifying debate whether ET is a clinically monosymptomatic disorder (as we discussed in the previous section), or ET can be maintained as a disease entity at all *versus* being a syndrome [6]. While we do not believe that this chapter is the appropriate forum to tackle such a grand scale debate, we do believe it is important to include some current perspectives relevant to it. It is important to address these perspectives as methodical problems partially stemming from diagnostic challenges and the lack of disease homogeneity, which -*as a consequence*- impede ET clinical trials and the successful development of better therapies of ET patients [44].

ET-PD SYNDROME

There is an ongoing discussion if ET and PD are related disorders or not. Besides both being age-related, there are several commonalities and other observations suggesting relatedness between ET and PD [45]. Population based studies suggest that there is at least a three and a half folds higher incidence of PD in ET compared to controls, and the risk of ET is increased for relatives of patients with PD [45]. Additionally, 20% of ET patients have a rest tremor which is characteristic for PD [13, 14]. PD patients can also present with action tremor, and both PD patients and ET patient have prolonged motor initiation compared to healthy controls [46]. Moreover, one study found mild striatal dopamine transporter loss as seen on a SPECT study in some patients with ET and a typical PD pattern of uptake loss [47]. Patients with tremor-dominant PD may benefit from medications (such as beta-blockers or primidone) that are traditionally used for the treatment of tremor in ET [48]. One could argue that there is no link between ET and PD because neuropathological studies did not find an elevated number of Lewy bodies in the substantia nigra in ET [49, 50], ET patients –*as a group*- do not have the neuro-chemical changes that are typically seen in PD, ET patients do not respond to Parkinson medications (such as dopaminergic medications), and neuroimaging studies such as ultrasound imaging of the substantia nigra have not revealed the changes of PD in ET patients as a group [51].

PD may present initially with postural and kinetic tremor [1]. These patients may be accidentally diagnosed as having ET and then mistakenly assumed to progress to PD, whereas they never actually had ET [1]. Older ET patients may have a small degree of bradykinesia and rigidity of the limbs as a non-specific sign of aging [1].

One way to resolve this apparent controversy is to suggest that there are patients with signs that are typical for ET and that there is a separate group of patients with an ET-PD syndrome (that is, patient with both illnesses). This is based on the observation that ET and PD are prevalent diseases and they may coexist in one patient [52]. Patients with typical ET are patients with predominantly action tremor (postural and kinetic tremor), "non-reemergent postural tremor" (that is, tremor which appears with no latency as soon as the arm muscles are activated), and subtle signs of cerebellar impairment such as tandem gait impairment [2, 15]. Separate from that are patients with an ET-PD syndrome who have a hypokinetic-rigid syndrome (that is, rigidity, bradykinesia, and postural instability), and new onset of rest tremor with previous action tremor, prominent asymmetry of symptoms and/or jaw or lip tremor [52].

ET PLUS DYSTONIA SUBTYPE

A recent review article suggests that clinical and neurophysiologic features of tremor in dystonia differ from those seen in ET [53]. In particular, in ET the tremor is regular in frequency and amplitude, while tremor in dystonia is irregular and varies in amplitude as well as presence based of position [53]. However, there are patients with dystonia who exhibit a tremor which appears similar to ET. There are two scenarios: (i) patients with clinical ET who develop dystonia years after onset of tremor and (ii) there are patients with an onset of dystonia at the same time as the tremor starts [54]. In either case, if the patient's tremor is relatively small in tremor amplitude and the tremor is irregular in the presence of a dystonia in a given patient, the tremor should be classified as dystonic tremor and not ET [55]. If the tremor in patients with dystonia meets the criteria of ET, these patients may be classified as "ET plus dystonia" as an ET subtype or they may be classified as "dystonia plus ET like arm tremor" as a separate syndrome [56].

In the context of ET *versus* cervical dystonia (CD) a long debated area of controversy is isolated head-neck tremor. We lack diagnostic markers for ET or CD and thus the debate on how to diagnose isolated head-neck tremor clinically between these 2 conditions remains academic, largely arbitrary, and perhaps fruitless. Some argue that CD should be diagnosed when the head-neck tremor is also associated with (1) some degree of asymmetry; and/or (2) head-neck posturing; and/or (3) cervical discomfort or pain; and/or (4) when the tremor is not entirely regular/rhythmic ("dystonic tremor"). Others argue that even in the absence of those phenotypical features isolated head-neck tremor, in the absence of tremor in any other body part (such as hand or larynx), should be diagnosed as CD. Family history, ethanol responsiveness, and even treatment response are generally not considered as useful distinguishing features due to great overlap between ET and CD. The long-standing debate regarding isolated head-neck tremor well illustrates the clinical heterogeneity of each disease, the arbitrariness of clinical diagnosis, and the need for diagnostic biomarkers.

Cerebellar Subtype of ET

There is a debate about the classification of ET into the "intention tremor" (or cerebellar) subtype and postural tremor subtype. In addition to such classification based on clinical grounds (*i.e.* the different characteristics of the main tremor component and the presence of associated cerebellar features), there are neurophysiological correlates [57] of such classifications.

Genetic Subtypes of ET

As noted above, past genetic studies in ET did not reveal a gene defect and universal genetic marker that may be used for diagnosing ET. In the future, with the presence of reliable gene testing in ET, the group of ET patients may get subdivided according to the respective genetic abnormality. At present, one may subdivide ET patients in (i) a hereditary ET if at least one relative is affected by

ET and the tremor started before age 65, (ii) sporadic ET if there is none in the immediate family with ET and if the tremor started before age 65, and (iii) senile ET if the tremor started after age 65 no matter if another family member has ET or not [58]. It is also possible that there are many different susceptibility genes or a combination of multiple genes that can play a role in ET etiopathogenesis, perhaps alongside with environmental factors or even in interaction with one another and/or environmental factors.

ETIOLOGY AND ENVIORNMENTAL FACTORS IN ESSENTIAL TREMOR

The etiology of ET is unknown. As outlined above, there is robust evidence of aging being a risk factor. How age increases the risk of ET is unknown. Additionally, genetics are important but further research is needed to reveal the precise genetic abnormality or abnormalities [40]. Moreover, studies in monozygotic twins showed that only in 60 to 63% of twins both are affected by ET [59, 60]. This suggests additional contributors being involved in the etiology of ET. Furthermore, the onset and severity of ET varies considerably between family members with ET [61, 62].

Comparable to other neurological illnesses such as PD, Alzheimer's dementia and Amyotrophic Lateral Sclerosis there is interest if environmental factors and life style habits are contributing or protective factors in the etiology of ET. Furthermore, there are many patients who have sporadic ET. That is, about half of ET patients do not have a family member with ET [63] and it is therefore of interest why they developed ET.

Toxins such as pesticides, lead, mercury, and beta-carboline alkaloids induce action tremor similar to ET [64]. Therefore, past research assessed if patients with ET had been exposed to these substances.

Pesticides

A case-control analysis with 142 ET subjects and 284 healthy subjects suggested that certain types of agricultural work (including the exposure to pesticides) and exposure to frosted glass are environmental factors associated with ET [37]. In contrast, two studies [65, 66] involving 136 and 79 ET patients, respectively, compared to 144 and 100 healthy controls, respectively, did not show that ET patients and healthy controls differ in this regard. Additionally, in one of these studies the concentration of organochlorine pesticides was measured and no difference was found between people with ET and controls [65]. Thus, exposure to pesticides may not be a proven contributing factor to ET.

Harmane

Harmane, found in meats such as chicken or pork, is a potent tremor-producing substance [63]. Therefore studies have been done to assess the role of harmane in the development of ET [67 - 69]. Blood concentration of harmane was elevated in hundred ET patients compared to hundred healthy subjects [67]. Additionally, magnetic resonance spectroscopic imaging showed cerebellar neuronal damage in relation to blood harmane concentration involving 12 ET patients [68]. On the other hand, a survey involving a detailed meat consumption questionnaire showed that "total meat consumption" was only higher in male patients with ET than in males without ET, but there was no correlation between meat consumption and presence of ET in women [69]. The specificity of a correlation between harmane ingestion and ET is made even less clear as its concentration was also elevated in patients with PD [70].

Lead

Like harmane, exposure to lead causes action tremor [71] and cerebellar impairment [72, 73]. Two studies found elevated lead concentrations in ET patients compared to controls [74, 75]. In contrast, a different research project which assessed exposure to environmental products containing lead did not reveal an association between ET and lead [37]. Therefore, no clear conclusion regarding lead and ET can be made at present.

Caffeine

There are reports on worsening of ET after caffeine intake [76, 77] and caffeine consumption does increase physiologic tremor, causing healthy subjects to have noticeable to functionally impairing tremor very similar to tremor in ET [78]. Thus, there were concerns that caffeine consumption may trigger the development of ET. A study involving 79 ET patients and 100 controls did not reveal any association between caffeine consumption and ET [66]. As a result, if someone with ET has the impression that caffeine worsens the tremor s/he may not consume it. On the other hand, people with ET do not necessarily need to abandon caffeine if they consume a reasonable amount of it and if their tremor is reasonably controlled with medication or surgical treatment.

Nicotine

Nicotine has shown to have protective properties against neuro-degeneration in the laboratory setting [79]. Several studies have shown a small, but statistically significant protective effect of nicotine on PD [80]. A few studies have looked if nicotine reduces the risk for ET. One study revealed that people who were smoking tobacco at some point in the past ("ever smokers") had a 50% higher likelihood to have ET opposed to people who never smoked tobacco ("never smokers") [81]. In contrast, two studies suggest that nicotine is not associated with ET [37, 66]. If there would be a small protective effect of tobacco use in ET, adverse effects of nicotine would not justify its use in ET.

Ethanol

A large number of people with ET have a reduction of tremor after using small amounts of ethanol (*i.e.* one glass of wine or beer) [82 - 84]. Additionally, when tested systematically, almost all ET patients have a reduction of tremor after ethanol ingestion [85]. Moreover, ethanol ingestion also improved tandem gait impairment in ET patient [86]. The effect of ethanol on ET may be related to a reduction of an overactivity of the deep cerebellar nuclei [83] as part of the cerebello-thalamo-cortical axis, which is believed to be involved in tremorogenesis in ET. The positive effect of ethanol is not limited to ET, as many other movement disorders improve after ethanol ingestions such as primary focal dystonia, task specific tremor or myoclonus-dystonia [83]. Thus, a positive effect of ethanol does not necessarily help to separate different movement disorders.

As chronic ethanol use is known to cause cerebellar impairment and the cerebellum is critically involved in the disordered function underlying ET, there are concerns that ethanol may trigger the development of ET [63]. Several large epidemiological research projects clearly indicated that ethanol intake is not a risk factor for getting ET [37, 63, 66, 87]. However, one study suggests that the chronic intake of very high doses of ethanol per day (more than 5 drinks per day) may promote the development of ET [88]. This is not surprising, considering the well-known cerebellotoxic effects of alcoholism. Ethanol's direct neurotoxic effect has been recently questioned and ethanol's mechanism of action in the cerebellum is not fully understood [89]. Also, a different study found that moderate wine drinking has a protective role for ET, possibly because of neuroprotective properties of its antioxidant components in the long run [87].

Since ethanol ingestion improves tremor in ET, there are concerns that ET patients may be at a higher risk to develop alcohol dependence throughout the course of ET. Additionally, movement disorder specialists recall anecdotal cases where patients used alcohol to self-medicate their tremor. There are many case-control studies that have assessed alcohol use in ET patients; some of them involved more than 2700 subjects [83]. None found a statistically increased risk for alcohol abuse in ET patients compared to controls. There is one observation in ET patients that may help to explain why there is no increased risk for alcohol abuse in ET patient. This may be the "tremor rebound" phenomenon after alcohol use that many ET patients describe [84, 90, 91]. That is, after drinking alcohol, the tremor first improves for hours, before the tremor is getting temporarily worse than before alcohol ingestion ("rebound tremor").

PATHOPHYSIOLOGY: THE "ESSENTIALS" OF ESSENTIAL TREMOR

The name "essential tremor" was introduced in 1847 by the Italian physician Dr. Alfredo Burresi because the cause of ET was completely unknown at that time [92]. We are still far away from completely understanding the disease, but we have some clues of where the tremor in ET is coming from. At present there are three different hypotheses aim to explain the pathophysiology of ET.

Oscillatory Network Hypothesis

There is agreement among experts that ET is centrally generated, which is in contrast to enhanced physiological tremor that looks very similar to ET but is caused by peripheral processes. A study of the coherence of tremor in different extremities of ET patients suggests that there are multiple cerebral oscillators, which are responsible for the observed rhythmic activity [93]. Additionally, these pacemakers interact with each other and some of the pacemakers become more or less active over time in the generation of tremor [94].

Very different findings give rise to the hypothesis that different spots in the brain are critical for tremor in ET to develop [95]. There are reports on "therapeutic strokes" in ET, where a single stroke in a single patient leads to a disappearance of tremor in ET [95]. The strokes were located along the olivo-cerebello-thalamc-cortical axis. Interestingly, a cerebellar stroke may also abolish the efficacy of thalamic stimulation [57].

Different observations point to the involvement of the inferior olive in the pathophysiology of ET [21]. Harmaline enhances the rhythmic bursting activity of neurons in the inferior olive which leads to an action tremor very similar to ET [96]. A PET study found an elevated metabolism for glucose in the inferior olive [97]. Lesions involving the Guillain-Mollaret triangle (which includes the inferior olive) lead to abnormal oscillation in the inferior olive that causes a tremor of the cranial nerves similar to limb tremor in ET [98]. Involvement of the inferior olive has been challenged, because increased glucose metabolism in the inferior olive were not found in other PET and MRI studies of ET [99 - 101] and a postmortem tissue study did not find any microscopic neural or glial changes in the inferior olive in ET [102].

Involvement of the cerebellum in the generation of tremor in ET has been shown in various imaging studies [101] in particular those of tremor-related cerebellar activation [99, 103] and morphological changes of the cerebellum in ET [104 - 108]. Also, oscillatory activity in the cerebellum in ET was found by coherent source analysis including simultaneously recorded magnetoencephalography (MEG) and peripheral tremor [56]. Additionally there are clear-cut signs of cerebellar impairment in ET such as intention tremor [109], gait [15] and stance impairment [2, 16], eye movement abnormalities [17] and motor learning deficits [21, 22].

Furthermore, neuropathological studies found decreased Purkinje cells and increased torpedoes (the remains of degenerated Purkinje cells) in the cerebellum [49, 50, 110], but there is a debate if these neuropathological changes are due to neurodegeneration or secondary to widespread oscillatory activity [58]. Likewise, there is a debate regarding the precise contribution of the cerebellum in the generation of tremor [58]. That is, it is still unclear whether cerebellar impairment is the consequence or the cause of tremor [58].

The main efferent projection target of the cerebellum is the ventrolateral thalamus [111]. Tremor related activity was indeed found in this thalamic area [57, 112]. The most important, clinically relevant argument of thalamic involvement in tremor of ET is the improvement of tremor with stereotactic procedures of the thalamus such as Deep Brain Stimulation (DBS) or thalamotomy [113].

The ventrolateral thalamus projects to the motor cortex, which has shown to be additionally involved in the network responsible for tremor generation. For instance, MEG and electroencephalography (EEG) studies showed cortical activity to be coherent with the tremor in ET [114 - 116], indicating rhythmic cortical output to the muscles [117, 118] which confirms the role of the cortex in tremor generation in ET [93]. Another piece of evidence for motor cortex involvement in ET is the improvement of tremor with electrical stimulation of the motor cortex in patients with ET [119].

GABA HYPOTHESIS

There is evidence that suggests a decrease of *gamma*-aminobutyric acid (GABA) in the pathophysiology of ET [120]. In the cerebrospinal fluid of patients with ET, reduced levels of GABA were found [120]. Medication such as primidone, topiramate and gabapentin that modulate GABA are used to treat tremor in ET [121]. Animal models such as the Harmaline model causes tremor by modulating GABAergic transmission in the inferior olive [122] and GABA receptor knockout mice exhibit tremor as seen in ET [123]. Additionally, imaging studies in patients with ET revealed altered binding of the GABA receptor [101]. Yet, diminished density of GABA receptors was only described in the dentate nucleus of the cerebellum [124]. Furthermore, a lower concentration of parvalbumin, a

GABAergic interneuron marker, was found in the brainstem but not in the cerebellum [50]. Studies assessing GABA receptor and transporter polymorphism in ET were negative [125, 126]. Although the GABA hypothesis in ET appears to be an important hypothesis to explain the pathophysiology of ET, more independent studies are needed to confirm present findings [127].

Neurodegenerative Hypothesis

Neuropathological assessment of brain tissue from ET patients did not reveal disease specific findings until 2005. The findings from the Essential Tremor Centralized Brain Repository initiative have led to a new discussion of neuropathological changes specific to ET. It is currently unclear whether these changes found are due to a neurodegenerative process that leads to tremor, or if tremor with its oscillatory brain activity causes these tissue changes [58]. Arguments that ET is a neurodegenerative process are (i) ET starts insidiously, (ii) it is slowly progressive, and (iii) it is associated with age and other degenerative diseases of the brain like PD or Alzheimer's dementia [128]. It is of note that there is no consensus regarding a disease stage (or duration) specific set of progressive pathological correlates.

One of the largest neuropathological studies of ET performed so far was conducted by Louis and associates which involved brains from 33 ET patients and 21 controls [110]. This study found a reduction of Purkinje cells by 25% in the cerebellum of ET patients. Additionally about 25% of the ET patients examined had Lewy bodies in the locus coeruleus. Several other publications were generated from this brain bank. One independent study replicated the findings of reduced Purkinje cells in the cerebellum of ET [50] while another did not [49]. An interesting finding from the "Essential Tremor Centralized Brain Repository" is that torpedoes (oval-shaped bulges of the axons of the Purkinje cell [129]), were elevated in different degenerative diseases of the brain such as Alzheimer's dementia, PD and in patients with action tremor and dystonia. Although the number of torpedoes was the highest in ET patients [130, 131], the specificity of these findings remains disputed. Further independent studies are warrant to support the neurodegenerative nature of ET [58]. For instance, if a neurodegenerative process underlies ET, a reduction of cerebellar Purkinje cells should be found in patients that are in the early stages of the disease [58]. Furthermore, younger patients must be assessed to rule out none-specific age-associated tissue changes [6]. Also, there have been repeat calls for independent and blinded confirmation of pathological changes in an ET sample.

Although there is a consensus that the thalamus is playing an important role in the pathophysiology of ET, knowledge regarding tissue changes in this part of the brain ET patients is very limited [58]. Neuropathological reports on postmortem studies in ET patients treated with thalamic DBS did not report any disease specific changes [132].

DIAGNOSTIC APPROACH

There are no tests to specifically diagnose ET [133 - 135]. The diagnosis is primarily based on the history provided by the patient and findings on clinical exam (see Table **1**) as well as the exclusion of mimicking diseases [133 - 135]. While it is believed that 30 to 40% of ET patient may be misdiagnosed, due to the lack of a diagnostic biomarker, *in vivo* or even on autopsy, there will be uncertainty regarding diagnostic errors. Clinical diagnostic errors occur in both directions [133 - 135].

Table 1. Diagnostic criteria for ET.

Inclusion criteria	Exclusion criteria
Bilateral, largely symmetric postural or kinetic tremor of hand and forearms but not at rest	Other neurological signs (*i.e.* dystonia)
Additional head tremor may occur	Causes of enhanced physiologic tremor
Absence of other neurological signs	Evidence for psychogenic tremor
	Sudden onset or stepwise deterioration
	Orthostatic tremor
	Isolated voice, tongue, chin, leg tremor
	Isolated position-specific or task-specific tremor

Data modified from [5].

History

The important elements in the history are age and location of tremor onset, course of progression as well as history of tremor and other movement disorders in the family (see Table **2**). In particular the history should include typical features of tremor that the patients perceive (*i.e.* "is the symptom rhythmic like the ticking of a clock", or "are the hands shaking when they are resting on the lap or when moving").

Table 2. Key elements of the history.

Age of tremor onset
Location of tremor onset
Course of tremor progression

(Table 2) contd.....

Age of tremor onset
Other impairments/concerns besides tremor
Family History of tremor and other movement disorders
Complete past medical and surgical history
Medication/drug history
Ethanol responsivity*

*Controversial (see text for details).

The questions regarding ETOH responsiveness of tremor may be asked in the clinical interview but alcohol responsiveness of tremor is not at all specific to tremor in ET [83, 84]. Additionally, many patients have not tried alcohol and can therefore not tell if alcohol improves their tremor.

Another important point besides the past medical history is current and past drug treatment to exclude enhanced physiological tremor, which can be clinically very similar to ET [5]. This is particularly important for people who are at risk of developing drug induced tremor such as elderly patients, multi-morbid patients and patients who require polypharmacy [136] (see Table **3**).

Table 3. Drugs known to cause tremor.

Antiarrhythmics
Antibiotics, Antivirals, Antimycotics
Anticonvulsives
Antipsychotics
Antiemetics
Antidepressants
Hormone supplements

Data modified from [1, 136, 137].

Drugs can lead to toxic tremor. Clinically, this is characterized by signs which point to an intoxication of the central nervous system (CNS) (*i.e.* eye movement abnormalities, gait disturbance, mental status changes) in addition to tremor [5].

Physical Exam of the Tremor

Besides the documentation of vital signs, the neurological examination should also include performance tests. It should be noted if the tremor is indeed rhythmic and whether or not the tremor amplitude changes with distraction. The main four characteristics that describe a tremor (localization, activation, frequency and amplitude) should be documented [138].

Assessment of Rest Tremor

First the patient is observed sitting in a chair [138, 139]. The patient is asked to rest the arms/hand on the armrest or the lap for at least 10 to 15 seconds. It is observed if there is a resting tremor of the any body part [138, 139]. Additionally, the patient is instructed to count back out loud or to name the months of the year to check if there is any tremor entrainment with distraction [138, 139]. A rest tremor may also appear when the patient is walking or lying on the exam table [140]. An arm, which is resting on the lap, could be still quite tense with actively contracted muscles. This could lead to an action tremor, which is mistakenly interpreted as resting tremor. To exclude this rare possibility, additional testing (such as surface EMG) may be helpful.

Assessment of Action Tremor

First postural tremor is assessed, which is typically present when the patient voluntarily maintains a position against gravity [5, 139]. The patient is instructed to stretch out the upper limbs (arms and hands) in front of his/her chest/body [5, 139]. It is important to note when the postural tremor starts: immediately as the arms are moved or after a few seconds holding the arms in this position. This helps to distinguish between the typical postural tremor in ET (which starts immediately when the arms are moved) and a re-emergent tremor as seen in PD. Next, there should be an evaluation of kinetic tremor, which includes "simple kinetic tremor" and "target directed tremor/intention tremor" [5, 139]. "Simple kinetic tremor" is observed with voluntary movements and which are not directed towards a goal (eg, tremor of the fingers or the hand while the elbow is flexed and extended or pronation/supination movements of the writs) [5]. Thereafter, the "finger-to-nose" or "finger-to-finger" maneuver is checked to look for target-directed tremor/intention tremor [5, 139]. Typically the amplitude of the tremor in ET increases with this maneuver, but may also fluctuate. The term "intention" tremor is not necessarily accurate as it has nothing to do with "mental intention" but with visually guided movement [5].

Additionally, performance tests are done such as writing a sentence (*e.g.* writing the sentence "This is an example of my handwriting"), drawing a Archimedes' spiral with at least five rounds and pouring water from one glass into another to assess action tremor while doing specific tasks [139]. There are several clinical rating scales to assess tremor in ET and its disability in daily life, which are primarily used for clinical trials [6, 141]. The "Fahn-Tolosa-Marin Scale" is the one used most frequently [141]. The "Essential Tremor Rating Assessment Scale" was recently introduced to overcome some of the limitations of the Fahn-Tolosa-

Marin Scale, such as the ceiling effect with tremor amplitudes greater than 4 cm [141].

Other Parts of the Neurological Exam

The neurological exam should evaluate the presence of other extra-pyramidal motor deficits such as dystonia, myoclonus, rigidity and bradykinesia. Moreover the physical exam should be directed to exclude a polyneuropathy or focal neurological deficit such as paresis. However, patients with ET often have subtle signs of cerebellar dysfunction such as tandem gait impairment [2, 86], dysarthria [2] or intention tremor [109] on detailed exam.

Although tremor in ET and tremor in PD differ, clinical overlap may occur on first glance. There are some hints that may help to distinguish tremor in the two diseases (Table **4**).

Table 4. Tremor characteristics and non-tremor motor symptoms in ET and Parkinson's disease.

Essential Tremor	Tremor in Parkinson's disease
Tremor at rest < action tremor	Rest tremor > action tremor
Mostly symmetric	asymmetric
postural tremor starts immediately	re-emergent tremor
Hands, head, voice	Hands, and legs
Head tremor "no-no" or multidirectional	Head tremor "yes-yes" tremor
additional non-tremor motor symptoms: gait/stance ataxia	additional non-tremor motor symptoms: hypomimia, asymmetric reduced arm swing, cogwheel rigidity, bradykinesia, impaired postural reflexes

Data modified from [5, 140, 142].

Technical Investigations/Diagnostic Workup

Although patients may present to have the typical clinical characteristics of ET, all should have their thyroid stimulating hormone (TSH) assessed to exclude enhanced physiological tremor due the impaired thyroid gland functioning [138]. Additionally, ceruloplasmin-levels should be assessed to rule out Wilson's disease in patients younger than 40 [142]. Although a considerable number of people with ET have a family history of tremor suggesting a genetic component of ET, there is no genetic test to diagnose ET at present. However, tremor in fragile X syndrome may look very similar to tremor in ET, but fragile X syndrome can be diagnosed using genetic testing [143].

Surface electromyography is helpful especially in clinically less clear cases, *e.g.* it may help to exclude enhanced physiological tremor or psychogenic tremor [5]. Another technical investigation to assess tremor in more detail is accelerometry [144]. An alternative to the classic accelerometry may be the use of the built in accelerometer of an iPhone to assess tremor at different positions (rest *versus* action), which may be helpful to more easily analyze the predominant tremor frequency in different types of tremor [145].

We suggest that ET patients should have brain Magnetic Resonance Imaging (MRI) (preferably at the time when the diagnosis is made) to exclude structural lesions causing tremor (*i.e.* vascular lesions). DaTscans can be helpful to assess presynaptic dopaminergic loss in the striatum, in particular when it is clinically difficult to differentiate a rest tremor in PD from tremor at rest in ET [5]. However, 10% of patients who have PD based on standard clinical criteria have normal DaTscans, leading to a new subgroup named as "SWEDDs (scans without evidence of dopaminergic loss)" [146].

TREATMENT OF ESSENTIAL TREMOR

At present, there is no protective or disease modifying agent proven or approved for ET [133, 134]. Additionally, life expectancy in ET is normal [147 - 149]. Thus, treatment for ET is solely symptomatic and determined by tremor related disability. Therefore, the need for prescribing any treatment in ET must be carefully evaluated in each case, determined by the patient's disability *versus* treatment related potential risks and adversities.

Only 7 to 8% of ET patients were found to have specific tremor treatment [150, 151]. There are several explanations for this perhaps surprising observation. Many ET patients are not aware that they have an illness because they think that the tremor is a family trait rather than a disease or they believe that tremor is a condition due to age [134]. For instance, epidemiological studies have shown that 85 to 93% of ET patient were unaware of their illness [150 - 153]. ET has a slow progression rate. Accordingly, ET patients may have time to modify and then gradually give up challenging activities [133]. Another reason why ET patient may not seek medical attention is that tremor in ET goes along with embarrassment as expressed by half of ET patients [154]. Many ET patients do not take medication because their tremor is functionally or psychologically not disabling and therefore do not require medication [6, 133, 134].

Additionally, a tendency to avoid harm as a personality feature in ET may a contributing factor that patient do not seek medical attention [23, 155]. This is particularly relevant when considering anti-tremor medication which has potential side effects in many cases. Quite a few patients have tried anti-tremor medication, but stopped taking because of side effects or experiencing very limited or no effect on their tremor [133].

Another dilemma of drug treatment for tremor in ET may be related to the design of drug trials. One such possible reason is that the ideal candidates for such trials were not enrolled [133]. That is, patients with longstanding ET entered the trials, who -*in contrast to de novo patients*- are more difficult to treat as they may have more pronounced changes in their brain [133]. Thus, medications that can have a meaningful effect for a subset of ET patients appear ineffective. Furthermore, a minimal meaningful effect of tremor treatment in ET needs to be defined to provide an understanding of what a new drug may offer to ET patients. Additionally, many trials did not rule out the possibility of enhanced physiologic tremor in their study subjects by simple measures such as tremor physiology studies.

The decision to start medication for tremor in ET is primarily determined by the patient's needs, as the patient's perception of quality of life impairment due to tremor varies greatly from patient to patient.

Various medications may be used for the tremor of ET, but propranolol and primidone have been found to be most effective (Table **5**).

Table 5. Drugs assessed for the treatment of tremor in ET.

effective	probably effective	possibly effective	insufficient evidence	ineffective
propranolol	topiramate	nimodipine	clonidine	trazadone
primidone	atenolol	clonazepam	gabapentin	acetazolamid
	alprazolam	metoprolol	glutethimide	amantadine
			L-tryptophane	carisbamte
			nicardine	isoniazid
			octanol	levetiracetam
			olanzapine	pinolol
			phenobarbital	3,4-diaminopyridine
			pregabaline	methazolamide
			quetiapine	mirtazapine
			theophylline	nifedipine
			tiagabine	verapamil
			zonisamide	
			oxybate	

Data modified from [121, 147].

Medications Used to Treat Tremor in ET

Beta-Adrenergic Blocker

This group of medication most likely works by blocking the muscle spindle and the deep cerebellar nuclei [156]. Over the past three decades, more than ten controlled studies assessed propranolol as a treatment option for ET [156]. About 50% of the patients will have a 50% improvement of their tremor as assessed with accelerometry and clinical rating scales [156]. Additionally, sotalol, metoprolol, and atenolol were tested, and compared with propranolol [157, 158]. These showed that these beta-blockers improved tremor but propranolol was most effective in tremor reduction [157, 158]. Side effects such as fatigue, impotence, bradycardia and lightheadedness occurred in 12 to 66% of ET patients [156]. Disadvantages of beta-adrenergic blockers include the many contraindications (*i.e.* COPD, asthma, congestive heart failure, diabetes mellitus, atrioventricular block), which are particularly present in elderly ET patients [156]. In the long run only a quarter of ET patients continued to have a good effect from beta-adrenegic blockers [159].

Anticonvulsants

The most effective drug in this group is primidone. Although metabolized to phenobarbital and other substances, primidone itself has the anti-tremor effect [134]. The precise mechanism of action by which primidone improves tremor is unknown [121]. As with propranolol, primidone improves tremor amplitude by 50% in ET patients [156]. However, primidone has many adverse effects, including sedation, drowsiness, fatigue, nausea, vomiting, ataxia, malaise, dizziness, unsteadiness, confusion, vertigo, and acute toxic reactions [156]. In contrast to propranolol, there are only a few contraindications for treatment with primidone (hypersensitivity to phenobarbital; porphyria, pregnancy, severe impairment of liver and kidney functioning). As side effects are very common in the titration phase, a slow increase was compared to a fast increase of primidone at the beginning of treatment (3x2.5 mg *versus* 25 mg per day) [160]. This study showed that a slow increase did not lead to fewer side effects than a faster increase. Another study showed that 250 mg of primidone taken per day is sufficient to improve tremor; very high doses such as 750 mg per day did not bear additional benefit [161].

Almost any new anti-seizure drug is tested for its effect on tremor in ET, but only a few such drugs are used in the routine care of patients with ET. Topiramate has been shown effective as a monotherapy or adjunct therapy to reduce tremor by about 29% while placebo improved tremor by 16% [162]. Another anticonvulsant for the treatment of tremor in ET is gabapentin. However, it appears only to be effective as a monotherapy (tremor reduction about 15% compared to placebo) [163].

Which Medication to Start With? Which Medications to Combine?

There are several different drugs that can be used to treat tremor in ET, with varying levels of effectiveness (see Table **5**). The question is which drug to try first. Propranolol and primidone appear equally effective at reducing tremor as outlined above, but is there a difference in the side effect profile?

One study compared treatment of primidone with propranolol in regard to side effects [159]. It showed that propranolol had fewer side effects than primidone in the titration phase (8% *versus* 32%), but in the long term follow up primidone caused fewer side effects than propranolol (0% *versus* 17%). Thus, the contraindications in a given patient will decide what drug should be tried first. If there are no contraindications for either drug, propranolol may be used first because of the better initial effect-side effect profile [147, 156].

Because of the many drugs than can be used to improve tremor in ET, the question arises what combination makes the most sense? As a first step the combination of primidone and propranolol makes the most sense because these drugs have been found to be most effective.

Interventional Treatments

Patients who have disabling tremor due to ET, who have tried adequate drug treatment, and are in good health are candidates for interventional treatment. The current literature specifies neither how many different drugs should be tried before considering an interventional treatment in ET nor what disabling implies [121, 147, 164, 165]. In our experience, at the very least propranolol, primidone and another drug should have been tried, keeping the contraindications in mind. Additionally, what a "disabling tremor" means depends very much on the patient: some patients are very disabled in their lives because of a relatively mild ET while others do not feel at all impaired with gross tremors. Thus, the decision regarding disability should be made on an individual basis.

Thalamotomy of the ventrolateral thalamus and the adjacent subthalamic area is a well known treatment for ET and other tremor disorders for more than a half century [113]. Because of the irreversible effects of thalamotomy in regard to side effect, DBS became widely used over the past 15 years. Another drawback of thalamotomy is that it is critical to perform thalamotomy on both sides because of adverse effects [113]. This is a particular limitation for a bilateral disease such as ET.

In regard to tremor, patients can expect a 60 to 75% improvement of their tremor by DBS [165, 166]. Side effects of DBS for tremor in ET occur in about one third of the ET patients (dysarthria 3-18%, paresthesias 6-36%, dystonia 2-9%, balance disturbance 3-8%, ataxia 6%, limb weakness 4-8%) and hardware complications such as skin erosions, infections and lead fracture in nearly 25% of patients [147]. Patient who have implants on both sides of the brain seems to be in particular prone for side effects [165].

There are reports which suggest that the brain may adapt to the DBS treatment and thus the tremor may reappear [167 - 169]. Reprogramming of the DBS [170] or on demand use of DBS may be a consideration for these rare cases [171]. Another reason for reduced efficacy of DBS on tremor in ET over time may be disease progression [172].

For patients who cannot undergo stereotactic brain surgery, thalamotomy by means of gamma knife is a consideration but outcomes and complications are less favorable than with DBS [147, 165]. A new development is thalamotomy by the use of focused ultrasound and guided by MRI for patients with ET [173, 174]. To further assess this treatment for ET, well controlled studies are warrant [173, 174].

Another interventional treatment option for tremor in ET is chemodenervation with injection of botulinum toxin into the tremulous muscles [175]. While formal studies demonstrating efficacy are lacking, many movement disorder specialists treating ET patients consider this off-label use of BoNT chemodenervation. There are two small published studies about the effects of botulinum neurotoxin treatment for hand tremor in ET [176, 177]. These revealed a mild to moderate effect on tremor, without significant functional improvement. The data were considered inadequate or conflicting and thus the recommendation for botulinum neurotoxin treatment in ET were rated as "given current knowledge, treatment is unproven" [175].

Treatment of the Non-tremor Deficits in ET

Tremor is the pre-dominant sign and source of disability in ET and therefore tremor is the focus of the treatment in ET patient. At present, knowledge about treatment for non-tremor deficits is limited [2].

Given that tremor and non-tremor deficits have a similar underlying pathophysiology in ET, one would expect that anti-tremor treatment would also affect the non-tremor deficits.

Surprisingly, there are no studies that looked at the effects of well-established drug treatments for tremor in ET such as propranolol or primidone on non-tremor deficits. There is one study that reported on the positive effects of ethanol on both tremor and gait impairment in ET [86], but ethanol is not a long term treatment option for ET.

In contrast, there are a few studies that looked at the impact of DBS on non-tremor changes in ET such as neuropsychological deficits and balance impairment [2, 178]. This may be related to the fact that DBS has immediate effects, making a DBS switch "OFF *versus* ON" comparison easy to perform [178]. Additionally, many ET patients with DBS are consistently followed from the work-up before DBS surgery and preparations for the surgical procedure to the post-surgical care for programming and monitoring of the DBS in the long term by the same team, making them accessible for such studies [178].

Findings from past neuropsychological studies in ET patients treated with DBS are inconsistent, which might be due to differing neuropsychological methods and experimental designs applied [*i.e.*, testing took place either pre- or postoperatively, or with the stimulator turned on (DBS-ON) or turned off (DBS-OFF)] [178]. Subtle impairment of verbal fluency was found 3 months post-surgery compared with pre-surgery [179, 180]. In contrast, improved verbal fluency was described in a single case report in the 3 month post-surgery follow up [181]. Interestingly, there were no group-wise declines in a one year follow-up study [182]. A more recent study comparing pre-surgical performance with DBS switched on and off at 1 year and 6 years post-surgery showed no group-wise changes [178]. Motor learning is impaired in ET patients and DBS at stimulation amplitudes up to 3 volts seem to improve this deficit [183, 184]. Thus, based on these different studies, DBS does not alter cognitive functioning in ET on a group level. But patients with disease onset after age 37 years and ET patients treated with DBS on an impulse width of DBS of more than 0.12 ms appear prone to neuropsychological decline after DBS surgery [180].

In regard to gait and stance, one study found no effects of DBS using clinical scores [185]. In contrast, two trials looked at the impact of DBS on stance in ET patients by the use of posturography method and found an improvement of stance with DBS [2, 186]. Likewise, DBS improved gait [2, 187] as revealed by detailed gait analysis technique. As a conclusion of these studies, DBS may have a beneficial effect on gait and stance performance in ET which is opposed to older, observational reports [167, 188].

DBS may impair the sense of smell [189], bladder functioning [190] and saccadic eye movements [191]. Further study is necessary to clarify these points.

Coping with ET

It is amazing how some patients with ET find ways to cope successfully with relatively gross tremors. However, there are also ET patients who do not cope well and for instance exhibit social withdrawal because of a much smaller tremor. There are no systematic studies at present to find explanations for these observations.

The International Essential Tremor Foundation (IETF) has released several tips for how to cope with ET (http://www.essentialtremor.org/coping).

CONCLUSION

ET is a highly prevalent movement disorder which can cause significant disability. The lack of diagnostic markers and questions about the heterogeneity of disease etiology and/or pathophysiology represent some of the current challenges that impede better understanding and treatment.

CONFLICT OF INEREST

The authors (editor) declare no conflict of interest, financial or otherwise.

ACKNOWLEDGEMENTS

We thank Rosalie L. Donaldson-Kronenbuerger for English language editing.

REFERENCES

[1]   Zeuner KE, Deuschl G. An update on tremors. Curr Opin Neurol 2012; 25(4): 475-82.
        [http://dx.doi.org/10.1097/WCO.0b013e3283550c7e] [PMID: 22772877]

[2]   Kronenbuerger M, Konczak J, Ziegler W, *et al.* Balance and motor speech impairment in essential tremor. Cerebellum 2009; 8(3): 389-98.
        [http://dx.doi.org/10.1007/s12311-009-0111-y] [PMID: 19452239]

[3]   Brennan KC, Jurewicz EC, Ford B, Pullman SL, Louis ED. Is essential tremor predominantly a kinetic or a postural tremor? A clinical and electrophysiological study. Mov Disord 2002; 17(2): 313-6.
        [http://dx.doi.org/10.1002/mds.10003] [PMID: 11921117]

[4]   Elble RJ. Essential tremor frequency decreases with time. Neurology 2000; 55(10): 1547-51.
        [http://dx.doi.org/10.1212/WNL.55.10.1547] [PMID: 11094112]

[5]   Deuschl G, Bain P, Brin M. Ad Hoc Scientific Committee. Consensus statement of the Movement Disorder Society on Tremor. Mov Disord 1998; 13 (Suppl. 3): 2-23.
        [http://dx.doi.org/10.1002/mds.870131303] [PMID: 9827589]

[6]   Elble RJ. What is essential tremor? Curr Neurol Neurosci Rep 2013; 13(6): 353.
        [http://dx.doi.org/10.1007/s11910-013-0353-4] [PMID: 23591755]

[7]   Louis ED. Clinical practice. Essential tremor. N Engl J Med 2001; 345(12): 887-91.
        [http://dx.doi.org/10.1056/NEJMcp010928] [PMID: 11565522]

[8]   Jankovic J, Fahn S. Physiologic and pathologic tremors. Diagnosis, mechanism, and management. Ann

Intern Med 1980; 93(3): 460-5.
[http://dx.doi.org/10.7326/0003-4819-93-3-460] [PMID: 7001967]

[9]   Raymond DO. Myoclonus-Dystonia.Source GeneReviews® Seattle, WA. Seattle: University of Washington 2003.

[10]   Rana AQ, Vaid HM. A review of primary writing tremor. Int J Neurosci 2012; 122(3): 114-8.
[http://dx.doi.org/10.3109/00207454.2011.635827] [PMID: 22050192]

[11]   Deuschl G, Raethjen J, Lindemann M, Krack P. The pathophysiology of tremor. Muscle Nerve 2001; 24(6): 716-35.
[http://dx.doi.org/10.1002/mus.1063] [PMID: 11360255]

[12]   Regragui W, Lachhab L, Razine R, Ait Benhaddou EH, Benomar A, Yahyaoui M. A clinical study of non-parkinsonian tremor in Moroccan patients. Rev Neurol (Paris) 2014; 170(1): 26-31.
[http://dx.doi.org/10.1016/j.neurol.2013.06.006] [PMID: 24321218]

[13]   Cohen O, Pullman S, Jurewicz E, Watner D, Louis ED. Rest tremor in patients with essential tremor: prevalence, clinical correlates, and electrophysiologic characteristics. Arch Neurol 2003; 60(3): 405-10.
[http://dx.doi.org/10.1001/archneur.60.3.405] [PMID: 12633153]

[14]   Louis ED. Essential tremor. Lancet Neurol 2005; 4(2): 100-10.
[http://dx.doi.org/10.1016/S1474-4422(05)00991-9] [PMID: 15664542]

[15]   Stolze H, Petersen G, Raethjen J, Wenzelburger R, Deuschl G. The gait disorder of advanced essential tremor. Brain 2001; 124(Pt 11): 2278-86.
[http://dx.doi.org/10.1093/brain/124.11.2278] [PMID: 11673328]

[16]   Bove M, Marinelli L, Avanzino L, Marchese R, Abbruzzese G. Posturographic analysis of balance control in patients with essential tremor. Mov Disord 2006; 21(2): 192-8.
[http://dx.doi.org/10.1002/mds.20696] [PMID: 16161140]

[17]   Helmchen C, Hagenow A, Miesner J, *et al.* Eye movement abnormalities in essential tremor may indicate cerebellar dysfunction. Brain 2003; 126(Pt 6): 1319-32.
[http://dx.doi.org/10.1093/brain/awg132] [PMID: 12764054]

[18]   Gamboa J, Jiménez-Jiménez FJ, Nieto A, *et al.* Acoustic voice analysis in patients with essential tremor. J Voice 1998; 12(4): 444-52.
[http://dx.doi.org/10.1016/S0892-1997(98)80053-2] [PMID: 9988031]

[19]   Ziegler W. Task-related factors in oral motor control: speech and oral diadochokinesis in dysarthria and apraxia of speech. Brain Lang 2002; 80(3): 556-75.
[http://dx.doi.org/10.1006/brln.2001.2614] [PMID: 11896657]

[20]   Ackermann H, Mathiak K, Riecker A. The contribution of the cerebellum to speech production and speech perception: clinical and functional imaging data. Cerebellum 2007; 6(3): 202-13.
[http://dx.doi.org/10.1080/14734220701266742] [PMID: 17786816]

[21]   Kronenbuerger M, Gerwig M, Brol B, Block F, Timmann D. Eyeblink conditioning is impaired in subjects with essential tremor. Brain 2007; 130(Pt 6): 1538-51.
[http://dx.doi.org/10.1093/brain/awm081] [PMID: 17468116]

[22]   Shill HA, De La Vega FJ, Samanta J, Stacy M. Motor learning in essential tremor. Mov Disord 2009; 24(6): 926-8.
[http://dx.doi.org/10.1002/mds.22479] [PMID: 19243062]

[23]   Chatterjee A, Jurewicz EC, Applegate LM, Louis ED. Personality in essential tremor: further evidence of non-motor manifestations of the disease. J Neurol Neurosurg Psychiatry 2004; 75(7): 958-61.
[http://dx.doi.org/10.1136/jnnp.2004.037176] [PMID: 15201349]

[24]   Gasparini M, Bonifati V, Fabrizio E, *et al.* Frontal lobe dysfunction in essential tremor: a preliminary study. J Neurol 2001; 248(5): 399-402.

[http://dx.doi.org/10.1007/s004150170181] [PMID: 11437162]

[25]   Lacritz LH, Dewey R Jr, Giller C, Cullum CM. Cognitive functioning in individuals with "benign" essential tremor. J Int Neuropsychol Soc 2002; 8(1): 125-9.
[http://dx.doi.org/10.1017/S1355617702001121] [PMID: 11843070]

[26]   Lombardi WJ, Woolston DJ, Roberts JW, Gross RE. Cognitive deficits in patients with essential tremor. Neurology 2001; 57(5): 785-90.
[http://dx.doi.org/10.1212/WNL.57.5.785] [PMID: 11552004]

[27]   Higginson CI, Wheelock VL, Levine D, King DS, Pappas CT, Sigvardt KA. Cognitive deficits in essential tremor consistent with frontosubcortical dysfunction. J Clin Exp Neuropsychol 2008; 30(7): 760-5.
[http://dx.doi.org/10.1080/13803390701754738] [PMID: 18608666]

[28]   Kim JS, Song IU, Shim YS, *et al.* Cognitive Impairment in Essential Tremor without Dementia. J Clin Neurol 2009; 5(2): 81-4.
[http://dx.doi.org/10.3988/jcn.2009.5.2.81] [PMID: 19587814]

[29]   Sahin HA, Terzi M, Uçak S, Yapici O, Basoglu T, Onar M. Frontal functions in young patients with essential tremor: a case comparison study. J Neuropsychiatry Clin Neurosci 2006; 18(1): 64-72.
[http://dx.doi.org/10.1176/jnp.18.1.64] [PMID: 16525072]

[30]   Applegate LM, Louis ED. Essential tremor: mild olfactory dysfunction in a cerebellar disorder. Parkinsonism Relat Disord 2005; 11(6): 399-402.
[http://dx.doi.org/10.1016/j.parkreldis.2005.03.003] [PMID: 16102998]

[31]   Shah M, Muhammed N, Findley LJ, Hawkes CH. Olfactory tests in the diagnosis of essential tremor. Parkinsonism Relat Disord 2008; 14(7): 563-8.
[http://dx.doi.org/10.1016/j.parkreldis.2007.12.006] [PMID: 18321760]

[32]   Balaban H, Altuntaş EE, Uysal IO, Sentürk IA, Topaktaş S. Audio-vestibular evaluation in patients with essential tremor. Eur Arch Otorhinolaryngol 2012; 269(6): 1577-81.
[http://dx.doi.org/10.1007/s00405-011-1801-x] [PMID: 22037719]

[33]   Benito-León J, Louis ED, Bermejo-Pareja F. Neurological Disorders in Central Spain (NEDICES) Study Group. Reported hearing impairment in essential tremor: a population-based case-control study. Neuroepidemiology 2007; 29(3-4): 213-7.
[http://dx.doi.org/10.1159/000112463] [PMID: 18073494]

[34]   Chandran V, Pal PK. Essential tremor: beyond the motor features. Parkinsonism Relat Disord 2012; 18(5): 407-13.
[http://dx.doi.org/10.1016/j.parkreldis.2011.12.003] [PMID: 22217558]

[35]   Lorenz D, Deuschl G. Update on pathogenesis and treatment of essential tremor. Curr Opin Neurol 2007; 20(4): 447-52.
[http://dx.doi.org/10.1097/WCO.0b013e3281e66942] [PMID: 17620881]

[36]   Louis ED, Ferreira JJ. How common is the most common adult movement disorder? Update on the worldwide prevalence of essential tremor. Mov Disord 2010; 25(5): 534-41.
[http://dx.doi.org/10.1002/mds.22838] [PMID: 20175185]

[37]   Jiménez-Jiménez FJ, de Toledo-Heras M, Alonso-Navarro H, *et al.* Environmental risk factors for essential tremor. Eur Neurol 2007; 58(2): 106-13.
[http://dx.doi.org/10.1159/000103646] [PMID: 17570916]

[38]   Louis ED, Ottman R. Study of possible factors associated with age of onset in essential tremor. Mov Disord 2006; 21(11): 1980-6.
[http://dx.doi.org/10.1002/mds.21102] [PMID: 16991147]

[39]   Jiménez-Jiménez FJ, Alonso-Navarro H, García-Martín E, Lorenzo-Betancor O, Pastor P, Agúndez JA. Update on genetics of essential tremor. Acta Neurol Scand 2013; 128(6): 359-71.
[http://dx.doi.org/10.1111/ane.12148] [PMID: 23682623]

[40]   Kuhlenbäumer G, Hopfner F, Deuschl G. Genetics of essential tremor: meta-analysis and review. Neurology 2014; 82(11): 1000-7.
[http://dx.doi.org/10.1212/WNL.0000000000000211] [PMID: 24532269]

[41]   Haerer AF, Anderson DW, Schoenberg BS. Prevalence of essential tremor. Results from the Copiah County study. Arch Neurol 1982; 39(12): 750-1.
[http://dx.doi.org/10.1001/archneur.1982.00510240012003] [PMID: 7138316]

[42]   Louis ED, Marder K, Cote L, *et al*. Differences in the prevalence of essential tremor among elderly African Americans, whites, and Hispanics in northern Manhattan, NY. Arch Neurol 1995; 52(12): 1201-5.
[http://dx.doi.org/10.1001/archneur.1995.00540360079019] [PMID: 7492295]

[43]   Putzke JD, Whaley NR, Baba Y, Wszolek ZK, Uitti RJ. Essential tremor: predictors of disease progression in a clinical cohort. J Neurol Neurosurg Psychiatry 2006; 77(11): 1235-7.
[http://dx.doi.org/10.1136/jnnp.2005.086579] [PMID: 17043291]

[44]   Carranza MA, Snyder MR, Elble RJ, Boutzoukas AE, Zesiewicz TA. Methodological issues in clinical drug development for essential tremor. Tremor Other Hyperkinet Mov (N Y) 2012; 2: 2.
[PMID: 23440401]

[45]   Fekete R, Jankovic J. Revisiting the relationship between essential tremor and Parkinson's disease. Mov Disord 2011; 26(3): 391-8.
[http://dx.doi.org/10.1002/mds.23512] [PMID: 21462256]

[46]   Montgomery EB Jr, Baker KB, Lyons K, Koller WC. Motor initiation and execution in essential tremor and Parkinson's disease. Mov Disord 2000; 15(3): 511-5.
[http://dx.doi.org/10.1002/1531-8257(200005)15:3<511::AID-MDS1014>3.0.CO;2-R]   [PMID: 10830417]

[47]   Isaias IU, Canesi M, Benti R, *et al*. Striatal dopamine transporter abnormalities in patients with essential tremor. Nucl Med Commun 2008; 29(4): 349-53.
[http://dx.doi.org/10.1097/MNM.0b013e3282f4d307] [PMID: 18317299]

[48]   Marjama-Lyons J, Koller W. Tremor-predominant Parkinson's disease. Approaches to treatment. Drugs Aging 2000; 16(4): 273-8.
[http://dx.doi.org/10.2165/00002512-200016040-00003] [PMID: 10874522]

[49]   Rajput AH, Robinson CA, Rajput ML, Robinson SL, Rajput A. Essential tremor is not dependent upon cerebellar Purkinje cell loss. Parkinsonism Relat Disord 2012; 18(5): 626-8.
[http://dx.doi.org/10.1016/j.parkreldis.2012.01.013] [PMID: 22306459]

[50]   Shill HA, Adler CH, Sabbagh MN, *et al*. Pathologic findings in prospectively ascertained essential tremor subjects. Neurology 2008; 70(16 Pt 2): 1452-5.
[http://dx.doi.org/10.1212/01.wnl.0000310425.76205.02] [PMID: 18413570]

[51]   Adler CH, Shill HA, Beach TG. Essential tremor and Parkinson's disease: lack of a link. Mov Disord 2011; 26(3): 372-7.
[http://dx.doi.org/10.1002/mds.23509] [PMID: 21284040]

[52]   Shahed J, Jankovic J. Exploring the relationship between essential tremor and Parkinson's disease. Parkinsonism Relat Disord 2007; 13(2): 67-76.
[http://dx.doi.org/10.1016/j.parkreldis.2006.05.033] [PMID: 16887374]

[53]   Defazio G, Conte A, Gigante AF, Fabbrini G, Berardelli A. Is tremor in dystonia a phenotypic feature of dystonia? Neurology 2015; 84(10): 1053-9.
[http://dx.doi.org/10.1212/WNL.0000000000001341] [PMID: 25663232]

[54]   Münchau A, Schrag A, Chuang C, *et al*. Arm tremor in cervical dystonia differs from essential tremor and can be classified by onset age and spread of symptoms. Brain 2001; 124(Pt 9): 1765-76.
[http://dx.doi.org/10.1093/brain/124.9.1765] [PMID: 11522579]

[55]   Shaikh AG, Jinnah HA, Tripp RM, *et al.* Irregularity distinguishes limb tremor in cervical dystonia from essential tremor. J Neurol Neurosurg Psychiatry 2008; 79(2): 187-9.
[http://dx.doi.org/10.1136/jnnp.2007.131110] [PMID: 17872981]

[56]   Schiebler S, Schmidt A, Zittel S, *et al.* Arm tremor in cervical dystonia--is it a manifestation of dystonia or essential tremor? Mov Disord 2011; 26(10): 1789-92.
[http://dx.doi.org/10.1002/mds.23837] [PMID: 21735481]

[57]   Zakaria R, Lenz FA, Hua S, Avin BH, Liu CC, Mari Z. Thalamic physiology of intentional essential tremor is more like cerebellar tremor than postural essential tremor. Brain Res 2013; 1529: 188-99.
[http://dx.doi.org/10.1016/j.brainres.2013.07.011] [PMID: 23856324]

[58]   Deuschl G, Elble R. Essential tremor--neurodegenerative or nondegenerative disease towards a working definition of ET. Mov Disord 2009; 24(14): 2033-41.
[http://dx.doi.org/10.1002/mds.22755] [PMID: 19750493]

[59]   Tanner CM, Goldman SM, Lyons KE, *et al.* Essential tremor in twins: an assessment of genetic *vs* environmental determinants of etiology. Neurology 2001; 57(8): 1389-91.
[http://dx.doi.org/10.1212/WNL.57.8.1389] [PMID: 11673577]

[60]   Lorenz D, Frederiksen H, Moises H, Kopper F, Deuschl G, Christensen K. High concordance for essential tremor in monozygotic twins of old age. Neurology 2004; 62(2): 208-11.
[http://dx.doi.org/10.1212/01.WNL.0000103236.26934.41] [PMID: 14745055]

[61]   Larsson T, Sjogren T. Essential tremor: a clinical and genetic population study. Acta Psychiatr Scand Suppl 1960; 36(144): 1-176.
[PMID: 14414307]

[62]   Louis ED, Ford B, Frucht S, Barnes LF, X-Tang M, Ottman R. Risk of tremor and impairment from tremor in relatives of patients with essential tremor: a community-based family study. Ann Neurol 2001; 49(6): 761-9.
[http://dx.doi.org/10.1002/ana.1022] [PMID: 11409428]

[63]   Louis ED. Environmental epidemiology of essential tremor. Neuroepidemiology 2008; 31(3): 139-49.
[http://dx.doi.org/10.1159/000151523] [PMID: 18716411]

[64]   Louis ED, Factor-Litvak P, Gerbin M, *et al.* Blood harmane, blood lead, and severity of hand tremor: evidence of additive effects. Neurotoxicology 2011; 32(2): 227-32.
[http://dx.doi.org/10.1016/j.neuro.2010.12.002] [PMID: 21145352]

[65]   Louis ED, Factor-Litvak P, Parides M, Andrews L, Santella RM, Wolff MS. Organochlorine pesticide exposure in essential tremor: a case-control study using biological and occupational exposure assessments. Neurotoxicology 2006; 27(4): 579-86.
[http://dx.doi.org/10.1016/j.neuro.2006.03.005] [PMID: 16620996]

[66]   Prakash KM, Fook-Choong S, Yuen Y, Tan EK. Exploring the relationship between caffeine intake and essential tremor. J Neurol Sci 2006; 251(1-2): 98-101.
[http://dx.doi.org/10.1016/j.jns.2006.09.007] [PMID: 17049563]

[67]   Louis ED, Zheng W, Jurewicz EC, *et al.* Elevation of blood beta-carboline alkaloids in essential tremor. Neurology 2002; 59(12): 1940-4.
[http://dx.doi.org/10.1212/01.WNL.0000038385.60538.19] [PMID: 12499487]

[68]   Louis ED, Zheng W, Mao X, Shungu DC. Blood harmane is correlated with cerebellar metabolism in essential tremor: a pilot study. Neurology 2007; 69(6): 515-20.
[http://dx.doi.org/10.1212/01.wnl.0000266663.27398.9f] [PMID: 17679670]

[69]   Louis ED, Keating GA, Bogen KT, Rios E, Pellegrino KM, Factor-Litvak P. Dietary epidemiology of essential tremor: meat consumption and meat cooking practices. Neuroepidemiology 2008; 30(3): 161-6.
[http://dx.doi.org/10.1159/000122333] [PMID: 18382115]

[70] Kuhn W, Müller T, Grosse H, Rommelspacher H. Plasma harman and norharman in Parkinson's disease. J Neural Transm Suppl 1995; 46: 291-5.
[PMID: 8821066]

[71] Valpey R, Sumi SM, Copass MK, Goble GJ. Acute and chronic progressive encephalopathy due to gasoline sniffing. Neurology 1978; 28(5): 507-10.
[http://dx.doi.org/10.1212/WNL.28.5.507] [PMID: 565491]

[72] Young RS, Grzyb SE, Crismon L. Recurrent cerebellar dysfunction as related to chronic gasoline sniffing in an adolescent girl. Lead poisoning from "leaded" gasoline as an attandent complication. Clin Pediatr (Phila) 1977; 16(8): 706-8.
[http://dx.doi.org/10.1177/000992287701600806] [PMID: 872480]

[73] McConnell P, Berry M. The effects of postnatal lead exposure on Purkinje cell dendritic development in the rat. Neuropathol Appl Neurobiol 1979; 5(2): 115-32.
[http://dx.doi.org/10.1111/j.1365-2990.1979.tb00665.x] [PMID: 471184]

[74] Louis ED, Jurewicz EC, Applegate L, *et al.* Association between essential tremor and blood lead concentration. Environ Health Perspect 2003; 111(14): 1707-11.
[http://dx.doi.org/10.1289/ehp.6404] [PMID: 14594619]

[75] Dogu O, Louis ED, Tamer L, Unal O, Yilmaz A, Kaleagasi H. Elevated blood lead concentrations in essential tremor: a case-control study in Mersin, Turkey. Environ Health Perspect 2007; 115(11): 1564-8.
[http://dx.doi.org/10.1289/ehp.10352] [PMID: 18007985]

[76] Jacobson BH, Winter-Roberts K, Gemmell HA. Influence of caffeine on selected manual manipulation skills. Percept Mot Skills 1991; 72(3 Pt 2): 1175-81.
[http://dx.doi.org/10.2466/pms.1991.72.3c.1175] [PMID: 1961665]

[77] Wharrad HJ, Birmingham AT, Macdonald IA, Inch PJ, Mead JL. The influence of fasting and of caffeine intake on finger tremor. Eur J Clin Pharmacol 1985; 29(1): 37-43.
[http://dx.doi.org/10.1007/BF00547366] [PMID: 4054205]

[78] Koller W, Cone S, Herbster G. Caffeine and tremor. Neurology 1987; 37(1): 169-72.
[http://dx.doi.org/10.1212/WNL.37.1.169] [PMID: 3796831]

[79] Zanardi A, Leo G, Biagini G, Zoli M. Nicotine and neurodegeneration in ageing. Toxicol Lett 2002; 127(1-3): 207-15.
[http://dx.doi.org/10.1016/S0378-4274(01)00502-1] [PMID: 12052660]

[80] Ritz B, Ascherio A, Checkoway H, *et al.* Pooled analysis of tobacco use and risk of Parkinson disease. Arch Neurol 2007; 64(7): 990-7.
[http://dx.doi.org/10.1001/archneur.64.7.990] [PMID: 17620489]

[81] Louis ED, Benito-León J, Bermejo-Pareja F. Neurological Disorders in Central Spain (NEDICES) Study Group. Population-based prospective study of cigarette smoking and risk of incident essential tremor. Neurology 2008; 70(19): 1682-7.
[http://dx.doi.org/10.1212/01.wnl.0000311271.42596.32] [PMID: 18458228]

[82] Lou JS, Jankovic J. Essential tremor: clinical correlates in 350 patients. Neurology 1991; 41(2 (Pt 1)): 234-8.
[http://dx.doi.org/10.1212/WNL.41.2_Part_1.234] [PMID: 1992367]

[83] Mostile G, Jankovic J. Alcohol in essential tremor and other movement disorders. Mov Disord 2010; 25(14): 2274-84.
[http://dx.doi.org/10.1002/mds.23240] [PMID: 20721919]

[84] Rajput AH, Jamieson H, Hirsh S, Quraishi A. Relative efficacy of alcohol and propranolol in action tremor. Can J Neurol Sci 1975; 2(1): 31-5.
[http://dx.doi.org/10.1017/S0317167100019958] [PMID: 1148953]

[85] Knudsen K, Lorenz D, Deuschl G. A clinical test for the alcohol sensitivity of essential tremor. Mov Disord 2011; 26(12): 2291-5.
[http://dx.doi.org/10.1002/mds.23846] [PMID: 22021159]

[86] Klebe S, Stolze H, Grensing K, Volkmann J, Wenzelburger R, Deuschl G. Influence of alcohol on gait in patients with essential tremor. Neurology 2005; 65(1): 96-101.
[http://dx.doi.org/10.1212/01.wnl.0000167550.97413.1f] [PMID: 16009892]

[87] Nicoletti A, Mostile G, Cappellani R, *et al.* Wine drinking and essential tremor: a possible protective role. Mov Disord 2011; 26(7): 1310-5.
[http://dx.doi.org/10.1002/mds.23603] [PMID: 21506162]

[88] Louis ED, Benito-León J, Bermejo-Pareja F. Population-based study of baseline ethanol consumption and risk of incident essential tremor. J Neurol Neurosurg Psychiatry 2009; 80(5): 494-7.
[http://dx.doi.org/10.1136/jnnp.2008.162701] [PMID: 19359288]

[89] Luo J. Effects of Ethanol on the Cerebellum: Advances and Prospects. Cerebellum 2015; 14(4): 383-5.
[http://dx.doi.org/10.1007/s12311-015-0674-8] [PMID: 25933648]

[90] Bain PG, Findley LJ, Thompson PD, *et al.* A study of hereditary essential tremor. Brain 1994; 117(Pt 4): 805-24.
[http://dx.doi.org/10.1093/brain/117.4.805] [PMID: 7922467]

[91] Koller WC, Biary N. Effect of alcohol on tremors: comparison with propranolol. Neurology 1984; 34(2): 221-2.
[http://dx.doi.org/10.1212/WNL.34.2.221] [PMID: 6538013]

[92] Louis ED, Broussolle E, Goetz CG, Krack P, Kaufmann P, Mazzoni P. Historical underpinnings of the term essential tremor in the late 19th century. Neurology 2008; 71(11): 856-9.
[http://dx.doi.org/10.1212/01.wnl.0000325564.38165.d1] [PMID: 18779514]

[93] Raethjen J, Deuschl G. The oscillating central network of Essential tremor. Clin Neurophysiol 2012; 123(1): 61-4.
[http://dx.doi.org/10.1016/j.clinph.2011.09.024] [PMID: 22055842]

[94] Helmich RC, Toni I, Deuschl G, Bloem BR. The pathophysiology of essential tremor and Parkinson's tremor. Curr Neurol Neurosci Rep 2013; 13(9): 378.
[http://dx.doi.org/10.1007/s11910-013-0378-8] [PMID: 23893097]

[95] Dupuis MJ, Evrard FL, Jacquerye PG, Picard GR, Lermen OG. Disappearance of essential tremor after stroke. Mov Disord 2010; 25(16): 2884-7.
[http://dx.doi.org/10.1002/mds.23328] [PMID: 20836089]

[96] Wilms H, Sievers J, Deuschl G. Animal models of tremor. Mov Disord 1999; 14(4): 557-71.
[http://dx.doi.org/10.1002/1531-8257(199907)14:4<557::AID-MDS1004>3.0.CO;2-G]   [PMID: 10435492]

[97] Hallett M, Dubinsky RM. Glucose metabolism in the brain of patients with essential tremor. J Neurol Sci 1993; 114(1): 45-8.
[http://dx.doi.org/10.1016/0022-510X(93)90047-3] [PMID: 8433096]

[98] Deuschl G, Toro C, Valls-Solé J, Zeffiro T, Zee DS, Hallett M. Symptomatic and essential palatal tremor. 1. Clinical, physiological and MRI analysis. Brain 1994; 117(Pt 4): 775-88.
[http://dx.doi.org/10.1093/brain/117.4.775] [PMID: 7922465]

[99] Bucher SF, Seelos KC, Dodel RC, Reiser M, Oertel WH. Activation mapping in essential tremor with functional magnetic resonance imaging. Ann Neurol 1997; 41(1): 32-40.
[http://dx.doi.org/10.1002/ana.410410108] [PMID: 9005863]

[100] Wills AJ, Jenkins IH, Thompson PD, Findley LJ, Brooks DJ. Red nuclear and cerebellar but no olivary activation associated with essential tremor: a positron emission tomographic study. Ann Neurol 1994; 36(4): 636-42.

[http://dx.doi.org/10.1002/ana.410360413] [PMID: 7944296]

[101] Bhalsing KS, Saini J, Pal PK. Understanding the pathophysiology of essential tremor through advanced neuroimaging: a review. J Neurol Sci 2013; 335(1-2): 9-13.
[http://dx.doi.org/10.1016/j.jns.2013.09.003] [PMID: 24060292]

[102] Louis ED, Babij R, Cortés E, Vonsattel JP, Faust PL. The inferior olivary nucleus: a postmortem study of essential tremor cases *versus* controls. Mov Disord 2013; 28(6): 779-86.
[http://dx.doi.org/10.1002/mds.25400] [PMID: 23483605]

[103] Jenkins IH, Bain PG, Colebatch JG, *et al.* A positron emission tomography study of essential tremor: evidence for overactivity of cerebellar connections. Ann Neurol 1993; 34(1): 82-90.
[http://dx.doi.org/10.1002/ana.410340115] [PMID: 8517685]

[104] Benito-León J, Alvarez-Linera J, Hernández-Tamames JA, Alonso-Navarro H, Jiménez-Jiménez FJ, Louis ED. Brain structural changes in essential tremor: voxel-based morphometry at 3-Tesla. J Neurol Sci 2009; 287(1-2): 138-42.
[http://dx.doi.org/10.1016/j.jns.2009.08.037] [PMID: 19717167]

[105] Cerasa A, Messina D, Nicoletti G, *et al.* Cerebellar atrophy in essential tremor using an automated segmentation method. AJNR Am J Neuroradiol 2009; 30(6): 1240-3.
[http://dx.doi.org/10.3174/ajnr.A1544] [PMID: 19342539]

[106] Daniels C, Peller M, Wolff S, *et al.* Voxel-based morphometry shows no decreases in cerebellar gray matter volume in essential tremor. Neurology 2006; 67(8): 1452-6.
[http://dx.doi.org/10.1212/01.wnl.0000240130.94408.99] [PMID: 17060572]

[107] Quattrone A, Cerasa A, Messina D, *et al.* Essential head tremor is associated with cerebellar vermis atrophy: a volumetric and voxel-based morphometry MR imaging study. AJNR Am J Neuroradiol 2008; 29(9): 1692-7.
[http://dx.doi.org/10.3174/ajnr.A1190] [PMID: 18653686]

[108] Louis ED, Shungu DC, Chan S, Mao X, Jurewicz EC, Watner D. Metabolic abnormality in the cerebellum in patients with essential tremor: a proton magnetic resonance spectroscopic imaging study. Neurosci Lett 2002; 333(1): 17-20.
[http://dx.doi.org/10.1016/S0304-3940(02)00966-7] [PMID: 12401550]

[109] Deuschl G, Wenzelburger R, Löffler K, Raethjen J, Stolze H. Essential tremor and cerebellar dysfunction clinical and kinematic analysis of intention tremor. Brain 2000; 123(Pt 8): 1568-80.
[http://dx.doi.org/10.1093/brain/123.8.1568] [PMID: 10908187]

[110] Louis ED, Faust PL, Vonsattel JP, *et al.* Neuropathological changes in essential tremor: 33 cases compared with 21 controls. Brain 2007; 130(Pt 12): 3297-307.
[http://dx.doi.org/10.1093/brain/awm266] [PMID: 18025031]

[111] Macchi G, Jones EG. Toward an agreement on terminology of nuclear and subnuclear divisions of the motor thalamus. J Neurosurg 1997; 86(1): 77-92.
[http://dx.doi.org/10.3171/jns.1997.86.1.0077] [PMID: 8988085]

[112] Hua SE, Lenz FA. Posture-related oscillations in human cerebellar thalamus in essential tremor are enabled by voluntary motor circuits. J Neurophysiol 2005; 93(1): 117-27.
[http://dx.doi.org/10.1152/jn.00527.2004] [PMID: 15317839]

[113] Speelman JD, Schuurman PR, de Bie RM, Bosch DA. Thalamic surgery and tremor. Mov Disord 1998; 13 (Suppl. 3): 103-6.
[http://dx.doi.org/10.1002/mds.870131318] [PMID: 9827604]

[114] Hellwig B, Häussler S, Schelter B, *et al.* Tremor-correlated cortical activity in essential tremor. Lancet 2001; 357(9255): 519-23.
[http://dx.doi.org/10.1016/S0140-6736(00)04044-7] [PMID: 11229671]

[115] Raethjen J, Govindan RB, Kopper F, Muthuraman M, Deuschl G. Cortical involvement in the generation of essential tremor. J Neurophysiol 2007; 97(5): 3219-28.

[http://dx.doi.org/10.1152/jn.00477.2006] [PMID: 17344375]

[116] Schnitzler A, Münks C, Butz M, Timmermann L, Gross J. Synchronized brain network associated with essential tremor as revealed by magnetoencephalography. Mov Disord 2009; 24(11): 1629-35.
[http://dx.doi.org/10.1002/mds.22633] [PMID: 19514010]

[117] Govindan RB, Raethjen J, Arning K, Kopper F, Deuschl G. Time delay and partial coherence analyses to identify cortical connectivities. Biol Cybern 2006; 94(4): 262-75.
[http://dx.doi.org/10.1007/s00422-005-0045-5] [PMID: 16453139]

[118] Schelter B, Timmer J, Eichler M. Assessing the strength of directed influences among neural signals using renormalized partial directed coherence. J Neurosci Methods 2009; 179(1): 121-30.
[http://dx.doi.org/10.1016/j.jneumeth.2009.01.006] [PMID: 19428518]

[119] Moro E, Schwalb JM, Piboolnurak P, et al. Unilateral subdural motor cortex stimulation improves essential tremor but not Parkinson's disease. Brain 2011; 134(Pt 7): 2096-105.
[http://dx.doi.org/10.1093/brain/awr072] [PMID: 21646329]

[120] Málly J, Baranyi M, Vizi ES. Change in the concentrations of amino acids in CSF and serum of patients with essential tremor. J Neural Transm (Vienna) 1996; 103(5): 555-60.
[http://dx.doi.org/10.1007/BF01273153] [PMID: 8811501]

[121] Zesiewicz TA, Shaw JD, Allison KG, Staffetti JS, Okun MS, Sullivan KL. Update on treatment of essential tremor. Curr Treat Options Neurol 2013; 15(4): 410-23.
[http://dx.doi.org/10.1007/s11940-013-0239-4] [PMID: 23881742]

[122] Stratton SE, Lorden JF. Effect of harmaline on cells of the inferior olive in the absence of tremor: differential response of genetically dystonic and harmaline-tolerant rats. Neuroscience 1991; 41(2-3): 543-9.
[http://dx.doi.org/10.1016/0306-4522(91)90347-Q] [PMID: 1870702]

[123] Kralic JE, Criswell HE, Osterman JL, et al. Genetic essential tremor in gamma-aminobutyric acidA receptor alpha1 subunit knockout mice. J Clin Invest 2005; 115(3): 774-9.
[http://dx.doi.org/10.1172/JCI200523625] [PMID: 15765150]

[124] Paris-Robidas S, Brochu E, Sintes M, et al. Defective dentate nucleus GABA receptors in essential tremor. Brain 2012; 135(Pt 1): 105-16.
[http://dx.doi.org/10.1093/brain/awr301] [PMID: 22120148]

[125] García-Martín E, Martínez C, Alonso-Navarro H, et al. Gamma-aminobutyric acid (GABA) receptor rho (GABRR) polymorphisms and risk for essential tremor. J Neurol 2011; 258(2): 203-11.
[http://dx.doi.org/10.1007/s00415-010-5708-z] [PMID: 20820800]

[126] Thier S, Kuhlenbäumer G, Lorenz D, et al. GABA(A) receptor- and GABA transporter polymorphisms and risk for essential tremor. Eur J Neurol 2011; 18(8): 1098-100.
[http://dx.doi.org/10.1111/j.1468-1331.2010.03308.x] [PMID: 21749575]

[127] Gironell A. The GABA Hypothesis in Essential Tremor: Lights and Shadows. Tremor Other Hyperkinet Mov (N Y) 2014; 4: 254.
[PMID: 25120944]

[128] Louis ED. Essential tremor: evolving clinicopathological concepts in an era of intensive post-mortem enquiry. Lancet Neurol 2010; 9(6): 613-22.
[http://dx.doi.org/10.1016/S1474-4422(10)70090-9] [PMID: 20451458]

[129] Louis ED, Kuo SH, Vonsattel JP, Faust PL. Torpedo formation and Purkinje cell loss: modeling their relationship in cerebellar disease. Cerebellum 2014; 13(4): 433-9.
[http://dx.doi.org/10.1007/s12311-014-0556-5] [PMID: 24590661]

[130] Louis ED, Faust PL, Vonsattel JP, et al. Torpedoes in Parkinson's disease, Alzheimer's disease, essential tremor, and control brains. Mov Disord 2009; 24(11): 1600-5.
[http://dx.doi.org/10.1002/mds.22567] [PMID: 19526585]

[131] Ma K, Babij R, Cortés E, Vonsattel JP, Louis ED. Cerebellar pathology of a dual clinical diagnosis: patients with essential tremor and dystonia. Tremor Other Hyperkinet Mov (N Y) 2012; 2: 2. [PMID: 23439731]

[132] DiLorenzo DJ, Jankovic J, Simpson RK, Takei H, Powell SZ. Neurohistopathological findings at the electrode-tissue interface in long-term deep brain stimulation: systematic literature review, case report, and assessment of stimulation threshold safety. Neuromodulation 2014; 17(5): 405-18. [http://dx.doi.org/10.1111/ner.12192] [PMID: 24947418]

[133] Louis ED. Treatment of Essential Tremor: Are there Issues We are Overlooking? Front Neurol 2012; 2: 91. [http://dx.doi.org/10.3389/fneur.2011.00091] [PMID: 22275907]

[134] Rajput AH, Rajput A. Medical treatment of essential tremor. J Cent Nerv Syst Dis 2014; 6: 29-39. [http://dx.doi.org/10.4137/JCNSD.S13570] [PMID: 24812533]

[135] Schrag A, Münchau A, Bhatia KP, Quinn NP, Marsden CD. Essential tremor: an overdiagnosed condition? J Neurol 2000; 247(12): 955-9. [http://dx.doi.org/10.1007/s004150070053] [PMID: 11200689]

[136] Morgan JC, Sethi KD. Drug-induced tremors. Lancet Neurol 2005; 4(12): 866-76. [http://dx.doi.org/10.1016/S1474-4422(05)70250-7] [PMID: 16297844]

[137] Block F, Dafotakis M. [Drug-induced tremor]. Fortschr Neurol Psychiatr 2011; 79(10): 570-5. [Drug-induced tremor]. [http://dx.doi.org/10.1055/s-0031-1281687] [PMID: 21989509]

[138] Alty JE, Kempster PA. A practical guide to the differential diagnosis of tremor. Postgrad Med J 2011; 87(1031): 623-9. [http://dx.doi.org/10.1136/pgmj.2009.089623] [PMID: 21690256]

[139] Hopfner F, Deuschl G. Examination of Patients with Essential Tremor. Mov Disord Clin Pract (Hoboken) 2014. [http://dx.doi.org/10.1002/mdc2.12012]

[140] Bhidayasiri R. Differential diagnosis of common tremor syndromes. Postgrad Med J 2005; 81(962): 756-62. [http://dx.doi.org/10.1136/pgmj.2005.032979] [PMID: 16344298]

[141] Elble R, Comella C, Fahn S, *et al.* Reliability of a new scale for essential tremor. Mov Disord 2012; 27(12): 1567-9. [http://dx.doi.org/10.1002/mds.25162] [PMID: 23032792]

[142] Benito-León J, Louis ED. Clinical update: diagnosis and treatment of essential tremor. Lancet 2007; 369(9568): 1152-4. [http://dx.doi.org/10.1016/S0140-6736(07)60544-3] [PMID: 17416247]

[143] Eye PG, Hawley JS. Pearls & Oy-sters: fragile X tremor/ataxia syndrome: a diagnostic dilemma. Neurology 2015; 84(7): e43-5. [http://dx.doi.org/10.1212/WNL.0000000000001267] [PMID: 25688154]

[144] Jankovic J, Frost JD Jr. Quantitative assessment of parkinsonian and essential tremor: clinical application of triaxial accelerometry. Neurology 1981; 31(10): 1235-40. [http://dx.doi.org/10.1212/WNL.31.10.1235] [PMID: 7202133]

[145] Joundi RA, Brittain JS, Jenkinson N, Green AL, Aziz T. Rapid tremor frequency assessment with the iPhone accelerometer. Parkinsonism Relat Disord 2011; 17(4): 288-90. [http://dx.doi.org/10.1016/j.parkreldis.2011.01.001] [PMID: 21300563]

[146] Schneider SA, Edwards MJ, Mir P, *et al.* Patients with adult-onset dystonic tremor resembling parkinsonian tremor have scans without evidence of dopaminergic deficit (SWEDDs). Mov Disord 2007; 22(15): 2210-5.

[http://dx.doi.org/10.1002/mds.21685] [PMID: 17712858]

[147] Deuschl G, Raethjen J, Hellriegel H, Elble R. Treatment of patients with essential tremor. Lancet Neurol 2011; 10(2): 148-61.
[http://dx.doi.org/10.1016/S1474-4422(10)70322-7] [PMID: 21256454]

[148] Elble RJ. Tremor: clinical features, pathophysiology, and treatment. Neurol Clin 2009; 27(3): 679-695, v-vi. [v-vi.].
[http://dx.doi.org/10.1016/j.ncl.2009.04.003] [PMID: 19555826]

[149] Rajput AH, Offord KP, Beard CM, Kurland LT. Essential tremor in Rochester, Minnesota: a 45-year study. J Neurol Neurosurg Psychiatry 1984; 47(5): 466-70.
[http://dx.doi.org/10.1136/jnnp.47.5.466] [PMID: 6736976]

[150] Louis ED, Ford B, Wendt KJ, Cameron G. Clinical characteristics of essential tremor: data from a community-based study. Mov Disord 1998; 13(5): 803-8.
[http://dx.doi.org/10.1002/mds.870130508] [PMID: 9756149]

[151] Sur H, Ilhan S, Erdoğan H, Oztürk E, Taşdemir M, Börü UT. Prevalence of essential tremor: a door-to-door survey in Sile, Istanbul, Turkey. Parkinsonism Relat Disord 2009; 15(2): 101-4.
[http://dx.doi.org/10.1016/j.parkreldis.2008.03.009] [PMID: 18474448]

[152] Barbosa MT, Caramelli P, Cunningham MC, Maia DP, Lima-Costa MF, Cardoso F. Prevalence and clinical classification of tremor in elderly--a community-based survey in Brazil. Mov Disord 2013; 28(5): 640-6.
[http://dx.doi.org/10.1002/mds.25355] [PMID: 23450620]

[153] Moghal S, Rajput AH, D'Arcy C, Rajput R. Prevalence of movement disorders in elderly community residents. Neuroepidemiology 1994; 13(4): 175-8.
[http://dx.doi.org/10.1159/000110376] [PMID: 8090259]

[154] Louis ED, Rios E. Embarrassment in essential tremor: prevalence, clinical correlates and therapeutic implications. Parkinsonism Relat Disord 2009; 15(7): 535-8.
[http://dx.doi.org/10.1016/j.parkreldis.2008.10.006] [PMID: 19028131]

[155] Lorenz D, Schwieger D, Moises H, Deuschl G. Quality of life and personality in essential tremor patients. Mov Disord 2006; 21(8): 1114-8.
[http://dx.doi.org/10.1002/mds.20884] [PMID: 16622851]

[156] Gironell A, Kulisevsky J. Diagnosis and management of essential tremor and dystonic tremor. Ther Adv Neurol Disorder 2009; 2(4): 215-22.
[http://dx.doi.org/10.1177/1756285609104791] [PMID: 21179530]

[157] Calzetti S, Findley LJ, Gresty MA, Perucca E, Richens A. Metoprolol and propranolol in essential tremor: a double-blind, controlled study. J Neurol Neurosurg Psychiatry 1981; 44(9): 814-9.
[http://dx.doi.org/10.1136/jnnp.44.9.814] [PMID: 7031187]

[158] Leigh PN, Marsden CD, Twomey A, Jefferson D. beta-Adrenoceptor antagonists and essential tremor. Lancet 1981; 1(8229): 1106.
[http://dx.doi.org/10.1016/S0140-6736(81)92276-5] [PMID: 6112475]

[159] Koller WC, Vetere-Overfield B. Acute and chronic effects of propranolol and primidone in essential tremor. Neurology 1989; 39(12): 1587-8.
[http://dx.doi.org/10.1212/WNL.39.12.1587] [PMID: 2586774]

[160] O'Suilleabhain P, Dewey RB Jr. Randomized trial comparing primidone initiation schedules for treating essential tremor. Mov Disord 2002; 17(2): 382-6.
[http://dx.doi.org/10.1002/mds.10083] [PMID: 11921128]

[161] Serrano-Dueñas M. Use of primidone in low doses (250 mg/day) *versus* high doses (750 mg/day) in the management of essential tremor. Double-blind comparative study with one-year follow-up. Parkinsonism Relat Disord 2003; 10(1): 29-33.
[http://dx.doi.org/10.1016/S1353-8020(03)00070-1] [PMID: 14499204]

[162] Ondo WG, Jankovic J, Connor GS, *et al.* Topiramate Essential Tremor Study Investigators. Topiramate in essential tremor: a double-blind, placebo-controlled trial. Neurology 2006; 66(5): 672-7.
[http://dx.doi.org/10.1212/01.wnl.0000200779.03748.0f] [PMID: 16436648]

[163] Gironell A, Kulisevsky J, Barbanoj M, López-Villegas D, Hernández G, Pascual-Sedano B. A randomized placebo-controlled comparative trial of gabapentin and propranolol in essential tremor. Arch Neurol 1999; 56(4): 475-80.
[http://dx.doi.org/10.1001/archneur.56.4.475] [PMID: 10199338]

[164] Sixel-Döring F, Benecke R, Fogel W, *et al.* German Deep Brain Stimulation Association. [Deep brain stimulation for essential tremor. Consensus recommendations of the German Deep Brain Stimulation Association]. Nervenarzt 2009; 80(6): 662-5. [Deep brain stimulation for essential tremor. Consensus recommendations of the German Deep Brain Stimulation Association].
[PMID: 19404603]

[165] Zappia M, Albanese A, Bruno E, *et al.* Treatment of essential tremor: a systematic review of evidence and recommendations from the Italian Movement Disorders Association. J Neurol 2013; 260(3): 714-40.
[http://dx.doi.org/10.1007/s00415-012-6628-x] [PMID: 22886006]

[166] Flora ED, Perera CL, Cameron AL, Maddern GJ. Deep brain stimulation for essential tremor: a systematic review. Mov Disord 2010; 25(11): 1550-9.
[http://dx.doi.org/10.1002/mds.23195] [PMID: 20623768]

[167] Benabid AL, Pollak P, Gao D, *et al.* Chronic electrical stimulation of the ventralis intermedius nucleus of the thalamus as a treatment of movement disorders. J Neurosurg 1996; 84(2): 203-14.
[http://dx.doi.org/10.3171/jns.1996.84.2.0203] [PMID: 8592222]

[168] Limousin P, Krack P, Pollak P, *et al.* Electrical stimulation of the subthalamic nucleus in advanced Parkinson's disease. N Engl J Med 1998; 339(16): 1105-11.
[http://dx.doi.org/10.1056/NEJM199810153391603] [PMID: 9770557]

[169] Hauser RA, Friedlander J, Smith DA, Nolan MF. Delayed stimulation-induced thalamic ataxia syndrome. Neurology 1998; 50(4): 1184-5.
[http://dx.doi.org/10.1212/WNL.50.4.1184] [PMID: 9566426]

[170] Barbe MT, Liebhart L, Runge M, *et al.* Deep brain stimulation in the nucleus ventralis intermedius in patients with essential tremor: habituation of tremor suppression. J Neurol 2011; 258(3): 434-9.
[http://dx.doi.org/10.1007/s00415-010-5773-3] [PMID: 20927533]

[171] Kronenbuerger M, Fromm C, Block F, *et al.* On-demand deep brain stimulation for essential tremor: a report on four cases. Mov Disord 2006; 21(3): 401-5.
[http://dx.doi.org/10.1002/mds.20714] [PMID: 16211619]

[172] Favilla CG, Ullman D, Wagle Shukla A, Foote KD, Jacobson CE IV, Okun MS. Worsening essential tremor following deep brain stimulation: disease progression *versus* tolerance. Brain 2012; 135(Pt 5): 1455-62.
[http://dx.doi.org/10.1093/brain/aws026] [PMID: 22344584]

[173] Elias WJ, Huss D, Voss T, *et al.* A pilot study of focused ultrasound thalamotomy for essential tremor. N Engl J Med 2013; 369(7): 640-8.
[http://dx.doi.org/10.1056/NEJMoa1300962] [PMID: 23944301]

[174] Lipsman N, Schwartz ML, Huang Y, *et al.* MR-guided focused ultrasound thalamotomy for essential tremor: a proof-of-concept study. Lancet Neurol 2013; 12(5): 462-8.
[http://dx.doi.org/10.1016/S1474-4422(13)70048-6] [PMID: 23523144]

[175] Hallett M, Albanese A, Dressler D, *et al.* Evidence-based review and assessment of botulinum neurotoxin for the treatment of movement disorders. Toxicon 2013; 67: 94-114.
[http://dx.doi.org/10.1016/j.toxicon.2012.12.004] [PMID: 23380701]

[176] Brin MF, Lyons KE, Doucette J, *et al.* A randomized, double masked, controlled trial of botulinum

toxin type A in essential hand tremor. Neurology 2001; 56(11): 1523-8.
[http://dx.doi.org/10.1212/WNL.56.11.1523] [PMID: 11402109]

[177]   Jankovic J, Schwartz K, Clemence W, Aswad A, Mordaunt J. A randomized, double-blind, placebo-controlled study to evaluate botulinum toxin type A in essential hand tremor. Mov Disord 1996; 11(3): 250-6.
[http://dx.doi.org/10.1002/mds.870110306] [PMID: 8723140]

[178]   Heber IA, Coenen VA, Reetz K, *et al.* Cognitive effects of deep brain stimulation for essential tremor: evaluation at 1 and 6 years. J Neural Transm (Vienna) 2013; 120(11): 1569-77.
[http://dx.doi.org/10.1007/s00702-013-1030-0] [PMID: 23649123]

[179]   Tröster AI, Fields JA, Pahwa R, *et al.* Neuropsychological and quality of life outcome after thalamic stimulation for essential tremor. Neurology 1999; 53(8): 1774-80.
[http://dx.doi.org/10.1212/WNL.53.8.1774] [PMID: 10563627]

[180]   Woods SP, Fields JA, Lyons KE, Pahwa R, Tröster AI. Pulse width is associated with cognitive decline after thalamic stimulation for essential tremor. Parkinsonism Relat Disord 2003; 9(5): 295-300.
[http://dx.doi.org/10.1016/S1353-8020(03)00014-2] [PMID: 12781597]

[181]   Lucas JA, Rippeth JD, Uitti RJ, Shuster EA, Wharen RE. Neuropsychological functioning in a patient with essential tremor with and without bilateral VIM stimulation. Brain Cogn 2000; 42(2): 253-67.
[http://dx.doi.org/10.1006/brcg.1999.1103] [PMID: 10744923]

[182]   Fields JA, Tröster AI, Woods SP, *et al.* Neuropsychological and quality of life outcomes 12 months after unilateral thalamic stimulation for essential tremor. J Neurol Neurosurg Psychiatry 2003; 74(3): 305-11.
[http://dx.doi.org/10.1136/jnnp.74.3.305] [PMID: 12588913]

[183]   Chen H, Hua SE, Smith MA, Lenz FA, Shadmehr R. Effects of human cerebellar thalamus disruption on adaptive control of reaching. Cereb Cortex 2006; 16(10): 1462-73.
[http://dx.doi.org/10.1093/cercor/bhj087] [PMID: 16357337]

[184]   Kronenbuerger M, Tronnier VM, Gerwig M, *et al.* Thalamic deep brain stimulation improves eyeblink conditioning deficits in essential tremor. Exp Neurol 2008; 211(2): 387-96.
[http://dx.doi.org/10.1016/j.expneurol.2008.02.002] [PMID: 18394604]

[185]   Earhart GM, Clark BR, Tabbal SD, Perlmutter JS. Gait and balance in essential tremor: variable effects of bilateral thalamic stimulation. Mov Disord 2009; 24(3): 386-91.
[http://dx.doi.org/10.1002/mds.22356] [PMID: 19006189]

[186]   Ondo WG, Almaguer M, Cohen H. Computerized posturography balance assessment of patients with bilateral ventralis intermedius nuclei deep brain stimulation. Mov Disord 2006; 21(12): 2243-7.
[http://dx.doi.org/10.1002/mds.21165] [PMID: 17078067]

[187]   Fasano A, Herzog J, Raethjen J, *et al.* Gait ataxia in essential tremor is differentially modulated by thalamic stimulation. Brain 2010; 133(Pt 12): 3635-48.
[http://dx.doi.org/10.1093/brain/awq267] [PMID: 20926368]

[188]   Limousin P, Speelman JD, Gielen F, Janssens M. Multicentre European study of thalamic stimulation in parkinsonian and essential tremor. J Neurol Neurosurg Psychiatry 1999; 66(3): 289-96.
[http://dx.doi.org/10.1136/jnnp.66.3.289] [PMID: 10084526]

[189]   Kronenbuerger M, Zobel S, Ilgner J, *et al.* Effects of deep brain stimulation of the cerebellothalamic pathways on the sense of smell. Exp Neurol 2010; 222(1): 144-52.
[http://dx.doi.org/10.1016/j.expneurol.2009.12.024] [PMID: 20051243]

[190]   Kessler TM, Burkhard FC, Z'Brun S, *et al.* Effect of thalamic deep brain stimulation on lower urinary tract function. Eur Urol 2008; 53(3): 607-12.
[http://dx.doi.org/10.1016/j.eururo.2007.07.015] [PMID: 17686571]

[191]   Kronenbuerger M, González EG, Liu LD, *et al.* Involvement of the human ventrolateral thalamus in the control of visually guided saccades. Brain Stimul 2010; 3(4): 226-9.
[http://dx.doi.org/10.1016/j.brs.2009.12.002] [PMID: 20965452]

CHAPTER 3

Natural Products for the Treatment of Alzheimer's Disease: Present and Future Expectations

Fernanda Rodríguez-Enríquez, Iria Torres and **Dolores Viña**[*]

Center for Research in Molecular Medicine and Chronic Diseases (CIMUS), University of Santiago de Compostela, Santiago de Compostela, Spain

Abstract: Improving living conditions and health care in developed countries has significantly increased life expectancy, which has led to an increase in age related disorders. In the population older than 60 years old the majority of cases of dementia are Alzheimer disease (AD). Progressive neurodegeneration in AD induces cognitive deterioration and constitutes a serious social problem. Currently, the drugs approved for the treatment of AD just slow down the progression of the disease or have a symptomatic effect. They are mainly acetylcholinesterase inhibitors (AChEIs) as donepezil, rivastigmine and galantamine or NMDA-receptor antagonist such as memantine. These drugs modestly improvement cognition, daily life activities and behavior in patients ranging from mild to severe stages of the disease. However, none of these agents has proven to be able to stop or reverse the underlying neurodegenerative process. Different studies point out that environmental factors and life style, such as diet and exercise have an important role in the biological mechanisms of the pathophysiology, considering them mutable. Actually, dietary compounds have been studied as therapeutics for neurodegenerative diseases and numerous studies have been focused on different nutritional approaches to benefit AD patients. On the other hand, during decades, medicinal plants have been studied as a potential treatment for dementia. This chapter includes a review of different natural products such as fatty acids, vitamins, alkaloids, amino acids, hormones and diverse groups of polyphenolic plant secondary metabolites, among others, which have a potential role in the prevention or treatment of AD.

Keywords: Alzheimer's disease, Dementia, Natural products, Neuroprotection, Pharmacological activity.

INTRODUCTION

Improving living conditions and health care in developed countries has significantly increased life expectancy, which has led to an increase in age related

[*] **Corresponding author Dolores Viña:** Center for Research in Molecular Medicine and Chronic Diseases (CIMUS), University of Santiago de Compostela, Santiago de Compostela, Spain; Tel/Fax: +348818115424; Email: mdolores.vina@usc.es

Atta-ur-Rahman (Ed.)

disorders. In the population older than 60 years dementia is mainly attributable to Alzheimer's disease (AD). Currently 35 million people has AD and by 2050 is estimated that this value will be, at least, triplicated [1, 2]. This disease is accompanied by a progressive neuropsychiatric disorder, which implies gradual cognitive decline and a decrease in the ability to accomplish normal daily living activities as well as progressive behavioral disturbances [3].

Although mild cognitive impairment (MCI) is considered a common stage both in aging and dementia, it is a risk state that does not necessarily lead to AD. Some symptoms such as depressive state and anxiety in patients with MCI have been linked with a major probability to progress from MCI to AD [4, 5]. Requirements suggested to introduce MCI drugs in the market by the European Medicines Agency as well as US Food and Drug Administration (FDA), include how to diagnose it with precise parameters and how to discriminate and recognize signs of future AD outbreak [6, 7]. Nowadays, the trials that have been carried out did not show any remarkable amelioration on AD dementia onset and evolution [8]. Thus, MCI lacks interest as a clinical target.

Currently approved treatments by FDA, includes five drugs for the cognitive manifestations of AD. Four of them are the AChEI such as rivastigmine (Exelon™), galantamine (Razadyne™, Reminyl™), tacrine (Cognex™) and donepezil (Aricept™) since in the course of AD there is a substantial loss in cholinergic neurons. Tacrine (Cognex™) is rarely prescribed because of its serious side effects related to liver damage. The fifth one is memantine (Namenda™), which acts as antagonist of N-methyl-D-aspartate (NMDA) receptor and it is the only drug authorized for more advances stages of the diseases that shows effectiveness [9]. All these market drugs slow down modestly the advance of cognitive symptoms, however they produce diverse side effects and do not stop the death of brain cells [10].

Therapeutic strategies for AD relying on one specific mechanism have not resulted to date due to the complexity of the pathophysiological mechanisms involved in this disease. Therefore, the research of new drugs which act in multiple pathomechanisms of AD is needed.

Diet is a relevant factor in maintaining the mental capacities during the aging process. Thus, malnutrition increases the risk of developing MCI and AD. Some epidemiological studies related to neurodegenerative disease factors highlight the influence of nutritional deficiencies on the risk of suffering these diseases [11]. Accordingly, a nutrition based on vegetables, antioxidants and polyunsaturated fats, such as the Mediterranean one, has been seen as an element which could decrease the risk of MCI and AD [12]. Through this chapter, we will describe some natural compounds whose therapeutic administration could possibly inhibit and prevent the AD outbreak or at least reduce the speed of its development. However, further research on the potential of these and other natural substances is needed to establish whether they actually constitute a remedy for AD and other neurodegenerative disorders.

Pathophysiology of AD

Structural modifications especially in cortex, basal forebrain nuclei and hippocampus have been identified as pathological hallmarks of AD [13]. These modifications which are mainly characterized by intracellular neurofibrillary tangles (NFT) containing hyperphosphorylated Tau proteins and extracellular senile plaques that are formed by aggregated β-amyloid peptide (Aβ), cause the loss of cholinergic synapses resulting in a harsh decrease of cholinergic tone [14].

In fact, Aβ deposits have been postulated as the elementary cause of AD and it has been experimentally demonstrated that neurophysiological modifications that are thought to be associated with AD's synaptic dysfunction can be initiated by neuronal exposure to very small concentrations of Aβ oligomers [15]. Therefore, the harmful effects on synapses and mitochondria are caused by the oxidative stress induced by $A\beta_{42}$ oligomers and Tau hyperphosphorylation [16, 17]. Aβ levels in the brain are regulated by several mechanisms: phagocytic clearance by microglia [18] and astrocytes [19], transport through the blood-brain-barrier (BBB) by various cell surface receptors, including the scavenger receptor RAGE (receptor for advanced glycation end-products) expressed on brain endothelium [20] and production or degradation by enzymes [21]. When β- and γ-secretases process the amyloid precursor protein (APP) cause an exaggerated storing of Aβ in the cortex and hippocampus in AD brain. However, APP can be alternatively cleaved by α-secretase pathway originating the soluble amyloid precursor α (sAPPα), which has some neuroprotective activities given that it facilitates axon growth and reduce Aβ formation [22]. APP is also one of the GABAergic transmission regulators. It has been demonstrated that GABA neurotransmission is involved in AD pathogenesis and has a remarkable impact on hippocampal adult neurogenesis, ranging from progenitor proliferation to newborn neuron maturation and integration [23].

NFT deposition, which is mainly misfolded hyperphosphorylated Tau (pTau), is another AD's typical feature. Tau belongs to microtubule-associated protein (MAP) family which is responsible for ensuring the microtubules integrity and stability. Microtubules act by preserving neurons conformation and, therefore, they maintain the contact among neurons. This contact is necessary for signal transduction but also for memory capacities. Furthermore, Tau contributes in the transport of essential nutrients, neurotransmitters and organelles along the axon. This transport is performed bidirectional, both from the nucleus to the periphery and from the periphery to the nucleus and it is necessary to maintain neuronal plasticity. Kinases and phosphatase are the enzymes which control Tau binding to microtubules [24]. However, in AD, there is an imbalance in kinases and phosphatases action resulting in accumulation and chronicle aggregation of pTau which has a lower affinity for microtubules [25]. Then occurs the depolymerization of the microtubules and consequently there is the interruption of axonal transport as well as the loss of the dendritic structure. In addition, given that protease act more easily on Tau than pTau, this one is accumulated [26] sequestering normal Tau and other MAPs and subsequently the microtubule function is even more compromised [27]. Hyperphosphorylated Tau originates amorphous oligomers and amorphous tangles that develop into pair helical filaments (PHF) which co-precipitate along with proteins such as various post-translationally modified and truncated Tau/pTau proteins in NFT [24].

Mutations in the gene for APP or some of the enzymes involved in its metabolism are responsible for only approximately 5% of AD cases [28]. Early-onset and family history cases are related to the genes that codify for presenilin 1 and 2 (PSEN1 and PSEN2) which correspond to chromosome 14 and 21, respectively. However, late-onset family AD is related to chromosome 19, because of the presence of allele e4 [29]. Although age constitutes the main risk factor for dementia, other risk factors such as environment and life style may be also implicated in AD [30]. Stress is currently one of the most known factors that influence sporadic and familial forms but the molecular mechanism for this risk factor is still unknown. However, caspase 6 (Casp6), a mediator of neuronal stress, has been identified as an important mediator in this process. It increases $A\beta$ production [31] and it is also present in NFTs [32]. Therefore, a possible strategy to prevent the progression of the disease could be Casp6 blockage in both AD or in aged individuals even though in elderly patients it is unadvisable to inhibit Casp 6 because of his potential role in physiological functions [33].

Several additional mechanisms such as oxidative stress, inflammatory pathways, metabolic dysfunction, dysregulation of metal ion homeostasis and hormonal influence have been proposed nowadays as alternative or combinatory mechanisms with $A\beta$ and NFT to explain the AD pathophysiology (Scheme **1**).

The brain consumes a large amount of oxygen to obtain adenosine triphosphate (ATP) generating oxidizing agents called free radicals [34, 35]. Free radicals, such as the reactive oxygen species (ROS) including hydroxyl radical (·OH) or superoxide radical (O_2^-) and the reactive nitrogen species (RNS) including nitric oxide (NO) are small molecules with unpaired electron which have high reactivity with macromolecules [36]. Also, they play a significant role as signaling molecules [37]. In addition, production and aggregation of Aβ and Tau phosphorylation can be aggravated by oxidative stress [38]. Nevertheless, it is still not demonstrated if one of the main causes of AD is the presence of free radicals or they are a result of other pathologic processes and act simply at the same time, resulting in more damage. Basal and inducible expression of a variety of antioxidant and detoxifying enzymes such as hemeoxygenase-1 (HO-1), nicotinamide adenine dinucleotide phosphate (NADPH), glutathione (GST) and glutathione synthetase is controlled by a transcription factor named the nuclear factor erythroid-derived 2 (NF-E2)-related factor 2 (Nrf2) [39].

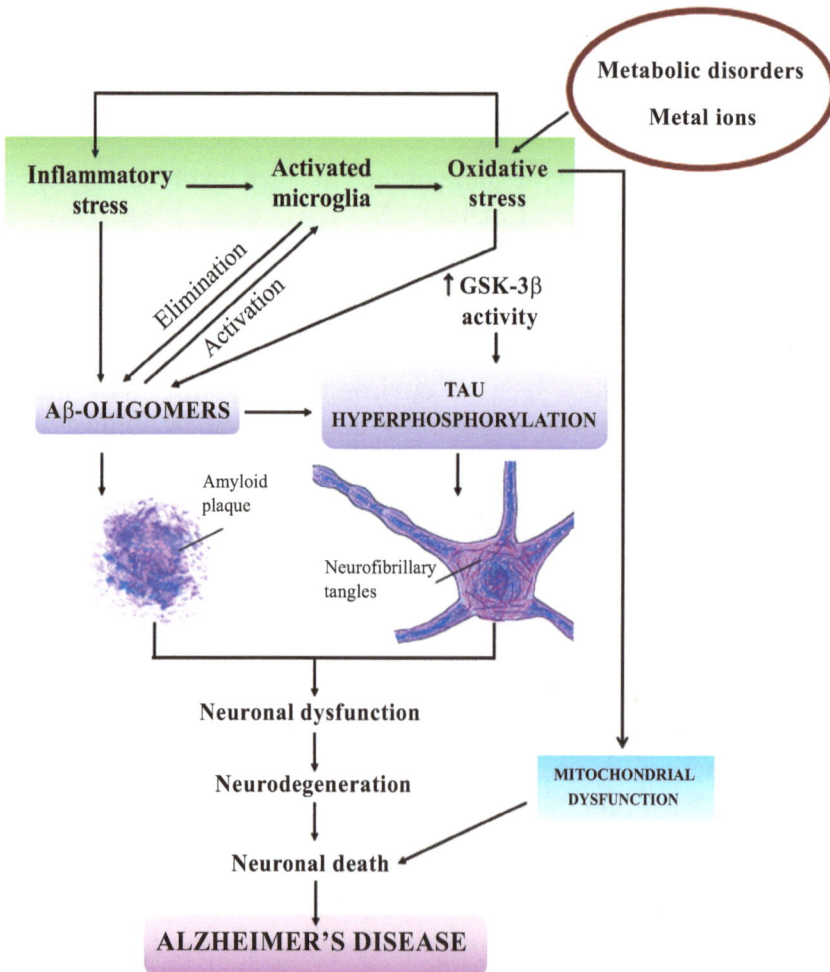

Scheme 1. Mechanisms involved in the pathophysiology of AD.

Different epidemiological studies indicate that neuroinflammation occurs in the early stages of AD. Neuroinflammation can be described as a cause and/or a result of chronic oxidative stress. Inflammation has an important role because it leads to modification in the microglia, from a ramified (resting) to amoeboid (active) morphology, as well as astrogliosis. Astrogliosis is detectable by the presence, around the senile plaques, of an increased number, size and motility of astrocytes [40]. In AD, Aβ may be the molecule that triggers microglial activation, but also molecules derived from degenerate neurons may increase microglial neurotoxicity. Activation of the microglial cells by both oligomeric and fibrous species of Aβ induces ROS production, including superoxide anions and hydroxyl radicals; proinflammatory cytokines such as interleukin-1 (IL-1), IL-6, and tumor necrosis factor-α (TNF-α); chemokines and prostaglandins. All these substances promote neuronal death, leading to exacerbation of AD pathogenesis [41]. Activated microglia also produces a large amount of glutamate. Glutamate is another factor that influences microglial neurotoxicity in AD. The discharged glutamate activates the extra synaptic NMDA receptors of astrocyte-associated neurons [42, 43]. As a consequence, there is an increase in the Ca^{+2} levels that triggers a series of events such as dysfunctional mitochondria pumping out ROS responsible for oxidative damage, Casp3 activation, Tau hyperphosphorylation, excess production of NO, ROS and vascular endothelial growth factor (VEG-F) thereby destroying dendritic spines and neuronal synapses and severing communications within the astrocyte's neurons and beyond [43].

On the other hand, microglial activation in AD may exert a protective effect by releasing neurotrophic factors to promote neuronal survival and seems to be crucial for Aβ phagocytosis and removal from the brain [44]. However, with the progression of the disease, due to the constant stimulation by cytokines and Aβ, microglia acquire a chronically activated phenotype which reduces their clearance function [45, 46] and produces neuronal damage. Thus, modulating microglial activity could slow down neurodegeneration even in the presence of Aβ [47].

The metabolic alterations known as "metabolic syndrome" can be often appreciated in brain aging and are associated with cognitive decline (Scheme **2**). Metabolic syndrome is characterized by insulin resistance or hypercholes-terolemia, among others, which are prodromal AD signs [48]. Even though it has not totally been identified, it has been thought that mitochondrial dysfunction is the link between the decrease of cognitive abilities and metabolic alterations. Brain constantly requires glucose because of the ionic gradients present in pre- and post –synaptic compartments which need to be maintained to sustain cellular excitability, as to ensure transmembrane lipid asymmetries [49]. Because of neuronal high energetic demand and its excitability, a correct mitochondrial activity is essential to brain's correct functioning and metabolism [50].

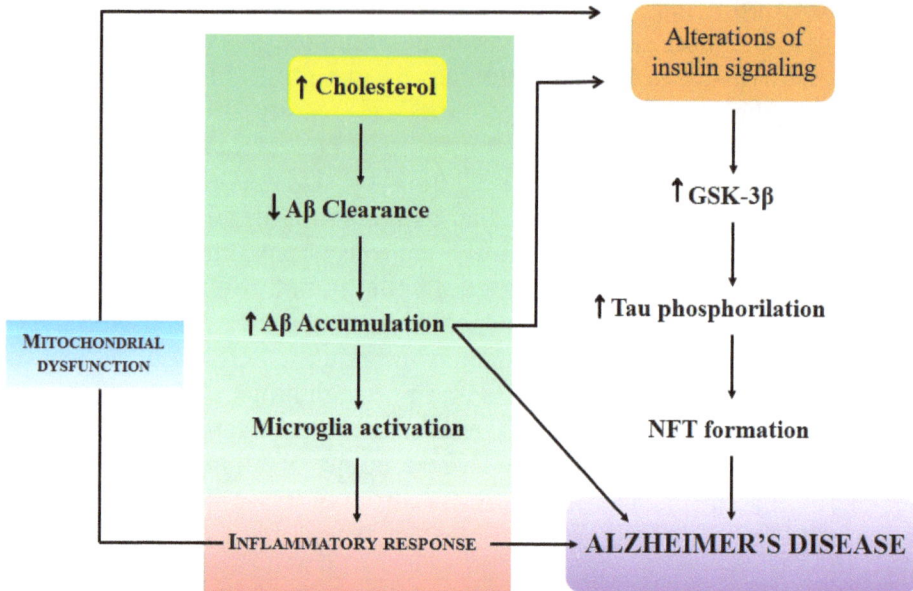

Scheme 2. Metabolic alterations involved in AD.

The analysis of senile plaques in AD patients has allowed the detection of high levels of metal ions (*i.e.*, $Cu^{I/II}$, Zn^{II}, $Fe^{II/III}$) [51]. Deregulation of these ions facilitates Aβ aggregation and allows stabilization of specific toxic conformations of the peptides [52]. In addition, redox-active metal ions can induce overproduction of ROS through Fenton-like reactions with and without Aβ, causing alterations in biological molecules that can lead to neuronal death [53]. Although these metal ions have also been found in the NFT [54], Tau's ability to mediate ROS production in the presence of metal ions remains a focus of study.

Also, hormonal changes during aging are thought to be related to AD pathogenesis. Men aged 69-79 with AD have lower serum total testosterone levels [55] and additionally it has been detected a reduction in the brain levels of this hormone. This reduction seems to play an important role because neuroprotective effects such as the reduction of oxidative stress and inflammatory processes that reduce Tau phosphorylation and Aβ production have been associated with testosterone [56]. Similarly, estrogens have shown a protective role against AD, because they are involved in the regulation of cognition, neuron viability, and Aβ accumulation [57]. Low brain levels of estrogens and androgens have been found in AD women patient aged 80 and older instead it is not a common pattern in those between 60-79 years old [58].

Aβ deposits and apoptosis justify synaptic loss in hippocampus and cerebral cortex found in AD patients. The protein p53 and transcription factor $FoxO_3$ induce neuron death in response to Aβ [59]. Apoptosis can be also mediated by oxidative stress, disorders of glucose metabolism or mitochondrial damage. The apoptotic process is mediated by the Bx/Bcl-2 system which is regulated by the cyclic adenosine monophosphate (cAMP) response element binding protein (CREB). CREB plays a critical role in neuronal survival [60] and its suppression or disruption results in a progressive neuronal apoptosis and neurodegeneration [61]. The rate of hydrolysis of cAMP is controlled by cyclic nucleotide phosphodiesterases [62].

Although cholinergic neuron loss and an overall decrease in choline acetyltransferase (CAT) activity seem to have a primary relevance in AD pathology, other neurotransmitters such as dopamine (DA), noradrenaline (NE), histamine (HA), and serotonin (5-HT) have also an important role in the alteration of cognitive processes in AD [63, 64].

Serotonin (5-hydroxytriptamine, 5-HT) has been recognized as a neurotransmitter with short- and long-term memory repercussions due to the action of activating 5-HT specific receptors, as well as the modulation of other neurotransmission systems such as cholinergic, glutamatergic, dopaminergic and GABAergic [65 - 67]. In the CNS of AD patients, 5-HT level is lower, and even a lower proportion of 5-HT receptors ($5HT_{1A}$, $5HT_{1B}$, $5HT_{1D}$, $5HT_{2A}$ and $5HT_6$) is observed [68]. This fact affects to a greater extent the raphe nuclei, as has been observed in postmortem studies. These neurons provide 5-HT projections to the septum and the hippocampus, which are severely affected in AD [69]. Also in these studies a degeneration of the monoaminergic neurons has been observed, which is produced by accumulation of neurotoxins like Aβ [70]. 5-HT has an inhibitory action of glutamatergic neurotransmission in the spinal cord as well as the hippocampus. Therefore, an increase in 5-HT levels in patients with AD, specifically around Aβ plaques, may play a beneficial role against the NMDA-induced neurotoxicity through the inhibition of calcium currents and membrane hyperpolarization [71]. These effects are mediated by the activation of $5\text{-}HT_{1A}$ and $5\text{-}HT_{1B}$ receptors [72]. Consequently, an increase in 5-HT fibers may be an intrinsic protective mechanism in response to Aβ-induced excitotoxic damage in AD [73]. In addition, low expression of serotonin transporter (SERT) has been noticed in the temporal cortex, frontal cortex and hippocampus of AD patients, which is responsible for the alteration of the 5-HT homeostasis/transmission in these areas and is related to the cognitive dysfunction in AD [74, 75].

In brains of AD patients, the levels of DA, its precursor; L-3,4-dihydroxyphenylalanine (L-DOPA), and its metabolite (DOPAC) are significantly lower in the cingulated gyrus, amygdale, striatum and raphe nuclei [76]. Moreover, also in the cerebrospinal fluid (CSF) [77] and urine [78] have been found DA lower levels. Likewise, D1 and D2 receptors in the striatum as well as DA transporter present lower levels in these patients [79, 80]. Thus, there seems to be a direct relationship between the deterioration of the DA-ergic system and severity of AD.

Also, lower levels of noradrenaline (NE) have been found in hippocampus, the frontal and temporal cortices but also in cingulated gyrus in AD [81]. Likewise, that DA, levels of NE determined in the CSF and urine are lower in AD patients [82]. It is important to highlight that in early AD onset it is more common to find NE decrease than in the late onset ones. DA is the precursor for NE synthesis and this is metabolized mainly by MAO-A and COMT. Both these and the enzymes involved in NE biosynthesis suffer alterations in AD [83]. NE and DA can protect neurons *in vivo* because they reduce Aβ production and therefore the effects of its accumulation [84].

Histamine (HA) is synthesized by neurons in the posterior hypothalamus, especially by hypothalamic tuberomamillary nucleus (TMN) [85]. In AD, these neurons show an important degeneration leading to cell death, cytoskeletal alterations and consequently a decline in HA levels in the hippocampus, temporal, occipital and frontal cortex [86]. Although overall HA levels appear to be decreased, a higher concentration of this neurotransmitter is observed in CSF and serum of AD patients. It is thought to be caused because cortical and subcortical neurons and inflammation-activated microglia in the central nervous system (CNS) and mast cells in the periphery [87] increase the HA production. Most H3 receptors which are dominant in TMN, are presynaptically located autoreceptors that modulate the release of HA but also of other neurotransmitters such as ACh, DA, and NE [88].

In addition to the previously described neurotransmitters, adenosine is an endogenous substance which is present in large amounts in the mammalian brain. It is responsible for modulating various synaptic functions in the CNS. Adenosine regulates and integrates neuronal activity, in addition it has an impact on relevant brain functions including sleep and arousal, cognition and memory and neuronal damage and degeneration [89]. The effects of adenosine are mediated by at least four adenosine receptor subtypes (A_1, A_{2A}, A_{2B} and A_3) that can inhibit (A_1 and A_3) or enhance (A_2) neuronal communication through neurotransmitter release [90]. Post-mortem brains with AD show decreased A_1 adenosine receptor density in the hippocampus [91]. Recent researches demonstrated that the activation of the adenosine receptors (specifically the A_{2A} subtype) can lead to the down-regulation of the inflammatory response and also the prevention of Aβ-induced synaptotoxicity by inducing the production of IL-10, the major anti-inflammatory cytokine [92]. Additionally, some direct and indirect data point to alterations of purine metabolism in AD. These alterations may break down the balance between intracellular nucleoside triphosphates and their metabolites and dependent products, modifying their release into the extracellular space as neurocrine factors in AD [93].

Recently epigenetic mechanisms controlling gene expression have been described to explain the late-onset in sporadic AD in addition to the multiple genetic and environmental risk factors above described. "Accordingly, abnormal patterns of histone acetylation and methylation, as well as anomalies in global and promoter-specific DNA methylation, have been documented in AD patients, together with a deregulation of noncoding RNA. Importantly, epigenetic deregulation interfaces with core pathophysiological processes underlying AD: excess production of $A\beta_{42}$, aberrant post-translational modification of Tau, deficient neurotoxic protein clearance, axonal-synaptic dysfunction, mitochondrial-dependent apoptosis, and cell cycle re-entry. Reciprocally, DNA methylation, histone marks and the levels of diverse species of microRNA are modulated by $A\beta_{42}$, oxidative stress and neuroinflammation" [94].

NATURAL PRODUCTS FOR THE TREATMENT OF AD

Plants and different organisms have been traditionally used as a source of bioactive compounds. In this chapter we review natural compounds with potential neuroprotective benefits and therefore useful in the prevention or/and treatment of AD (**Table 1**). Some of these compounds have been isolated from medicinal herbs and health food supplements, many of them used in traditional Chinese medicine. Also some extracts from kitchen species have shown neuroprotective properties. In other cases, the plant itself is used to achieve the therapeutic benefit.

Fatty Acids (FA)

Alterations in the composition of brain lipid profile including phospholipids, sphingomyelin, ceramides and gangliosides appear to play an important role in AD. Moreover, studies suggest that lipid composition in plasma and CSF could serve as AD biomarkers as well as prognostic indicators for AD therapies [95].

Cholesterol is a sterol lipid which is involved in the structure and functions of cellular membrane. Cholesterol is mainly present in the brain, being the richest organ in this fatty acid; 25-30% of the total in the body. The homeostasis of this fatty acid in the brain is tightly regulated, so a low level of cholesterol induces an increase of $A\beta$ and high level of cholesterol is a risk factor for amyloidogenesis [96].

Table 1. Different categories of natural products discussed in this review.

1. Fatty acids	1.1. Docosahexaenoic acid (DHA)	
	1.2. Arachidonic acid (ARA)	
	1.3. AC-1202	
	1.4. AC-1204	
2. Polyphenols	2.1. Flavonoids	2.1.1. Flavones
		2.1.2. Flavonones
		2.1.3. Flavanols
		2.1.4. Flavonols
		2.1.5. Anthocyanins
	2.2. Curcumin	
	2.3. Resveratrol	
	2.4. Chalcones	
	2.5. Xantones	
3. Alkaloids	3.1. Isoquinolines	
	3.2. Physostigmine	
	3.3. Huperzine A	
	3.4. Caffeine	
	3.5. Rhynchophylline and isorhynchophyline	
	3.6. Indirubin	
	3.7. Harmine	
	3.8. Morphine	
	3.9. Nicotine	
4. Vitamins	4.1. Vitamin E	
	4.2. Folate, Vitamins B2, B3, B6 and B12	
	4.3. Vitamin B1 (thiamine)	
	4.4. Vitamin D	
	4.5. Vitamin A and other retinoids	
	4.6. Vitamin C	
5. Coenzyme Q10		

(Table 1) contd.....

	6.1. Acetyl-L-carnitine
	6.2. Glutamine
	6.3. L-arginine
6. Amino acids	6.4. S-adenosylmethionine
	6.5. L-tryptophan
	6.6. S-allyl-L-cysteine
	6.7. Homotaurine
	7.1. Melatonin
	7.2. Testosterone
	7.3. Pregnenolone and Dehydroepiandrosterone
7. Hormones	7.4. Estrogen and Phytoestrogen
	7.5. Progestogens
	7.6. Insulin
	8.1. *Ginkgo biloba* L.
	8.2. *Panax ginseng* Meyer
	8.3. *Withania somnifera* Dunal
8. Plants and herbal extracts	8.4. *Melissa officinalis* L.
	8.5. *Crocus sativus*
	8.6. *Polygonum multiflorum* Thunb.
	8.7. *Salvia spp.*

It has recently been described how lipids determine the APP proteolysis. α-Secretase acts on non-lipid rafts, whereas β-secretase (BACE) is responsible for APP cleavage in lipid rafts. Binding of APP to lipid rafts is promoted by the increase in membrane cholesterol [97]. There is previous evidence in the BACE-1 inhibitor's role modifying the BACE-1 location related to raft, given that this secretase appears in non-lipid raft after the action of the BACE inhibitor, instead of linked to lipid raft. Ceramides, which are central molecules of sphingolipid biosynthesis and catabolism, contribute to the increase of Aβ levels from APP because the stabilize BACE-1 [98]. In turn, the oxidative stress associated to the cellular membrane is induced by Aβ, which in this way disrupts the ceramide and cholesterol metabolism and therefore causes their accumulation [99]. In addition, the increase in ceramide concentration stimulates ROS generation, establishing a relationship between oxidative stress and metabolism of sphingolipids with detrimental consequences for neuronal survival [98].

Additionally, the increase of 4-hydroxynonenal (HNE) which is produced by oxidative degradation of lipids and it is particularly harmful, is accompanied with polyunsaturated fatty acids (PUFAs) depletion [100]. Epidemiological studies suggest that increased intake of PUFAs, primarily n-3 PUFAs, such as docosahexaenoic acid (DHA) and eicosapentaenoic acid (EPA), contribute to maintain brain function in adults [101]. In AD patients, levels of DHA such as arachidonic acid (ARA) and other n-3 PUFAs are lower in blood or raft and brain phospholipids in comparison to those shown by control subjects [102, 103]. CSF profile of n-3 FA is modified with oral n-3 PUFAs supplementation, suggesting that these FAs can cross the BBB in adults [104]. The n-6 PUFA, ARA, is related to memory. Consequently, the supplementation of aged rats with ARA has demonstrated to relieve spatial cognition deficits related to aging [102].

In vitro studies have also shown the role in APP proteolysis of triglycerides combined with PUFAs and/or caprylic acids [105].

Docosahexaenoic Acid (DHA)

DHA is an omega-3 fatty acid that belongs to the broader category of PUFAs. It can be obtained from alpha-linoleic acid or isolated directly from breast milk or fish oil [106]. DHA can be found in high proportion in oceanic fish oils from cold-water. In these and other multicellular organisms comes from photosynthetic and heterotrophic microalgae, and will be more concentrated depending on their content in the food chain. DHA can also be obtained commercially from microalgae such as *Crypthecodinium cohnii* and another of the genus *Schizochytrium* [107, 108].

Research has shown that DHA is particularly important for the brain at different stages of life. It is involved in neurological fetal growth, synthesis and degradation of lipids during childhood and mid-age, and cognition in the elderly. In brain neural membranes, DHA acts by favoring both fluidity and permeability of the membrane as well as the receptor structure and quantity. Therefore, DHA will be responsible for vehicle-mediated transport of nutrients through the cell membrane, enzymatic activities and intercellular communication [109]. In aged animals, memory impairment is accompanied by a decrease of cerebral DHA levels. Moreover, the lowering of brain DHA levels in Alzheimer's patients is linked to memory loss and learning difficulties [110].

DHA is involved in the antioxidant defense, subserves the oxidative stress diminution and prevents memory loss. These actions can be explained because lipid peroxides (LPO) inversely correlate with the DHA/ARA ratio and in rats with long-term administration of DHA has been shown a lower correlation [111]. Additionally, low levels of DHA in plasma have been linked in cross-sectional analysis with dementia and particularly AD. Moreover, patients with highest blood DHA concentrations are less likely to develop dementia than those with lower levels of DHA in the blood, according to a study in which patients were followed up for 9 years [112]. Up to now the clinical trials carried out with DHA supplements in patients with mild to moderate Alzheimer's disease have not achieve the expected results. Nevertheless, it is not ruled out that they are effective for those cases in which cognitive symptoms have not appeared yet [113]. The association between n-3 PUFA consumption and Alzheimer's risk has been studied in several epidemiological studies and these have determined that they are inversely correlated. In particular, experience has demonstrated that a fish-rich diet as a source of this omega-3 fatty acid is associated with a lower probability of cognitive function impairment [111]. More recently, krill oil (KO) has been highlighted as one of the main sources of n-3 PUFA [114]. This oil comes from *Euphausia superba,* an Antarctic microcrustacean which is rich in n-3 PUFAs, such as EPA and DHA, as well as phospholipids, and astaxanthin (ASTA). KO has a good bioavailability, which makes it a better choice as a source of n-3 PUFA than fish oil (FO). On the other hand, it stands out as a good option because phospholipids in KO are mainly EPA and DHA, while FO contains mainly triglycerides [115]. KO produces slight activation of astrocytes in motor cortex and corpus callosum of Wistar rats. However, its long-term administration does not significantly alter the redox or inflammatory index. Likewise, it has not been experimentally demonstrated that other sources of PUFA such as FO, ASTA-rich algal biomass or them in combination, had an impact in the baseline redox or neuro-inflammation in brain regions related to neuromotor in experiments with animals [115].

Arachidonic Acid (ARA)

ARA is with DHA a major component of the cellular membranes and thus also it has an impact in the preservation of physiology and psychological functions, as well as in the development and maintenance of the CNS [116]. Eicosanoids, as prostaglandins, leukotrienes, and thromboxanes that are obtained from ARA metabolization, are potent mediators of inflammation. Prostaglandin E2 (PGE2) which is synthesized by cyclooxygenase (COX) action stimulates the Aβ production through internalization of PGE2 receptors [117]. AD patients seem to present higher levels of PGE2 in the brain than healthy subjects [118], whereas a lower level of phospolipase A2 improves the cognitive deficits in studies carried out with animal models with AD [119]. Despite the above described, some studies have determined that ARA could have protective effects. Therefore, a rich diet in ARA and DHA could stimulate synaptic plasticity in aged rats [120] and improve cognitive function [121]. On the other hand, *in vitro*, neuronal exposure to ARA and phospholipase A2 increases secretion of sAPPα and decreases Aβ levels in SH-SY5Y cells. It results in an increase in membrane fluidity, which impedes proper enzyme functionality related to Aβ production [122, 123]. In addition, an

ARA-enriched diet increases ARA levels in the mouse brain cortex and decreases levels of insoluble Aβ in mouse brain Tg2576 [124].

Although these data may suggest that ARA supplements can improve the cognitive dysfunction related to brain disorders or ageing, there are not significant results achieved in AD patients [125].

AC-1202

AC-1202, as known as Ketasyn or Axona, is a proprietary formulation of processed coconut or other oils and glycerin, which contains medium chain triglycerides (MCTs). This formulation allows safe elevation of serum ketone bodies, even when the diet provides carbohydrates. The basis of MCT use is due to a theory linking brain glucose hypometabolism with an increased risk of AD [126]. In long fasting periods, due to the low glucose levels the main source of energy for the neurons become the ketones or ketone bodies, obtained by metabolizing fatty acids by the liver.

The brain uses two ketones named β-hydroxybutyric acid and acetoacetate as the only significant back-up fuel for glucose [102]. Although it would be more logical to think of direct supplementation of glucose to improve the low metabolism in the brain, the consequence could be levels of glucemia too high leading to an old-age diabetes [101].

Studies carried out with AD transgenic mouse models in which lower Aβ levels are appreciated as well as protective effects on neuronal cells also support the use of ketone bodies as an alternative energy source. However, the neuroprotective effect of MCTs in humans remains to be determined. Clinical trials performed with MCT and their following post hoc analysis indicate that it could be a treatment for patients who are apolipoprotein E4 (ApoE4) (-) and/or patients with single-nucleotide polymorphisms of insulin-degrading enzyme or IL-1β genes that modified the function, gaining and losing it, respectively [127]. Conversely, no beneficial effects were observed in ApoE4 (+) patients. Therefore, it has been proposed to carry out a genomic study prior to the administration of MCT [128]. However, as low level of neuronal metabolism is one of the primary events in this disease, it has been propounded to administer MCTs in potentially at-risk populations (before AD strikes). In this way, MCT could produce greater benefits regarding neurodegeneration, principally in patients with ApoE4.

AC-1204

AC-1204 is the follow-up formulation to AC-1202 medical food product. AC-1204 is a caprylic triglyceride supplement approved for cognitive function improvement in early stages of AD. It increases the ketone bodies synthesis because it improves the mitochondrial mechanism and therefore provides an alternative energy source for brain cells. Therefore, its aim is to correct the metabolic dysfunction relative to AD. The rationale behind AC-1204 is to boost cellular metabolism in Alzheimer's by providing a fuel alternative to glucose as previously described for AC-1202. Caprylic acid is also metabolized into acetoacetic acid and β-hydroxybutyric acid, which can be transformed to acetyl-CoA to produce energy *via* the citric acid cycle [129].

A new formulation as medical food similar to AC-1204 is nowadays in Phase II/III of clinical evaluation with the aim of studying its effects on participants' cognitive status compared to AC-1204 or a placebo. The drug is a powder that will be mixed with water, other liquid, or soft food [130].

Polyphenols

Recently, research has been growing importance in the study of polyphenols because of their multiple effects as Aβ-inhibitors, metal chelators, and because they are able to prevent mitochondrial dysfunction and apoptosis. Additionally, they show interesting antioxidant and anti-inflammatory properties [131, 132]. However, nowadays it is not clear whether neuroprotection of natural polyphenols is related to the described mechanisms or they could affect novel targets. Epidemiological studies suggest that the incidence of AD may be inversely related to the consumption of polyphenols in the diet [131, 133]. Although the mechanism of action is not enough known to understand the action of polyphenols against AD, the results for the most studied polyphenols that could support their possible interest are presented below. Other polyphenol derivatives such as salvionolic acid B, tannic acid and rosmarinic acid have also shown to inhibit Aβ fibrillization through different mechanisms [132, 134].

Flavonoids

Flavones

Apigenin (4',5,7-trihydroxyflavone) is a non-mutagenic flavone with low toxicity. This compound is found in varieties of fruits and vegetables such as pepper and parsley as well as in medicinal plants such as *Elsholtzia rugulosa* Hemsley and *Carduus crispus* L [135]. Specifically, in the plant of camomile can be found up to 68% apigenin of total flavonoids in its flowers, and it can be also found in parsley and peppermint [136].

Apigenin is considered very safe since even when it is administered at high doses, toxicity has not been described. It crosses the BBB and shows effect against oxidation and inflammation [136]. The mechanism of action is based on the inhibition of pro-inflammatory mediators in LPS-activated murine cell lines, as well as decreasing of IL-6 and TNF-α levels in interferon gamma (IFN-γ) activated murine. Also, apigenin has been shown to suppress inflammatory mediators such as NO and prostaglandin in both microglial and macrophage mouse cells in a dose-dependent relation. These actions can be explained because apigenin is able to inhibit inducible nitric oxide synthase (iNOS) and COX-2 [137]. Furthermore, it has been shown to suppress phosphorylation by kinases responsive to stress stimuli such as p38MAPK and c-Jun N-terminal kinase (JNK) without repercussion on the activity of extracellular signal-regulated kinase (ERK) [136].

Regarding AD, apigenin protects against damage induced by $A\beta_{25-35}$, as has been determined in microvascular endothelial cells of rat models. In $A\beta_{25-35}$ intracerebroventricularly injected mice, apigenin manages to improve learning and memory disabilities. In these mice a decrease in oxidative damage and an improvement in cholinergic neuronal transmission have also been observed in addition to it preserves the BBB integrity [138]. In terms of its *in vitro* activity, apigenin has been shown to protect against stress-induced apoptosis of endoplasmic reticulum in the HT_{22} murine hippocampal neuronal cells [139], quinoloinic acid-induced excitotoxicity in humans and neurotoxicity induced by glutamate in murine cerebellar and cortical cell cultures [140]. Its neuroprotective action could be due, at least partially, to its antagonism of NMDA and GABA receptor channels in neurons. Furthermore, apigenin protects against copper mediated Aβ toxicity in a human AD neuroblastoma cell model. Mechanisms to explain the neuroprotection exerted by apigenin against Aβ toxicity may include regulation of redox imbalance, preservation of mitochondrial function, inhibition of p38MAPK and JNK pathways, and apoptosis depression [135, 141].

Despite the above, there are some limitations to explain the neuroprotective effect of apigenin against Aβ peptides [138, 141 - 143] and others of their actions. However, apigenin represents a potential therapy against AD

Baicalein (5,6,7-trihidroxyflavone) is found in the root of *Scutellaria baicalensis* Georgi, a medicinal plant native to China. It is believed that baicalein by reducing oxidative stress can inhibit Aβ-induced neurotoxicity in PC12 cells [144]. In addition, Aβ production has been reduced by baicalein in different animal models. It elevates sAPPα promoting the nonamyloidogenic α-secretase proteolytic pathway. This action results in an increase in cognitive performance of treated animals compared to untreated [145]. However, further research would be needed for its use in humans against AD.

Luteolin (3',4',5,7-tetrahydroxyflavone) is a flavonoid present in a wide range of vegetables, fruits and medicinal plants. Luteolin shows very strong antioxidant properties by inhibition of ROS generation and scavenging activity. It also inhibits both types of ChEs (AChE and BuChE) and BACE-1 [146].

Flavonones

Hesperidin, a common flavonone glucoside of citrus fruits, has shown antioxidant activity and free radical scavenging properties *in vitro* as it has been described previously for other flavonoids [147]. Hesperidin treatment acts on ROS and LPO achieving a decrease in their levels, while it also promotes the activity of HO-1, oxide super-dismutase (SOD), catalase, and GSH peroxidase (GSH-Px). Therefore, the oxidative stress is reduced. On the other hand, hesperidin exhibits an anti-inflammatory action by activating Akt / Nrf2 signaling and inhibiting RAGE / NF-κB [148]. Also, hesperidin has been shown to improve glucose utilization and protect cultured cortical neurons against the Aβ-induced neuronal damage [149]. Long treatment with hesperidin results in the reduction of learning and memory deficits and improves locomotor activity in a transgenic mouse model of AD. The increased in the cognitive function could be explain due to a reduction in mitochondrial dysfunction through the inhibition of glycogen synthase kinase-3β (GSK-3β) activity coupled with an increase in anti-oxidative defense [150].

Naringenin is also a natural flavonone abundantly found in citrus fruits and grapes. Naringenin has anti-inflammatory and neuroprotective effects and improves the cholinergic function due to its antioxidant property and through inhibition of ChE activity in the hippocampal region. Pretreatment with naringenin has been shown that, by decreasing lipid peroxidation and apoptosis, attenuates the impairment of learning and memory induced by Aβ. This positive action may be partially carried out by an estrogenic pathway [151].

Flavanols

Epigallocatechin (EGCG) is the main polyphenolic constituent (over 60% of the total catechins) of green tea. EGCG stimulates α-secretase activity and therefore the Aβ aggregation is prevented in animals. In addition, it modulates neuronal survival and apoptosis by regulating gene expression and signal transduction pathways and mitochondrial function [152]. It has now reached phase II/III of clinical trials for patients with early AD [153]

Epigallocatechin but also catechin are capable of soothe microglia and/or astrocyte mediated inflammation through inhibition of iNOS and COX-2 expression, NO production, cytokine release and NADPH oxidase activation

[154]. The properties of this compound are mainly due to its capacity of modulate various pathways related to protein and lipid kinase signaling, as tyrosine kinase, protein kinase C and MAPK. As far as these favanols prevent the activation of Casp3, it explains their potent anti-apoptotic action [155].

Epicatechin protects neurons against oxidative damage through JNK suppression. Also it is active against excitotoxicity induced by NMDA and stimulates the activation of CREB in cortical neuronal cells. However, in transgenic mice lacking HO-1 or Nrf2 it has no neuroprotective effect [156].

Flavonols

Myricetin (3,3',4',5,5',7-hexahydroxyflavone) is a common flavonol present in a variety of plants such as berries, fruits, vegetables and medicinal plants. Myricetin acts as a metal chelator which may lead to a complex with a 1:1 or 1:2 metal / myricetin binding. Possible positions for metal chelation include positions between the 4-oxo and 3 or 5-hydroxy groups [157]. Myricetin is able to modulate Aβ aggregation by binding to fibrillar Aβ reversibly although it does not bind to the monomeric species. Moreover, it is more effective to modulate metal-induced Aβ aggregation than in the absence of metal [158]. Additionally, myricetin inhibits glutamate-induced neuronal toxicity by different biochemical pathways. It affects modulation of NMDA receptors by phosphorylation and therefore decreases intracellular Ca^{2+} overload, inhibits ROS production and activation of Casp3 glutamate-induced [159].

Quercetin (3,3',4',5,7-pentahydroxyflavone) is a flavonol which could be found in a variety of fruits and vegetables such as onions and apples and red wine. This flavonol is characterized because of its potential pharmaceutical uses. Among those, it may have neuroprotective effects and slow down the progression of degenerative diseases. These effects can be explained because quercetin shows high oxygen radical scavenging activity as well as inhibition of lipid peroxidation [160]. *In vivo*, long term administration of quercetin to different AD mice models was able to improve learning and memory impairments. Therefore, quercetin represents a potential effective therapy against cognitive impairment in AD [161, 162].

Kaempferol (3,4',5,7-tetrahydroxyflavone) is also a flavonol which can be found in the same sources of quercetin. As myricetin and quercetin, it inhibits BACE-1 [163]. Also, as above described for others flavonoids, it suppresses lipid peroxidation in PC12 cells and thus might be beneficial in protecting cells against oxidative damage [164]. *In vivo*, also kaempferol improves Aβ-induced memory impairment in mice [165].

Anthocyanins

Anthocyanins are flavonoids that give color to leaves, fruits and flowers whose pigments are blue, purple or red. These compounds stand out for their anti-inflammatory and neuroprotective effect linked to their antioxidant activity. Its antioxidant activity is a consequence of its free radical scavenging activity reducing ROS. In addition, anthocyanins produce an increase of antioxidant enzymatic activity [166].

Anthocyanins have shown to inhibit the neurotoxic effects of Aβ *in vitro* and *in vivo* preserving normal cellular functionality. They are able to reverse protein expression of mitochondrial apoptotic pathway (Bax, cytochrome C, Casp9 and Casp3) induced by Aβ as well as important AD markers such as APP, p-Tau and BACE-1 [167].

Curcumin

Curcumin is a diaryl heptanoid polyphenol isolated from *Curcuma longa* L., a herbaceous plant, belonging to the family Zingiberaceae. It has been widely cultivated in countries of Asia such as India and China because of its link with traditional Chinese and Ayurvedic medicine. Furthermore, in tropical climate countries it is also widely cultivated. The medicinal properties of the plant are in the rhizome and they characterized by their yellow color, characteristic color of the curry. In addition to its color, its flavor is also characteristic so it is widely used for culinary purposes [168]. Curcumin, the active compound, presents two substituted planar aromatic rings linked by a chain with α,β-unsaturated carbonyl groups. Stable enols from diketones could be easily deprotonated to form enolates [169]. A wide variety of pharmacological properties has been attributed to curcumin, related to the inflammation, oxidative stress, cancer, besides antimicrobial activities and wound healing [170]. It addition, a beneficial role has been described against several neurodegenerative diseases such as AD, multiple sclerosis, Parkinson disease, epilepsy, cerebral injury, neurodegeneration related to aging, and neurological diseases such as schizophrenia, (Creutzfeldte-Jakob disease), neuropathic pain and depression and others such as spongiform encephalopathies [171]. It is thought that in India, the widespread consumption of curcumin used for culinary purposes, is related to lower prevalence of AD [172].

Low doses of curcumin decrease oxidative stress levels exacerbated by Cu[II] and significantly increase the levels of matured APP. An increased of concentration is related to a lower level of matured APP. Therefore, curcumin treatment markedly increases the ratio of matured/immature APP at lower concentrations whereas the ratio decreases significantly at higher concentrations of curcumin [173]. Curcumin supplementation has been shown to significantly alleviate cognitive deficits resulting from the intracerebroventricular (icv)-streptozotocin (STZ) administration. In addition, curcumin has protective properties against oxidative damage in the hippocampus and cerebral cortex [174]. Curcumin has been found to be effective in reversing behavioral and biochemical cognitive parameters induced by icv-STZ. As a consequence, it is under study that the consumption thereof can be protect against sporadic dementia related to aging of Alzheimer type.

In *in vivo* studies, curcumin administration reduces both Aβ accumulation and oligomerization in the animals [175]. It has also been seen that it is involved in the upregulation of BACE-1 by suppressing it, so that it has a protector effect against the mechanism of Aβ [176]. In relation to the activity of Tau, properties against aggregation of this have been attributed to curcumin [177].

Curcumin is able to cross the BBB and thus exerts anti-inflammatory effects, however a point against curcumin is its low bioavailability [178 - 180]. Several clinical trials using normal curcumin even in high doses did not show any difference in plasma and CSF levels of $A\beta_{40}$ and $A\beta_{42}$ or Tau among treatment groups. This may be explained because the plasma levels reached were lower than 1mM in all cases. This concentration may result ineffective taking into account the data obtained from cell culture and animal assays [181, 182]. In order to solve bioavailability problems, several formulations of curcumin have been developed, such as nanoparticles, liposomes or inclusion complexes [183]. Among the curcumin formulations developed so far, the formulation "LongVida" (VS Corp) (650 mg) has increased bioavailability, reaching a maximum of 22 mg /l in plasma. This is a good result since the first formulations of curcumin did not reach the detection threshold of the same. A study carried out in humans, being randomized, double-blind and placebo controlled with a daily dose of this formulation has shown an improvement in memory and mood after one month treatment [184, 185]. However, some studies suggest that curcumin can produce major benefits in people without immediate disease states [186].

Resveratrol

Resveratrol (3,4',5-trihydroxy-*trans*-stilbene) is a polyphenol which is found in different plants, especially in the skin and seeds of the grapes, and a phytoalexin produced against bacteria or fungi [187].

Resveratrol displays antioxidant and anti-inflammatory activities as well as anti-amyloidogenic effects, therefore several hallmarks for AD are well recognized target for resveratrol modulation [188].

Resveratrol inhibits pro-oxidative genes such as NADPH and myeloperoxidase while favoring the activity of several enzymes such as SOD, catalase, thioredoxin and GSH-Px, which are related to antioxidant properties. Therefore, resveratrol acts in two pathways. Firstly, it suppresses the production of oxygen free radicals, but it also acts as direct scavenger of free radical in tissues [189, 190]. In addition, resveratrol has proven to protect the rats from Aβ-induced neurotoxicity because it suppresses iNOS production, which is involved in the Aβ-induced lipid peroxidation and HO-1 down regulation [191]. Also it protects against Aβ-metal complexes due to its beneficial effects against metal ion dyshomeostasis through different mechanism [192]. Other mechanisms that explain the anti-amyloidogenic activity are related to SIRT1 activation [193] as well as reduction of poly [ADP-ribose] polymerase 1 (PARP-1) cleavage by resveratrol [194].

Resveratrol is effective in reducing the inflammatory status in *in vitro* and *in vivo* neuroinflammation models [195, 196]. Although mechanisms by which resveratrol attenuates neuroinflammation are still not completely clear, it has been described that resveratrol can reduce the release of inflammatory mediators such as IL-1β, IL-6 and TNFα through the inhibition of cellular cascade signaling pathways involving NF-κβ and activator protein 1 (AP-1) [197]. Likewise, resveratrol can inhibit PGE formation by activated microglial cells [198].

Because of its a lipophilia, resveratrol can readily diffuse through BBB [199]. Evidences about therapeutic effects of resveratrol in AD are increasing, however it has some limitations as its low bioavailability, low aqueous solubility and chemical instability [200]. Recently a phase II trial of resveratrol in individuals with early stages of AD has been performed. Although resveratrol was administrated in high dose due to its low bioavailability, it resulted safe and well-tolerated. In addition, resveratrol alters some AD biomarker trajectories in AD patients [201]. Although results suggest CNS effects, more studies are needed to determine whether resveratrol is actually beneficial.

Nowadays, different drug delivery systems are being developed to improve the bioavailability and solubility of resveratrol. Some examples are encapsulation in liposomal formulations, resveratrol-protein complexes, solid-lipid nanoparticles, pectinate delivery systems, chitosan microspheres or polyethylene glycol (PEG) derivatives [202, 203].

Chalcones

Among chalcones, the chalcone glycoside hydroxyl-safflor yellow A (HSYA) has been particularly studied. It is the active compound that forms part of the yellow pigments of *Carthamus tinctorius* L. It has been demonstrated that HSYA has a strong anti-inflammatory function which may contribute to the neuroprotection.

HSYA inhibits the expression of pro-inflammatory mediators whilst restores the expression of anti-inflammatory mediators [204, 205]. Additionally, HSYA suppresses $A\beta_{42}$-induced activation of microglia and astrocytes and ameliorates the memory deficits in Aβ-induced AD mice [205].

Xantones

Mangiferin is a glucosylxantone majorly present in leaves and bark of *Mangifera indica* L. *In vitro*, mangiferin has shown to protect oligodendrocytes and brain neurons from excitotoxicity [206]. *In vivo*, mangiferin improves the learning and memory. Administration of mangiferin in young mice decreases oxidative stress. Therefore, this compound could be used as a treatment against neurodegenerative and natural aging processes as well as a treatment against AD [207].

Other xantones isolated from *Gentiana campestris* L. have exhibited significant inhibition of AChE activity [208].

Alkaloids

Alkaloids constitute a large group of compounds; their common feature is the presence of nitrogen atom(s) in a cyclic ring. Plants, particularly flowering ones, are abundant of these active ingredients. Many pharmacological applications have been described for alkaloids, however only few of them are nowadays sold in the drug market. Despite that, the importance of these compounds in the treatment of AD can be easily understood by the fact that FDA has approved two alkaloids, galantamine and rivastigmine (a synthetic derivative of physostigmine), as treatment for AD, due to their capacity of acting as cholinesterase inhibitors [209, 210].

Galantamine

Galantamine was firstly isolated in the 1950s from plants which belong to the Amaryllidaceae such as *Galanthus nivalis* L. and *Galanthus woronowii* Losinsk. Later, it was also found in *Narcissus spp.* (daffodil), *Leucojum spp.* (snow flake), and Lycoris including *Lycoris radiate* Herb (red spider Lily). These plants are the ones from which nowadays are extracted for industrial production.

It has been found that galantamine works following two mechanisms of action, both involving the cholinergic system. It inhibits AChE but also acts as an allosteric modulator of nicotinic acetylcholine receptors (nAChR) [211]. Galantamine inhibits more potently AChE than butyrylcholinesterase (BuChE) [211, 212]. Physostigmine, other natural alkaloid blocks both AChE and BuChE activities [213]. In addition to AChE inhibition, galantamine allosterically potentiates pre- and post-synaptic nAChR [214]. In the presynaptic neuron, galantamine acts by elevating the liberation of ACh and other neurotransmitters which are important to ensure the correct function of brain such as glutamate and dopamine [215]. In addition, galantamine has been described to show inhibition of Aβ aggregation and it can scavenge ROS protecting neurons against oxidative stress [216]. Consequently, it has been thought that it could promote neuroprotection and neurogenesis which facilitate a reduction of neurodegeneration in AD [217]. Furthermore, as clinical studies demonstrated [218], the administration of galantamine at 8-32 mg/day shows a considerable improvement of symptoms in both the cognitive functions and daily living activities in persons who suffer mild to moderate AD over 3-6 months [219] and these effects are sustained for 12 months using 24 mg/day dose [220].

Berberine

Berberine is an isoquinole alkaloid which can be found in roots, bark and stems of a few medicinal plants such as *Berberis spp.*, *Hydrastis canadensis* L., *Phellodendron amurense* Rupr. and *Rhizome coptidis*. Traditionally it has been used in Indian and Chinese medicine and as a natural yellow dye, this application is due to the fact that it has a yellow fluorescence. Diverse pharmacological effects have been reported for berberine [221], for example it protects neural cells against cerebral ischemia, psychological depression, schizophrenia, anxiety, and AD [222, 223]. As anti-inflammatory agent, berberine acts by suppressing the liberation of proinflammatory cytokines such as IL-6 and monocyte chemotactic protein (MCP)-1 in both Aβ-stimulated primary microglia and in a murine microglial cell line (BV-2). In addition, expression of COX-2 and iNOS is decreased by the action of this compound. Even though mechanisms are not totally clear, what has been discovered at the moment suggests that anti-inflammatory effects of berberine could be due to inhibition of the NF-κB, phosphoinositide 3-kinase, and MAPK signaling pathways [224]. Berberine has been also demonstrated being an effective anti-oxidant, increasing the action of anti-oxidative enzymes such as SOD and GSH-Px to counter oxidative stress [225, 226]. It is also a good scavenger of NO, peroxynitrite (ONOO), H_2O_2, and 1,2-diphenyl-2-picryl hydrazyl (DPPH) radicals and decreases lipid peroxidation [227].

In recent years, the anti-AD effects of berberine have been attributed to its inhibitory activity of AChE and BuChE, indoleamine 2,3-dioxygenase inhibition, improvement of cognitive deficit induced by $Aβ_{1-40}$ and inhibition of $Aβ_{1-42}$ fibrillation [228].

Berberine also reduces the neurotoxicity induced by homocysteic acid in HT-22 hippocampal neuronal cells and significantly reduces the levels of ROS which are induced by homocysteic acid as well as the lactate dehydrogenase release, and subsequent cell death [229]. Also it rescues calyculin-induced cell death in neuroblastoma-2a cells by decreasing oxidative stress [230].

Taking into account the above described, berberine is able to reduce the oxidative stress which characterizes AD. Furthermore, berberine can also decrease Tau hyperphosphorylation in several *in vitro* systems among which we can count HEK293 cells, calyculin treated neuroblastoma-2a cells, and neuroblastoma-2a cells stably expressing human Swedish mutant APP by modulating phosphatase 2A activity. In HEK293 cells berberine acts by recovering protein phosphatase-2A activity and inhibiting GSK-3β activation and therefore it reduces Tau hyperphosphorylation upon calyculin-A treatment [230].

Berberine blocks the development of pathogenic Aβ and Aβ aggregation in different cellular models such as neuronal cells which show a stable expression of human Swedish mutant APP [231], H4 neuroglioma (APPNL-H4) cells [194] and human embryonic kidney 293 (HEK293) cells [232]. Moreover, it has been noticed that it improves learning and long-term spatial memory of TgCRND8 transgenic AD mice in comparison to controls. During an experimentation of 4 months in which mice have been treated with berberine, they exhibited a reduction of plaque load as well as soluble and insoluble Aβ levels in the brain by decreasing the GSK-3β activity [230].

Even if berberine has several proved beneficial results on AD and it has been always considered innocuous, the outcome of deepen investigations suggest that it could produce neuronal toxicity, induction of bradycardia, and increased neurodegeneration, particularly as a result of its accumulation in the CNS when it is administrated in high doses [233]. Accordingly, supplementary experimentations are essential to assess if berberine can be considered neurotoxic or can be used without serious damaging effects in the treatment of AD.

Physostigmine

Physostigmine is also a reversible AChEI under clinical trial which effectively increases the concentration of ACh and thus the stimulation of both nicotinic and muscarinic receptors due to the increased in available ACh at the synapse. However, physostigmine appears to have no advantage over some other anticholinesterase drugs [213]. It displays a short half-life representing a serious disadvantage and requires complex forms of administration [234]. Nevertheless, two separate and chirally pure physostigmine derivatives, (−)- and (+)-phenserine seem to improve some of the main features of AD. (−)-Phenserine, has reached phase III clinical trials, but now is being reformulated to enhance its pharmacological actions [235], increases the ACh levels and, thereby, reinforces cognition [236]. Moreover, both isomers support the reduction of Aβ levels APP because of their role as inhibitors of APP synthesis [237].

Huperzine A

Huperzine A has been employed from old in Chinese folk medicine to treat several medical conditions. It is obtained from the Chinese moss *Huperzia serrata* Thunb, also known as *Lycopodium serratum* Thunb and its function is to reduce the neuronal deterioration and apoptosis in the hippocampus and cortex generated by Aβ and free radicals and suppress the glutamate toxicity by acting as a NMDA receptor antagonist [238, 239].

In addition to that, researchers have demonstrated that Huperzine A is a potent, reversible and selective AChEI, has good oral bioavailability, has the capacity to pass the BBB and acts with an extended half-life. Its action is about 1000 times more intense on AChE than the one on BuChE and it has a calculated IC_{50} of 0.1 mM [240]. Research has shown that Huperzine A has a far effective action in rising the levels of ACh in the cortex if related to either donepezil or rivastigmine and it also increases the *in vivo* AChE inhibition [239, 241].

Several studies carried out in adult and aged rodents, monkeys and in various experimental cognitive impairment models have demonstrated that huperazine A has an important function in improving the learning and memory capacities [242, 243]. In confirmation to this, clinical trials at phase IV, performed in China, reported similar results showing substantial improvement of memory deficits in geriatric patients with benign senescent forgetfulness and in persons who suffer AD. At the same time, no remarkable peripheral cholinergic side effects and toxicity are shown [239, 244]. Therefore, in China, huperzine A has encountered the authorization to be used as a treatment in AD and vascular dementia [245].

In other phase II clinical trial performed in USA, patients with mild and moderate AD who received huperzine A (200 mg twice daily) did not show any significant cognitive effect, but if the patients were administered an increased dose (400 mg twice daily), they exhibit an enhancement of cognitive efficiency [209, 246, 247], daily living activities, and global clinical assessment [248]. Despite the fact that larger studies on long period treatment are needed to confirm this theory, currently it seems to be that either Huperzine A or Huperzine A combined with memantine have a higher efficacy and are better tolerated than FDA-approved AChEIs [249].

Caffeine

Caffeine is mainly known as a component of coffee, tea, cola, and cocoa [250] and as a short-term stimulating agent on the CNS because of its ability to antagonize adenosine A2A receptors [251, 252]; however, its action on the long-term has not been totally clarified yet.

It has been established that caffeine reduces *in vitro* and *in vivo* levels of Aβ and also the Aβ-promoted neurotoxicity, moreover it raises the cognitive performance of Aβ-induced AD mouse models [253, 254] possibly by decreasing Aβ levels in plasma and brain [255]. This preserving activity is partway exercised by reducing Casp3 expression. Furthermore, caffeine works by decreasing the levels of PSEN1 and β-secretase [253, 256]. The reduction of β-secretase induced by caffeine in the brains of AD transgenic mice is mediated by the downregulation of the cRaf-1/NF-κB pathway [257]. Another mechanism of action of this compound is the stimulation of PKA activity and phospho-CREB levels while at the same time suppressing the expression of phospho-JNK and phospho-ERK in the striatum of the Swedish mutant APP transgenic mouse model leading to a promotion of the survival pathway of neurons in the AD pathogenesis [258].

Caffeine also acts as anti-inflammatory and anti-oxidative drug in AD transgenic mice by decreasing significantly the mRNA levels of CD45, toll-like receptor 2, CCL4, and TNF-α [259]. As anti-oxidant, caffeine acts by scavenging OH and OCH3 free radicals [260] and minimizes the mRNA levels of several oxidative stress markers [259]; it also diminishes the hippocampal Tau phosphorylation and the respective proteolytic Tau fragments [259]. It has also been proved that there is relationship between caffeine anti-oxidative effect and the reduction of Aβ and phosphorylated Tau in the cholesterol-induced sporadic AD model in rabbits [261].

Also evaluation of the therapeutic activity of caffeine in AD has been performed by clinical trials. One of them examined the relationship between caffeine consumption and the risk of developing AD, showing an inverse correlation between daily caffeine intake during the 20 years that preceded the diagnosis of AD [262]. A "Cardiovascular Risk Factors, Aging and Dementia" (CAIDE) study describes also that people in their midlife who drink three to five cups of coffee daily has a 65% lower risk of dementia [262 - 264] and eventually it is possible to state that caffeine exercises a significant function in slowing down AD.

Nevertheless, the latest meta-analysis of observational epidemiological studies has not found a correlation between caffeine consumption from coffee or tea and the risk of cognitive disorders [265]. However, to achieve a more certain data on the longer period further studies are needed.

Rhynchophylline and Isorhynchophyline

Rhynchophylline and isorhynchophylline are alkaloids with tetracyclic oxindole structure which can be isolated from the herb *Uncaria rhynchophylla* Miq [266]. These alkaloids exert a variety of pharmacological effects and they recently have been investigated in AD. Both show neuroprotective effects and the capacity to defend PC12 neuronal cells from Aβ-induced death [267].

Tau protein hyperphosphorylation and the reduction of Ca^{2+} overload are just some of key features of the neuroprotective pattern of these compounds [268].

Additionally, isorhynchophylline has anti-oxidant activity and inhibits Casp3, increases the ratio of Bcl-2/Bax protein expression, and acts as a stabilizer of mitochondrial membrane potential [269]. Isorhynchophylline can also reverse the attenuated phosphorylation of Akt, cAMP response element binding protein, and GSK-3β signaling proteins Aβ [267].

AD rats treated with isorhynchophylline against Aβ showed the reversal of the cognitive and behavioral damage, demonstrating the efficacy of this compound as neuroprotective agent [270].

Meanwhile, rhynchophylline has also been classified as an inhibitor of the ephrin type-A receptor 4 (EphA4), which means that it improves hippocampal synaptic dysfunctions in AD. It has been seen in cultured hippocampal neurons that EphA4 could be stimulated by soluble Aβ oligomers, and EphA4 signaling is improved in the hippocampus of APP/PSEN1 transgenic mouse model of AD. This EphA4 activation by Aβ has a leading part in the synaptic dysfunction observed in AD. Rhynchophylline was identified to inhibit EphA4 activation in *in vitro* assays and when orally administrated works by reducing EphA4 activity which leads to a reestablishment of the damaged long-term potentiation in the hippocampus of APP/PSEN1 transgenic mice, therefore, rhynchophylline may reverse the synaptic dysfunctions in AD [271].

Indirubin

Indirubin, a bis-indole alkaloid isolated of plants such as *Baphicacavthus cusia* (Nees) Bremek, *Polygonum tinctorium* Aiton, *Isatis indigotica* L. and *Indigo feratinctoria* L., and its analogues are mostly known because of their activity as kinase inhibitors, however they have other properties that make them suitable candidates for the treatment of AD [272]. Some of these analogues have shown inhibition of GSK-3β both *in vitro* and *in vivo*; prevention of abnormal Tau phosphorylation and improvements of spatial memory deficits in AD mouse models [272 - 274].

Harmine

Harmine is a β-carboline alkaloid which has been found in the plant *Peganum Harmala* L [275]. Its action is thought to be the inhibition of dual specificity tyrosine phosphorylation regulated kinase1A protein. This enzyme acts phosphorylating Tau protein on sites that, in the AD pathology are usually hyperphosphorylated [276, 277]. Therefore, one can conclude that this compound may be interesting for AD treatment.

Morphine

Morphine is a well-known active ingredient obtained from the seedpod extract or opium found in the poppy plant, *Papaver somniferum* L. Opioids such as morphine through the centuries have been mostly recognized as pain killers. Recently, several research studies describe morphine medications as a possible therapeutic approach in the treatment of AD because of its action as neuroprotector against microglia-mediated neuroinflammation and oxidative stress [278, 279]. Additionally, morphine promotes the release of estradiol and the activation of μ-opioid receptors, resulting in a protection against intracellular Aβ toxicity and identifying opioid receptors as a possible target for AD treatment [280]. Nevertheless, the use of morphine and its derivatives has to be strictly regulated because of their high chance to induce addiction.

Nicotine

The therapeutic potential of tobacco alkaloid nicotine has been extensively investigated because it can enhance cholinergic function. Aβ is believed to bind nAChR with high affinity [281]. Nicotine exerts a neuroprotective effect because it links Aβ and prevents its aggregation [282] and it also reduces AD pathology in animal models [283]. Despite that, currently, clinical studies show no evidence of remarkable signs of memory improvements related to nicotine [284, 285], although an evident positive effect on attention in AD patients has been noticed [284]. However, from it does not to improve memory in AD patients, associated to its short half-life, inherent toxicity, induction of tachyphylaxis, and addictiveness explain why nicotine is not truly considered for clinical use.

Vitamins

According to a report by the European Nutrition and Health Survey, the individual needs of vitamin D and folate are not covered in most countries, while in 50% of countries the average consumption of vitamins E and C is lower than recommended [286]. In USA the dietary intake is similar, however this situation is partially offset by the intake of dietary supplements [287]. For vitamin B12, even when the intake is adequate, malabsorption causes low serum levels [288]. Considering the role of vitamins, structural nutrients, and minerals in regulating the proper functioning of neural tissue, an adequate intake of them throughout the life, can contribute to delay the onset of AD [289].

Vitamin E

Vitamin E (α-tocopherol) and its analogues show in their structure a 6-membered aromatic chromanol ring and an aliphatic side chain. Vegetable oils are natural sources of vitamin E, being α, β, and γ-tocopherol derivatives the most abundant. However, in human tissues α-tocopherol is the most common and bioavailable antioxidant form [290].

They are lipophilic compounds which can be found in fat deposits or cellular membranes, associated with lipoproteins, where they protect the PUFA from peroxidation reactions [291]. Owing to its antioxidant properties, vitamin E (α-tocopherol) was tested in clinical trials in AD patients in order to prevent or delay the cognitive decline. This is because the important roles of vitamin E in the CNS described at the beginning of the 20th by Evans *et al.*, who observed paralytic offspring in rats deprived of dietary vitamin E [292]. Epidemiological and experimental studies have demonstrated a reduced risk of developing AD in population with high plasma levels of vitamin E, as well as a diet rich in fruit and vegetables, which is plentiful in antioxidants [293, 294].

It addition, it was tested throw neuronal incubation with Trolox (water-soluble analogue of vitamin E) that this compound inhibits p38MAPK, responsible for Tau hyperphosphorylation [295].

The use of vitamin E as an antioxidant strategy for the treatment of AD has advanced in the last decade. However, not all studies performed have consistent and uniform outcomes, and their translation to clinical trials have not given the hypothesized results [296, 297]. Consequently, the usefulness of vitamin E in neurodegenerative disorders is still under debate.

Folate and Vitamins B2, B3, B6 and B12

Folate and B12 as methylcobalamin facilitate cellular replacement of S-adenosyl methionine (SAM). SAM is obtained from methionine which is afforded from homocysteine in one-carbon metabolism. For this reaction, it is necessary a methyl group from 5-methyltetrahydrofolate which has been recognized as B-vitamin folic acid most biologically effective structure. Therefore, in order to regenerate 5-methyltetrahydrofolate, B12 results essential [298]. SAM donates methyl groups to DNA methylation. This is important in the control of gene expression and determines chromosomal conformation. The bioavailability of SAM depends on diet [299]. Green leafy vegetables and supplemented foods act as a source of folate that cannot be synthesized de novo. Although the activity of microbes through the digestion generates folate, the quantity generated is not enough to satisfy the metabolic requirements [300]. The active forms of folic acid need the contribution of a considerable amount of enzymes and adequate supplies of riboflavin (B2), niacin (B3), pyridoxine (B6), zinc, vitamin C, and serine to be synthesized. Accumulation of homocysteine (HCY) and S-adenosylhomocysteine (SAH), and therefore, SAM reduction is produced by deficiency in folate and vitamin B12 [301]. SAH inhibits DNA methyl transferase and thus induces DNA hypomethylation [302].

Epigenetic diet could represent a valid strategy to prevent pathological mechanisms of AD with early onset, but it may also be a strategy to treat AD already developed [303]. It has been observed that target genes in AD as APP, PSEN1, ApoE and BACE-1 show changes in methylation patterns. Thus, methylation in particular sections of the promoter region of the APP gene is slowed in aging. Promoter CpG of the APP gene are hypomethylated in the post mortem brains of patients with AD. Therefore, modifications in DNA methylation could result in modifications in APP gene expression, and lead to a progressive increase of Aβ deposition [303]. However, these results are controversial, Wang *et al.* stated that it has not been noticed any variation in the methylation status of the APP gene promoter in the AD [304]. By contrast, genes like those encoding ApoE and methylenetetrahydrofolate reductase are hypermethylated in the cortex of AD brains [305].

It has been noticed that HCY high levels are the most significant markers of folate and vitamin B12 insufficiency, and they also biochemically reveal functional impairment. Several studies performed since the 1970s, try to correlate cognition and folic acid [306]. The high plasma levels of HCY found in patients at risk of being affected by AD brought to the thesis that there could be a risk reduction it they received folic acid supplementation. In addition, the results obtained in AD patients with temporal lobe atrophy, which correlates with elevated plasma levels of HCY could support the hypothesis that an extra doses of folic acid may be beneficial [305]. However, further studies are needed to determine the epigenetic regulation of AD-related genes by diet, as well as the quantity required to obtain a therapeutic effect [303].

Vitamin B1 (thiamine)

The cerebral metabolism of glucose that plays a critical role in normal brain functioning has shown to be regulate by thiamine-dependent processes. Although a decrease in thiamine-dependent enzymes is related to an increase of Aβ and NFT in AD patients' brains, this association does not appear to be clear. It seems logical to think that if the decline in brain glucose utilization is a consequence of the reduction of thiamine-dependent enzymes, an increase of these could improve the symptomatology. However, in the few clinical trials performed, thiamine has been shown to be ineffective [307]. Also, thiamine derivatives have been administrated in clinical studies. Fursultiamine and sulbutiamine seem to be more promising to increase the cognition in AD patients [308, 309].

Vitamin D

Vitamin D refers to more than one member of a group of steroid molecules: Vitamin D3 (cholecalciferol) and vitamin D2 (ergocalciferol). Vitamin D3 is available in a limited number of foods such as flesh and fatty fish and fish liver oils. Vitamin D2 is synthesized by irradiating plants and fungi. Vitamin D3 is more efficacious than D2 at raising serum 25(OH)D, which is the active form [310].

Recently, low levels of vitamin D have been associated with increased odds of cognitive dysfunction and AD in geriatric patients. Numerous evidences indicate a direct correlation between serum vitamin D levels and cognitive ability [311]. Increase of vitamin D levels has been shown to have positive effects on cholinergic, dopaminergic and noradrenergic neurotransmitter systems [312]. Serum vitamin D levels are inversely proportional to TNF-α levels and moreover, vitamin D administration reduces TNF-α levels and other inflammatory markers. Therefore, vitamin D protects neurons by preventing apoptosis [313].

Vitamin D is also reported to exert a positive influence on glucose homeostasis avoiding Aβ deposition [314]. Soluble and insoluble Aβ levels in mice's brains decrease after long-term vitamin D treatment [315]. Additionally, higher serum vitamin D levels correlates with increased Aβ in the CSF suggesting increased clearance from the brain [316].

Although several studies highlight the link between low vitamin D and AD in adults, more research is necessary in order to determine whether vitamin D serum levels are only a biomarker of the disease or supplementation with vitamin D would be preemptive of AD [317, 318].

Vitamin A and Other Retinoids

Preformed vitamin A can be obtained from various sources of animal origin such as liver, fish liver oil, eggs and milk products; where it is consumed mainly esterified form palmitate. It can also be consumed in plant foods as its most important chemical precursor β-carotene which is converted into retinoids after cellular metabolism. The term retinoid is applicable to vitamin A derivatives including its biological precursors such as carotenoids, metabolites that activate nuclear receptors (RAR) and compounds that activate retinoid X receptors (RXR) [319]. All these receptors mediate retinoid signaling. Retinoid signaling is highly expressed in brain areas related to neuronal plasticity like hippocampus, prefrontal cortex and retrosplenial areas. Several experimental reports describe the involvement of retinoid signaling in amyloid pathology. RARα signaling induces the expression of α-secretase, hence reducing the Aβ formation [320]. Additionally, RARα signaling system up-regulates neprilysin mediated proteolytic cleavage of Aβ in both neurons and microglia [321].

Retinoids regulate the expression of AD-related genes such as BACE, PSEN1 and PSEN2 which are engaged in the Aβ formation [322]. Administration of retinoids to APP/PSEN1 or APP23 transgenic mice produces attenuation of Aβ accumulation and Tau hyperphosphorilation [323, 324]. Intracellular cholesterol concentration and its distribution are modulated by retinoids. Thereby retinoids modulate APP processing and ultimately reduce Aβ secretion [325].

Retinoids have also shown anti-inflammatory effects. They significantly suppress the LPS/Aβ-induced TNF-α formation, expression of iNOS and production of IL-6 from the activated microglia. Suppression of inflammatory response by retionoids may be attributed *via* inhibition of the NF-κβ signaling pathway [319]. Retinoids have also shown antioxidant potential in human neuroblastom cells, inducing the expression of the manganese superoxide dismutase (MnSOD2) gene, an antioxidant enzyme located in mitochondria [319].

Retinoids improve the cholinergic transmission by up-regulating the CAT expression [326]. Furthermore, it has also been reported that retinoids may regulate the expression of tyrosine hydroxylase, DA-β-hydroxylase and D2 receptor expression [327].

In view of the above described results, retionoids might reverse memory-related deficits linked with AD. However, further *in vivo* studies are needed to elucidate the potential of RAR and RXR to fully understand the precise mechanisms of retinoids in AD.

Vitamin C

Vitamin C also named as L-ascorbic acid, or ascorbate (the anion of ascorbic acid), is an essential nutrient which acts as a cofactor for different enzymatic reactions in humans and other animal species. However, a large portion of the population is deficient in vitamin C. Although the daily intake is adequate, transport of vitamin C into the blood is performed by a sodium-dependent saturable transporter (SVCT1). This means that not always the high intake or supplementary administration serve to cover the needs of vitamin C [328]. Different evidence suggests that adequate levels of vitamin C may play a protective role against cognitive decline related to age and AD, which does not involve taking supplements over a normal and healthy diet [329, 330]. In the brain parenchyma, ascorbic acid levels are controlled through the sodium dependent vitamin C transporter (SVCT2), which transports vitamin C into the choroid plexus from blood to cerebral spinal fluid, and subsequently from extracellular fluid to neurons. This transfer process allows higher levels of vitamin C to be achieved in brain than in blood, except when there is insufficient consumption over a prolonged period [331].

It is unknown which is the optimal level of vitamin C intake needed to preserve brain function, but vitamin C is necessary for brain development and protection throughout life [331]. Due to its antioxidant action, vitamin C protects against oxidative stress [332]. Low vitamin C levels available with a low dietary intake are sufficient for the patients have not symptoms of scurvy such as lethargy, hemorrhage and hyperkeratosis, but they cannot combat the oxidative stress of AD [331, 333]. The latest research with respect to vitamin C has proved that it is able to dissolve neurotoxic senile plaques. Equally effective in the treatment of AD is the oxidized form of vitamin C, *i.e.* dehydroascorbic acid [334]. Furthermore, Vitamin C may be one of several factors that could help prevent the formation of amyloid plaques. However, these results need to be confirmed by more research before can be make recommendations that may change feeding people [335].

Seizures in AD patients have been reported to accelerate cognitive and neuropathological dysfunction. In APP/PSEN1 mice by crossing them with sodium vitamin C transporter 2 (SVCT2) heterozygous knockout mice, it was observed that cerebral ascorbic acid levels are decreased. In these mice there is an increase in oxidative stress in brain, an increase in mortality, a faster latency of seizure onset when treated with kainic acid, and more ictal events following pentylenetetrazol treatment. Furthermore, it was shown that low levels of ascorbic acid aggravate the seizures produced by kainic acid and pentylenetetrazol. Therefore, it is particularly important to meet the needs of ascorbic acid in populations at increased risk for epilepsy and seizures, such as AD [336].

Stem cell therapy is also a promise for the treatment of neurodegenerative disorders. Vitamin C has been shown to improve the generation of pluripotent stem cell in mouse and human cells, while accelerating changes in gene expression. The vitamin C appears to act, in part, by slowing cell senescence [337].

Coenzyme Q10

The Coenzyme Q10 (CoQ10), the most abundant form of coenzyme Q found in humans, is a quinone derivative with 10 isoprenoid units in its structure. This compound is synthesized intracellularly from tyrosine and the first step of the reaction, requires vitamin B6 as a cofactor. It's distributed in all cells of the human body, including brain cells. Those cells or tissues which are metabolically active show the highest levels of CoQ10 and they are very susceptible to its deficiency. Due to its structure and ubiquity, this compound is also named as ubiquinone [338].

In addition, CoQ10 is a lipophilic compound that acts as cofactor in the electron transport system of the cell's mitochondria, and also as intracellular antioxidant. Therefore, adequate levels of ubiquinone are required for cellular respiration and ATP production, and deficiency states of this biomolecule could cause negative effects in several pathological processes. Recent research in this field show that CoQ10 levels decline with age, and it contributes in part to some of the manifestations of aging, one of the most important risk factors for developing AD [338, 339].

In addition, deficiency of CoQ10 may be related to various factors such as nutritional deficiencies of vitamin B6 that lead to a decrease in CoQ10 levels or an increase in its needs caused by genetic or acquired factors. In fact, it has been described that CoQ10 supplementation could have beneficial effects in neurodegenerative process, reducing oxidative stress. In addition, studies carried out in a transgenic mouse model of AD demonstrated that the use of CoQ10 also improves the cognitive function and decreases the Aβ levels [340].

Although the *in vitro* results and preclinical studies performed show CoQ10 as a potential neuroprotective agent, it is necessary to carry out clinical trials that reveal its neuroprotective activity [339].

Amino Acids

At the present, numerous studies using amino acid for the treatment of AD are being conducted. However, there is not yet sufficient correlation between their efficacy and increased cognitive function. Therefore, these products are marketed as medical food.

Acetyl-L-Carnitine

L-carnitine is widely distributed in foods such as red meat, fish and dairy products. It is also available as a dietary supplement in the form of acetyl-L-carnitine (ALC). ALC is an endogenous substance involved in maintaining mitochondrial bioenergetics while decreasing oxidative stress associated with aging in mitochondrial membrane [341]. Neurobiological properties have been described for both ALC present in the brain at high concentrations and for its carnitine and acetyl moieties [342].

ALC is involved in ACh production [343], membrane stabilization and potentiation of mitochondrial function in cholinergic neurons [344, 345]. Through its action on the cholinergic system, ALC could protect against oxidative stress mediated by Aβ and neurotoxicity and thus improve the course of AD [346]. These pharmacological actions have been shown in primary cortical neurons cultures where pretreatment with ALC significantly decreases Aβ-induced cytotoxicity, protein oxidation and lipid peroxidation in a dose dependent manner. Also, ALC may modify the APP physiological metabolism by enhancing the α-secretase pathway [347].

Double-blind controlled studies performed in recent years have shown that ALC has beneficial effects on AD. Compared to placebo, ALC administered for one year seems to decrease the impairment associated with AD. However, the effectiveness of ALC appears to be greater in the early stages of the disease [348 - 350].

Glutamine

Glutamine is the most abundant free amino acid in the human blood stream. In brain, glutamine is synthesized from glutamate and ammonia by glutamine synthetase (GS) [351, 352].

In adult brains GS protects against oxidative stress and inflammation [353], suggesting that it plays a neuroprotective role [354, 355]. Changes in GS levels and its activity have been appreciated in patients with AD [356].

GS activity and glutamine uptake regulate endogenous levels of glutamine. Additionally, it has been shown that in the cultures of primary cortical neurons, the reduction of exogenous glutamine slightly reduces neuronal viability. Moreover, reductions in exogenous and endogenous glutamine by decreasing of GS activity are synergistic. Although a healthy neuron needs glutamine endogenously obtained by the action of GS, the supply of exogenous glutamine is beneficial for the neurons. Low levels of glutamine lead to a greater sensitivity of the cell against stress factors such as H_2O_2 or Aβ [352].

The *in vitro* findings correlate with data obtained *in vivo* on dietary glutamine administration in two different mouse models of familial AD (phenotypes of

R1.40 and 8.9 transgenic mouse strains). The results obtained strongly support neuroprotective effect for glutamine supplementation. Taking into account that adequate concentrations of glutamine are neuroprotective against the oxidative stress associated with aging, glutamine supplementation could avoid neurodegeneration in AD [352].

L-Arginine

L-arginine is a semi-essential proteinogenic amino acid that plays a role in two major metabolic pathways: NOS and arginase pathways [357].

L-arginine is widely distributed in animal and vegetable foods such as dairy products, beef, pork, poultry, seafood, grains, nuts, seeds and soy. Dietary patterns characterized by intakes of these foods are strongly associated with lower AD risk [358]. In addition, some clinical trials have shown that 3-month supplementation with L-arginine increases cognition in patients with cerebrovascular disease but it decreases 3-month after the end of the treatment [359]. Several reasons could explain the effects of L-arginine on AD and some of them are described below.

L-arginine acts directly and indirectly on human vasculature through the NO production [360]. In patients with high cholesterol levels, L-arginine supplements may have beneficial effects by decreasing platelet aggregation and mononuclear cell adhesiveness [361, 362]. Also, in both healthy individuals and patients with essential hypertension, it causes a reduction in blood pressure [363]. Therefore, L-arginine improves the vascular factors associated to AD [364, 365].

It has been described that in cell cultures NO provides protection against ROS. This effect may be explained because NO directly reacts with O_2^- to give ONOO$^-$ which rapidly and before interacting with cells yields nitrate at physiological pH [366]. Thus, L-arginine modulates neuroinflammation through NO production [367]. In addition, it has been reported that NO protects cells in the CNS against damage induced by Aβ [368].

On the other hand, L-arginine and NO can modulate the glucose metabolism and insulin activity related to AD [369].

L-arginine has shown effects on neurogenesis through arginase and NOS pathways. Polyamides produced from L-arginine through the arginase pathway can improve the neuronal survival. However, neurogenesis may be both improved and impaired by L-arginine metabolism *via* NOS pathway [370].

Despite above described, excess of L-arginine in the CNS may not be entirely beneficial. Excessive NO production can result in nitroxidative stress which is implicated in the pathogenesis of AD [371].

S-Adenosylmethionine

S-adenosylmethionine (SAM) is an endogenous substance in the human body that has numerous roles in metabolic reactions, some of them have been previously described in this chapter (to see section *Folate and Vitamins B2, B3, B6 and B12*). SAM deficiency may alter the expression of genes involved in APP metabolism and thus, lead to Aβ-accumulation [372]. Therefore, it has been suggested that a SAM dietary supplementation could achieve neuroprotective effects in AD patients [373]. SAM appears to be a relatively benign medication with few side effects [374].

Nowadays, research on SAM has been mostly performed by using AD mouse models [375, 376]. Few results are available in humans. SAM has been shown to increase GST activity in APOE -/- mice but not in normal mice [377, 378]. However, in animals fed a nutrient deficient diet, SAM supplementation suggests evidence of improving cognition in both healthy and transgenic mice [379].

Clinical studies related to SAM supplementation in humans, show brief duration and small sample size. However, they suggest the possibility that SAM may improve function in AD, particularly in cognition [373].

L-Tryptophan

L-Tryptophan (TrP) is an essential amino acid necessary for 5-HT obtention [380]. Basal extracellular levels in multiple brain regions including the hippocampus depend on the contribution of tryptophan in the diet. Therefore, in patients with AD, the dietary TrP deficit could further impair cognition [381].

In 3xTg-AD animals it has been shown that chronic increase of TrP in the diet reduces the density of intraneuronal Aβ compared to those fed a normal diet [382].

The mechanisms involved in the TrP-mediated decrease in the intraneuronal accumulation of Aβ have not yet been well established, however, some hypotheses suggest that 5-HT (obtained from TrP) favors the non-amyloidogenic processing of APP by stimulating the cell liberation of sAPPα [383]. Also, increased central 5-HT could decrease β-secretase activity [384]. Additionally, supplementation with TrP enhances extracellular 5-HT and induces antidepressant effect similar to selective serotonin reuptake inhibitors (SSRI) administration in rats [385].

On the other hand, it has been described that the mice fed with a TrP deficient diet show a significant decrease in nerve growth factor (NGF) in the hippocampus as well as brain derived neurotrophic factor (BDNF) in the cerebral cortex. Both, NGF and BDNF are responsible for stimulating differentiation and/or growth of basal forebrain cholinergic neurons. However, feeding with a TrP excess diet produces no further increase in the brain levels of any of these neurotrophins [386].

S-Allyl-L-Cysteine

S-Allyl-L-cysteine (SAC) is an organosulfur compound naturally formed in Allium plants such as garlic and onion [387]. Presumably, despite its hydrophilic it is still sufficiently hydrophobic to cross the BBB and access to the CNS. SAC has shown a possible anti-amyloidogenic effect both *in vitro* and *in vivo* studies and even it is able to destabilize preformed Aβ fibrils *in vitro* [388].

Treatment of APP-Tg mice with SAC decreases level of Tau phosphorylation. Likely, this fact can be explained due to the decline of GSK-3β activity [389]. Another possible mechanism is that SAC decreases secreted APP which produces microglia activation releasing IL-1, which has been described to be involved in Tau pathology [390]. Because of its properties as ROS-scavenger and decreased APP [391], SAC can ameliorate neuroinflammation and therefore to be effective in preventing excessive Tau phosphorylation [392].

SAC modulates intracellular levels of GSH which has been repeatedly cited in this chapter because it protects against ROS [393]. Further, SAC administration to APP-Tg mice significantly decreases IL-1β in immunoreactive microglia compared to untreated APP-Tg mice [394]. SAC also suppresses NF-κβ activation in cultured cells when they are separately treated with ROS and TNF-α [395, 396].

Also, increasing of affinity choline uptake and CAT activity has been appreciated in SAC and ROS cotreated cholinergic human SK-N-SH cells [392].

SAC is also beneficial for improving cardiovascular health which may be a risk factor for AD as evidenced by epidemiological and clinical studies [397, 398].

Similar results have been obtained after aged garlic extract supplementation [399]. Besides SAC, aged garlic extracts contain di-allyl-disulfides, ajonjoe, allixin, flavonoids/polyphenols and many others thiosulfinates which are known to exert multiple benefits [400].

Therefore, SAC and aged garlic extract could be developed as a potential therapeutic agent for AD.

Homotaurine

Homotaurine also known as tramiprosate or 3-APS (aka 3-amino-1-propanesulfonic acid), Alzhemed™ or Vivimind™ is a small aminosulfonate that can be obtained from different species of red seaweed [401].

It was the first anti-amyloidogenic drug to enter in phase III trial. The inhibition of amyloidogenic aggregation and subsequent deposition as well as its capability to binds soluble Aβ make this orally-administered compound an interesting option for AD patients. Therefore, it can act as a neuroprotector agent in both *in vitro* and *in vivo* models [401]. *In vitro* Alzhemed™ has shown a neuroprotective effect against Aβ-induced neurotoxicity in mouse organotypic neuronal hippocampal cultures whereas *in vivo* studies have shown dose-dependent Aβ reductions in brain of transgenic mice (hAPP-TgCRND8) [402]. Besides, in rat hippocampus, Alzhemed™ activates GABA-A receptors [403] and therefore, it reverses the effect in the inhibition for induced long-term potentiation [402].

Clinical studies have shown that it is safe and well tolerated [404]. The post-hoc analyses of a Phase III clinical study have shown positive and significant effects for Alzhemed™ in subgroups of patients. These include a decrease in hippocampal volume loss and lower memory impairment in the overall cohort, as well as a decrease in global cognitive impairment in a population of patients with AD who are ApoE4-ε4, a variant related to an increased likelihood of AD, suggesting disease-modifying effects [405].

In mild to moderate AD patients it also reduced $A\beta_{42}$ levels in CSF [404]. However, Neurochem Inc., the drug company developing Alzhemed™ considered the results inconclusive in reference to its clinical efficacy [406]. Therefore, the compound was subsequently rebranded and marketed as a nutraceutical, unregulated by the FDA. Later, the development of a tramiprosate pro-drug, named ALZ-801, that claims achieves higher levels in the brain has been started.

Phase I trials of ALZ-801 have shown an extended half-life, which leads to an efficient once-daily dose since it posses extensive bioavailability. Also, the reduction of nausea and vomiting incidents indicates gastrointestinal tolerability and safety. Also ALZ-801 shows low intersubject pharmacokinetic variability and thus solving some of the problems associated to tramiprosate, the parent molecule [407].

Hormones

Both aging women and men, are affected by hormonal problems. Respectively, during menopause women are afflicted by a reduction of estrogen and progesterone [408], while in men there is a drop of the testosterone levels [55]. Recently this aged-related depletion of hormones has been taken into consideration as an underlying factor of AD which led to a new therapeutic approach to the disease [409].

Also melatonin, which is secreted by the pineal gland declines with age [410]. In the same way, age, but also alterations in metabolism as a result of a sedentary lifestyle and inadequate nutrition contribute to a decrease in insulin production and insulin sensitivity [411, 412]. Alterations in insulin signaling pathways and the resulting cerebral hypometabolism are also associated with AD and cognitive impairment [413, 414]. Since hormones have fundamental roles in neural health, hormone replacement therapy is studied as a solution to treat AD.

Melatonin

Melatonin is the major hormone secreted by the pineal gland into the CSF and circulatory system during the night. Its role is to maintain the normal circadian rhythms, it means that it controls the biological clock and functions such as the sleep-wake cycle and the induction of physiological sleep [415].

Its production has been seen reduced as the body ages and, because of that, there has been a raised concern about the fact that it could be one of the principal predisposing factors of neurodegenerative diseases. To confirm this, it has been noticed a decrease of CSF melatonin levels in AD accompanied by early neuropathological modifications. Furthermore, the severity of mental and sleep impairments in demented patients has a clear relationship with the decrease of melatonin levels which can be already observed in preclinical stages [416, 417].

Research on melatonin showed that it acts as a potent suppressant of oxidative and nitrosative stress-induced damage within the nervous system. It seems to be more valid than many other natural antioxidants, implying that it could be a favorable option as a treatment of several diseases characterized by oxygen radical-mediated tissue damage [418]. Melatonin is a potent scavenger of free radicals and ROS, including the hydroxy and peroxyl radicals, the singlet oxygen and NO, it also stimulates the antioxidant enzymes SOD, GSH-Px and catalase. *In vivo*, either vitamins C or E, which are known as valuable antioxidants, demonstrated lower efficacy in reducing lipid peroxidation than melatonin [415].

Although evidence from preclinical animal experiments as well as open label studies in AD and in MCI patients have reported on the beneficial effects of exogenous melatonin on cognitive decline and sleep, much more research is needed to conclusively establish this connection [419, 420].

Testosterone

As mentioned previously, androgens play an important role in improving mood and promoting selective aspects of cognition, inclusive of spatial abilities and verbal fluency. For instance, a better performance of visual, verbal and long-term memory as well as visuospatial processing and visuomotor scanning, have been noticed in men whose free testosterone levels were higher [421]. Several studies have shown that low levels of free testosterone in blood in men younger than 80 years of age may be related to AD [55, 422]. Despite of the relationship found, it is still not possible to conclude whether the reduced levels of testosterone are a cause or consequence of the disease. However, several complementary studies have revealed that low testosterone levels can be detected before or at the onset of AD, so it appears to be a risk factor [423 - 425].

Androgens could exert their neuroprotective action by different mechanisms of action: directly by producing the activation of androgen pathways or indirectly by their aromatization to estradiol and the activation of protective estrogen signaling pathways. The reduction of bioavailable testosterone has been associated with an increase of Aβ peptides, Tau protein hyperphosphorylation, and neuronal death [422]. In fact, it has been shown that androgens regulate Aβ levels through androgen receptor (AR) and estrogen receptor (ER) in cell cultures and rodent AD models [426]. Moreover, hippocampal neurons cultured have been demonstrated to increase neuronal viability *via* AR-dependent MAPK/ERK signaling pathway when they are treated with testosterone [427]. Testosterone is also able to facilitate Aβ clearance by increasing the levels of neprilysin [428]. Furthermore, dihydrotestosterone (DHT), an androgenic hormone obtained from testosterone, has also been found to have neuroprotective properties through the activation of the AR-dependent AMPc-CREB signaling pathway in PC12 cells and cultured hippocampal neurons [427]. Recently it has been discovered that in addition to the mechanisms previously described, testosterone promotes Aβ degradation through a non-estrogenic route [429]. Because of the all above mentioned physiological level of testosterone appears to be a suitable factor which can decrease the possibility of developing AD [421].

It has been proved that castration, which leads to an almost complete loss of testosterone, results in elevation of Aβ in guinea pigs, rats and 3xTg-AD mice [430 - 432]. It has also been suspected that the real contribution to the regulation of Aβ levels is played by estrogen and androgen because testosterone is only a prohormone that is soon enzymatically converted within tissues to both the active DHT and the estrogen E2. Between the two pathways, researchers believe that androgen has a major role, because the elevated levels of Aβ induced by castration were prevented by supplementation with DHT but not with E2 [431]. In addition to that, when testosterone conversion to E2 is inhibited by genetically limiting aromatase activity it results in elevated testosterone levels, low E2 levels, and reduced Aβ accumulation in male APP23 mice [433]. However, castrated male 3xTg-AD mice happened to positively respond to testosterone and DHT, but also to E2, that suggested an active role of estrogens in the reduction of Aβ burden in male brain [432].

Studies evaluating the results of testosterone treatment in subjects with AD have achieved controversial results. While in a small clinical study including men newly diagnosed with AD, testosterone administration for up to one year contributed to improvement of both general cognitive ability and spatial visual ability [434]; in a randomized, placebo-controlled and crossover study including men with subjective memory complaint and low testosterone levels, the findings did not show significant improvements on global cognition [435]. In addition to that, other studies have not found significant improvements in men with mild cognitive impairment and AD who were treated with testosterone [436 - 438].

Several factors such as the age and characteristics of the patients, the type of treatment and its duration as well as the cognitive domains could explain the contradictions found among the studies with testosterone. Therefore, larger clinical studies are required, with longer-term follow-up and including blood samples and brain imaging markers to be able to establish whether testosterone is actually relevant in the treatment of AD.

Pregnenolone and Dehydroepiandrosterone

Pregnenolone (PREG) and dehydroepiandrosterone (DHEA) and their sulfate forms (PREGS and DHEAS) are the most abundant neurosteroids produced in the neural tissues of rats and humans [439, 440]; their action is thought to be an excitatory one on cells. PREG is synthesized from cholesterol and converted to DHEA *via* cytochrome P450c17, and, as many other hormones, humans, especially those affected by AD, suffer from a significant decrease of PREG and DHEA and their sulfate forms with age [441 - 445]. PREGS and DHEAS display GABA-antagonistic properties [446], DHEAS acts as sigma-1 receptor antagonist and it protects hippocampal neurons against glutamate-induced neurotoxicity [447]. The activity of PREG, at low doses is to be a protective agent which prevents cell damage induced by Aβ peptide in rat PC12 cells [448].

Although it seems that the administration of PREG, DHEA or their sulfate esters in both male and female adult rodents may improve memory, the physiological significance of these effects is still to verify considering the fact that contradictory results were found in human trials [449, 450].

Estrogen and Phytoestrogen

Among the hormones studied in AD research, estrogens are by far the most investigated because of results from epidemiological and clinical studies indicating that risk of AD in postmenopausal women is increased due to depletion of ovarian hormones [451]. However, recent studies which have been carried out both in animal and humans revealed that the development of the disease is more related to the depletion of estrogen derived from the brain than to the circulating estrogen [452, 453]. Therefore, female AD transgenic mice that genetically eliminate aromatase (an enzyme involved in the production of estrogen) are more prone to develop early and more severe AD pathology compared with control mice [452, 454].

Because of estrogen well-known action as neuroprotective agent, its loss, during menopause, could, at least partially, result in the brain metabolism deficits (mitochondrial impairment) which are recurrently seen in AD [455]. Estrogen also plays leading role in enhancing the generation of new neurons in the dentate gyrus of hippocampus and other brain regions and thus facilitates specific learning and memory functions [456]. Recent studies have confirmed that estrogens have potent effects on the morphology and plasticity of neurons in certain areas of hippocampus demonstrating its specific regional effect [452, 457].

Another acknowledged neuroprotective property of estrogen in AD is to decrease Aβ levels, as well as to control their increase against pathological triggers [458]. Activation of non-amyloidogenic APP processing by activation of MAPK/ERK pathway and the reduction of BACE-1 levels are the ways by which estrogen decreases Aβ production, while Aβ clearance is promoted by facilitating microglial phagocytosis and degradation; furthermore as it was described for testosterone, estrogen also regulates the levels of major enzymes involved in Aβ degradation [459].

At the moment, the effects of hormone replacement therapy are being taken into examination because its possibility to improve cognition and lower the risk of AD. However, the effectiveness of estrogen therapy seems to decrease with age and consequently it becomes almost useless in old women [460]. In addition, progestogens have a function in regulating estrogen neuroprotective actions; it means that if the brain tissue is constantly exposed to progestogen there will be a suppression of estrogen actions whereas cyclic exposure could potentiate estrogen-induced neural benefits [451]. Although the results of basic science, observational, and clinical studies indicate that estrogen therapy in premenopausal women may reduce the risk of developing AD, its effectiveness remains unclear [461].

Doubts about the cognitive benefits of hormone therapy in AD are based on controlled and uncontrolled clinical trials that have limitations on the sensitivity of cognitive test performed. In addition, in these studies, predominantly conjugated equine estrogens (CEE) instead of estradiol are administered as replacement therapy. CEE is composed of estrone sulfate and at least ten other steroid hormones whose neurobiological effects are unknown.

A research realized by Women's Health Initiative (WHI) and WHI Memory Study (WHIMS), showed that postmenopausal women older than 65 were more exposed to develop dementia when treated with CEE or CEE conjugated with medroxyprogesterone acetate (MPA) [462]. As valid alternative of CEE, 17-β estradiol, the most potent and natural human form of estrogen, could be also used. In consequence of the relationship established between oral formulations, increased thrombotic risk and cognitive deficits, transdermal estradiol formulations are usually preferred [463]. In this regard, a three months study of hormone therapy administration on postmenopausal women affected by AD, with transdermal 17β-estradiol showed significant improvements on semantic memory and visual memory [463]. These findings, in accordance with others earlier reports [464], indicate that short-term hormone therapy with transdermal 17β-estradiol could enhance some cognitive abilities in older postmenopausal women with AD [463, 464]. Despite that, it has not totally been established the clinical relevance of hormone therapy, thus larger clinical trials over the extended period of time are needed.

Further investigations are also required to find the appropriate parameters which can provide a successful therapeutic approach in aging women to delay, prevent, and or treat AD; in addition to that, it is also essential to develop new drugs which interact selectively with estrogen and androgen receptors.

Soy-derived phytoestrogen compounds have shown similar results; they seem to have neuroprotective effects both in animal-models and cell cultures [465, 466]. In addition to their antioxidant properties, soy products could be a potential tool against AD because of their interaction with estrogen receptors which affects cognition and eventually can alleviate the risk of AD progression. However, the results of observational studies and randomized controlled trials in humans have not been conclusive and they are consistent with estrogen treatment studies above described [467].

Progestogens

Progesterone, also known as pregn-4-ene-3,20-dione, is a steroidal hormone which plays an important role in the female menstrual cycle, also promotes pregnancy and embryogenesis of humans and other species [468]. It works as neurosteroid and moreover it is also an essential intermediate in the synthesis of several endogenous steroids, such as sex hormones and corticosteroids [469].

It has been found a correlation between increased level of progesterone in the prefrontal cortex and hippocampus and a regression of both cognitive impairment and cell loss [470] and, because of that, progesterone is starting to be considerate as a neuroprotective agent. *In vitro*, progesterone has shown protective effects on hippocampal neurons against toxic agents; it has a similar efficacy to estradiol against glutamate toxicity and glucose deprivation, but it is less effective against Aβ-induced toxicity [471, 472].

However, it has been reported that the concentration of progesterone decreases in the media of $A\beta_{25-35}$-induced primary rat cortical neurons, and progesterone treatment inhibits $A\beta_{25-35}$-induced cell toxicity as well as apoptosis [473, 474]. The effect of progesterone administration against $A\beta_{25-35}$-induced impairment has been also studied *in vivo*. $A\beta_{25-35}$ impairs learning and memory abilities of rats, accompanied by reduced levels of progesterone. Treatment of these AD rats with progesterone reverses cognitive impairment [474]. Researches showed that progesterone decreased the expression of TNF-α and IL-1β in the hippocampus of $A\beta_{25-35}$-treated rats. Because progesterone is converted into its metabolites, allopregnanolone and estradiol, further studies are necessary to know whether progesterone itself or its metabolites are responsible for the protective effect [475]. Progesterone also potentiates the stimulation of mRNA expression for CAT and BDNF in the basal forebrain of the rat produced by estradiol [476].

In humans, an oestro-progestagen treatment combined with rivastigmine in menopausal women suffering from AD did not provide further improvement during mild to moderately severe stages than single rivastigmine. The evaluated parameters such as cognitive function, functioning and neuropsychiatric symptoms were similar in both groups [477].

Insulin

Insulin resistance at the brain level causes alterations in insulin signaling pathways which are closely associated with increased oxidative stress and mitochondrial dysfunction, neuroinflammation and overexpression of Aβ [478]. Variations in glucose and insulin levels regulate GSK-3β activity in the hippocampus and cerebral cortex by phosphorylation/dephosphorylation reactions as well as its expression levels. Inhibition of GSK-3β decreases the processing of APP and inhibits Tau phosphorylation, preventing the formation of both Aβ and hyperphosphorylated Tau oligomers and thereby avoiding the neurotoxicity and neuroinflammation phenomena that lead to deterioration of the cognitive function characteristic of AD [479].

The involvement of insulin in the neurodegeneration associated to AD has been demonstrated in mouse models administered with streptozotocin (STZ), which is a compound that inhibits neuronal insulin receptor function. Adult rats develop long-term and progressive deficits in learning, memory and cognitive behavior when they are intracerebroventricularly injected with STZ [480]. In transgenic mice, STZ-induced insulin-deficient hyperglycemia increases the advanced glycation end products (AGE) in brain tissue and further enhances the expression levels of RAGE and NF-κB in the brain and accelerates the senile plaque formation [481].

Animal studies have shown that intranasal insulin is able to cross the cribriform plate, dispersing readily throughout the brain, and is capable of improving learning and memory in a rodent model of AD [482]. Similarly, intranasal insulin administrated to patients with AD or mild cognitive impairment has demonstrated a delay in memory impairment [483] and modulation of Aβ formation because it increases $A\beta_{40}/A\beta_{42}$ ratio [484]. However, not all patients with AD respond similarly to insulin given intranasally and differences related to APOE allele can be noticed. Insulin therapies have proved less effective in Apoε4 carriers, but these individuals had positive results when treated with fast acting insulin [485, 486].

Therefore, it is also important to study the therapeutic relationship of co-administration of insulin and other hormones that modulate insulin sensitivity and function such as leptin which is also a promise therapy for AD [487].

Plants and Herbal Extracts

In addition to the isolated natural compounds and phytochemicals described in this chapter for treating AD, we below describe plants and herbal extracts which have shown some interest in clinical trials for the treatment of this disease.

Ginkgo Biloba L.

Ginkgo biloba L. has long been linked to Eastern and Western medicine. In recent years, several compounds have been extracted and purified from this plant including terpenoids, diterpene lactones, and polyphenols [488]. These compounds could be responsible for the different activities of *Ginkgo biloba* L. Ginkgo-specific acylated flavonols [489] improve availability dopamine and acetylcholine in the prefrontal cortex and neuroplasticity is partly related to isorhamnetine [490]. On the other hand, it is thought mitochondrial properties are related to flavonoids, bilobalide, and the ginkgolides B and J [491]. Many studies have demonstrated that ginkgolides can enhance cognitive function [492, 493].

The quality of the starting material, the extraction method used, and the solvents employed are decisive for the extraction of pharmacodynamically active compounds from the plants. Therefore, it is well known that each *Ginkgo biloba* extract has to be evaluated in studies individually since each extract is different and possesses its own properties [493, 494]. Because of their different composition, the data obtained from the EGb 761® trials, the standardized formulation, cannot be extended to other preparations of other Ginkgo extracts, as well as data from them could not be extrapolated to EGb 761®. Taking into account the preclinical effects of *Ginkgo biloba* [495], EGb 761® may be considered as a multitarget compound, because of its capability of act in different targets involved in AD and natural cognitive impairment [496 - 498]. EGb 761® is a dried acetone extract (60% weight/weight) from *Ginkgo biloba* leaves (35-67:1). The extract is adjusted to contain 22.0%-27.0% ginkgo flavonoids, calculated as ginkgo flavone glycosides, 5.0%-7.0% terpene lactones consisting of 2.8%-3.4% ginkgolides A, B, and C, 2.6%-3.2% bilobalide, and less than 5 ppm ginkgolic acids. (EGb 761® is a registered trademark of Dr Willmar Schwabe GmbH & Co. KG, Karlsruhe, Germany.) [494].

EGb 761® modifies several of the molecular mechanisms related to AD. It improves neuronal energy supply by restoring altered mitochondrial function [499], enhances compromised hippocampal neurogenesis and neuroplasticity [500], prevents the deposition and toxicity of Aβ [501], promotes blood flow, and improves microperfusion [502].

In the clinical trials carried out with EGb 761®, it was found that it is a compound with no tolerance issues at daily doses of 240 mg which was revealed to be appropriate to improve cognitive function as well as activities of daily living and clinical parameters outstanding [503].

Panax Ginseng Meyer

Panax ginseng Meyer belongs to the family Araliaceae. Ginseng, the root of this perennial plant, has been linked for 2000 years to Asian medicine, particularly in China, Korea, and Japan [504]. It is currently popularly used worldwide as a natural medicine and it is also used as a functional healthy food [505, 506]. Due to its composition of ginsenosides, the saponins of Ginseng, has a number of effects on the CNS. It acts as stimulant and depressant, anticonvulsive, analgesic, also it has anti-fatigue activity, and improves cognitive performance [507]. Mountains of evidence highlight the physiological and pharmacological interest of *Panax ginseng* in neurodegenerative disorders [508] due to the action of ginsenosides on the maintenance of homeostasis and immunity and their properties against inflammation, stress oxidative and apoptosis [509]. In aged mice long-term administration of ginseng total saponins significantly prevents memory loss. This is related to lower levels of oxidative stress and an improvement of neuronal plasticity in the hippocampus [510]. In human neuroblastoma cells SH-SY5Y, neuroprotection exerted by ginseng Panax is believed to be due to the regulation of the phosphatase activity of purified calcineurin and Tau phosphorylation [511]. Also, *Panax ginseng* could enhance significantly the learning and memory ability of AD rats and attenuate oxidative stress damage. A possible mechanism may be associated with its blockade of RAGE/NF-κB activation [512].

To data, around 31 ginsenosides have been extracted of *Panax ginseng*, and many others remain unidentified [513, 514]. All ginsenosides are characterized by a common four-ring hydrophobic, steroid-like structure which has linked sugar moieties [515]. Differences in their pharmacological actions may lie in the diversity of sugar components as well as the number and position of sugar residues [505, 516]. Although each ginsenoside has different properties, as they are multitarget compound, they can act by different pathways in the same tissue [509]. Based on their structure, the ginsenoside monomers can be dammarane or oleanolate alkyl type.

The first ones described before, are divided into two groups: the Rb1 group, which have a protopanaxadiol in their structure (Rb1, Rb2, Rc and Rd) and the Rg1 group which have protopanaxatriol (Rg1, Re, Rf, Rg2 and Rg3) [516].

- It has been observed that ginsenoside Rb1 and Rb2 are related to an improvement of learning and memory ability in Aβ dementia mice. These effects may be explained because these ginsenosides cause an increase in the number of axons of the cerebral cortex and hippocampus and improve the expression of synaptophysin [517 - 519].
- Rg1 has a wide variety of effects including its neuroprotective action in AD mouse model [519]. Probably because of its antioxidant action it protects neurons against the $A\beta_{42}$ toxicity [406], leading to improved cognition. It further modulates APP hydrolysis [520], and activation of the PKA/CREB signaling pathway [521]. *In vitro* it has been observed that ginsenoside Rg1 decreases Tau phosphorylation in SKN-SH cells related to Aβ and the involvement of p38 pathway activation [522]. In cultured primary rat neurons, Rg1 decreases apoptosis due to its mitochondrial action [523]. It has also been shown that Rg1 favors neuronal plasticity, as well as stimulates the cognition and induces the mossy fiber sprouting and the expression of growth associated protein 43 (GAP-43 protein) in the hippocampus [517, 524]. These properties are dose dependent. Furthermore, other mechanisms that can explain their actions are its ability to prevent the activation of NF-κB/p65, Akt and the ERK1/2 in H_2O_2-induced PC12 cells [525, 526]. In brain slices from AD model rats it has been observed

that inhibits the expression of Casp3 to obviate apoptosis by decreasing PKA RIIα level (isoform IIα of the regulatory subunit of PKA) and increasing p-CREB and BDNF levels in the hippocampus [521]. In addition, its protective effects on the toxicity of $A\beta_{25-35}$ [523] and/or interferon (IFN)-γ in microglia, may be due to the inhibiting microglial respiratory burst activity and decreased of the NO accumulation [509].

- Rg2 is found mostly in the root and stem leaves of *Panax ginseng*. Research indicates that Rg2 acts through different mechanisms. Rg2 protects PC12 cells against glutamate-induced neurotoxicity by blocking the Ca^{2+} channels, also it inhibits Casp3 that mediates Aβ formation and apoptosis of neuronal cells [517, 527].

- Rg3 promotes the Aβ clearance by stimulating of microglial phagocytosis through the activation of the macrophage scavenger receptor type A (MSRA) [528]. In addition, Rg3 increases neprilysin gene expression in the SK-N-SH cells which present Swedish mutant APP, implanted by transfection. It also decreases $A\beta_{40}$ and $A\beta_{42}$ levels [509, 529].

- Rh2 and Re: Rh2 attenuates Aβ-induced toxicity by stimulating the gene expression of neurotrophic factor and promotes the proliferation and survival of rat brain type I astrocytes [530]. On the other hand, it has been observed that Re significantly reduces both *in vitro* and *in vivo* the $A\beta_{42}$ levels [531]. In addition, the increased administration of Re leads to increased DA and ACh levels at the hippocampus and prefrontal cortex in rats [532].

- Gintonin promotes non amyloidogenic pathway of APP in a concentration and time-dependent manner and thus decreases $A\beta_{42}$ formation and $A\beta_{1-40}$-induced cytotoxicity in SH-SY5Y cells. Gintonin also improves cognitive impairment induced by $A\beta_{1-40}$ in mice. Furthermore, in studies carried out with a transgenic mouse model with AD, it has observed that prolonged oral administration of gintonin reduces Aβ aggregation and cognitive impairment slowed [509, 533, 534].

Clinical trials have shown that treatment with a high dose (4.5 g/d) of *Panax ginseng* powder improves cognitive function in AD patients [535]. However, further large scale and long-term research is needed to ensure the efficacy observed in clinical trials with AD patients carried out with *Panax ginseng*.

Cereboost™, has been marketed as a functional food that improves cognition in humans. It is an extract obtained from American ginseng root (*Panax quinquefolius* L.) containing many saponins in common with *Panax ginseng* [536].

Withania Somnifera Dunal

Withania somnifera, also called ashwagandha or Indian ginseng is a small evergreen shrub which has been traditionally used in India with different medicinal applications [537]. It has a neuroprotective effect that may be explained because of its anti-inflammatory and antioxidant properties. It also protects against Aβ [538], decreases calcium levels and inhibits AChE activity [539]. Many of these actions lead to a lower apoptosis [540 - 542]. *In vitro* it has been observed that ashwagandha regenerates injured neuron cells. Mice oral treatment with ashwagandha manages to decrease neurite atrophy and restore synapses in the hippocampus and cortex and thus improves memory [543]. To date, at least 18 active components, known as withanolides, have been identified in ashwagandha. Each one has a variety of neuroprotective effects, acting in different areas of the neuron for example such as withanolide-A whose target is the axon, while withanolides IV and VI have dendrites as a target [544].

Melissa Officinalis L.

Melissa officinalis L. traditionally used for treatment of neurological disorders has also been identified as a memory-enhancing herb. The extract of *Melissa officinalis* has shown cholinergic properties. It inhibits AChE activity and is active on nicotinic receptors [545, 546]. Also it has shown neurotropic action [547] and protects PC12 cells against Aβ induced toxicity [548].

In vivo, *Melissa officinalis* L. extract can improve memory in healthy rats and scopolamine induced memory impairment because of its cholinergic activity [546]. It also has been proven to ameliorate mild to moderate AD patients and decreases agitation in these patients [549].

Crocus Sativus L.

Crocus sativus L, also identified as saffron, is cultivated in a variety of countries of Europe and Asia. It is characterized because of its small size and it is a perennial plant belonging to the family of Iridaceae. Saffron properties are linked to it stigmas because on them we can find a variety of compounds such as carotenoids, α-crocetin and glycoside crocin (responsible for saffron yellow color) and picrocrocin, the aglycone safranal (responsible for saffron aroma), the antioxidant carotenoids lycopene and zeaxanthin and vitamin B2 [550].

Preclinical and clinical studies have revealed that saffron or its active components show pharmacological activity against convulsion, depression, inflammation, cancer, oxidative stress and cognitive impairment [551, 552]. Additionally, it has been carried out a variety of studies about antioxidant properties related to this plant [553]. Studies carried out with crocin 10 µM have shown that this compound inhibits lipid peroxidation in cultured PC12 cells [554], moderately restores SOD activity and preserves neurons integrity. At some concentrations, crocin's properties against oxidation are even more pronounced than that shown by α-tocopherol [550]. Moreover, this compound activates SOD and glutathione peroxidase (GSH-Px) and significantly reduces the levels of malondialdehyde (MDA) found in the ischemic cortex in rat model of ischemic stroke [555]. Saffron extract has a moderate inhibitory activity of AChE, decreasing the ACh breakdown [556]. One-week treatment with saffron to normal and aged mice is linked to an improvement in learning and memory [557] whilst crocin treatment could significantly prevent the cognitive impairments following icv injection of STZ suggesting the therapeutic potential of this component in aging and age-related neurodegenerative disorders [558].

In a 16-week study, saffron administration significantly enhanced cognitive function compared to placebo whilst no significant differences related to adverse events were noticed in the two groups [559]. Saffron was also found to show a similar result to donepezil in the treatment of early stages of AD after 22 weeks in a Phase II study. The frequency of vomiting was lower with saffron extract than donepezil. These results suggest a possible application of saffron extract in early stages of AD [560].

Polygonum Multiflorum Thunb

Polygonum multiflorum Thunb., also called He shouwu in regions of Asia and Fo-ti in North America, is a perennial vine linked to natural Chinese medicine [561] and officially included in the Chinese Pharmacopoeia [562].

Pharmacological studies with *polygonum multiflorum* extract claimed that this medicinal plant may be beneficial in preventing AD [561, 563] showing significant neuroprotective activity on mild AD clinical cases [564]. It could improve the fluidity of mitochondria membrane and the activity of mitochondrial COX in a model of AD [565].

One of its compounds, tetrahydroxystilbene glucoside, has been identified to be responsible for improving spatial learning-memory impairment in Morris water maze as well as the object recognition in 16 months old Tg mice [566, 567]. The effects of this compound on learning-memory improvement need further investigation [568].

Salvia spp

Salvia miltiorrhiza Bunge, plant which belongs to the family of Lamiaceae, has been linked to treatments against cardiovascular and cerebrovascular diseases, specifically its roots [569, 570]. Chemical compounds found in *Salvia miltiorrhiza* can be classified into polyphenolic compounds which are water-soluble such as salvianolic acids and danshensu and non-polar compounds such as tanshinones [571]. All of them are responsible for the biological activities of *Salvia miltiorrhiza*.

Salvia miltiorrhiza shows antiamyloidogenic, antioxidant, anti-apoptotic and anti-inflammatory properties. Also it enhances cholinergic signaling and stimulates neurogenesis of neuronal cells *in vitro* and *in vivo*. Therefore, this plant is useful for AD treatment [572].

Different constituents show different activities. Both salvianolic acids and tanshinones prevent Aβ deposition, protecting SH-SY5Y cells against cytotoxicity induced by Aβ [573]. In addition, cryptotanshinone promotes the non-amyloidogenic pathway of APP [574]. Regarding its antioxidant activity, tanshinone IIA can inhibit lipid peroxidation and ROS generation [575]. Similar mechanisms are described for salvianolic acids which protect SH-SY5Y cells against H_2O_2-induced toxicity [576 - 578]. Tanshinone IIA reduces Casp3 expression and thereby attenuating apoptosis while enhancing Bcl-2 protein expression in ischemic cortex [579]. Salvianolic acids protect SH-SY5Y cells inhibiting mitochondria-dependent apoptotic pathway [576, 580].

Tanshinone derivatives are also AChE inhibitors [581, 582] which attenuate learning and memory impairments induced by scopolamine in mice [583]. Tanshinone IIA, salvianolic B and PF2401-SF (a standardized fraction of *Salvia miltiorrhiza*) increase HO-1 expression that leads to suppress iNOS expression and NO production. Additionally, tanshinone IIA but also cryptotanshinone inhibit NF-κB and activate ERK signaling pathways [584].

Salvianolic B activates the PI3K/Akt signaling pathway in a concentration and time-dependent manner and thereby favors the proliferation of neural stem cells from mouse and rat embryonic cortex [585] but also *Salvia miltiorrhiza* promotes differentiation of rat mesenchymal stem cells into neurons [586].

The numerous components of *Salvia miltiorrhiza* are thought to have synergistic effects [587]. However, the low oral bioavailability of some of them has a detrimental effect on the clinical application of its extract [588].

Clinical studies have been performed with herbal preparations of others Salvia species such as *Salvia officinalis* L. and *Salvia lavandulaefolia* Vahl. Results show that they may also improve the cognitive function and memory in both healthy subjects and those suffering from dementia or cognitive decline. It is necessary to standardize the composition and to know the information on the different extracts to carry out clinical studies with greater reliability [589].

FUTURE EXPECTATIONS IN ALZHEIMER'S DISEASE

AD is a devastating disease both for patients and their families. In spite of the many efforts made to know the ultimate causes of the disease and to develop an effective treatment, the advances achieved are limited.

Therefore, the first step to afford effective drugs should be to fully understand the causes of the disease at the molecular level. Because of that it is necessary to develop basic and clinical research with close collaboration and improve the preclinical studies by developing more appropriate animal models [590]. The improvement of recent methodologies based on pluripotent stem cells to access neural models may be a good starting point to obtain a valid model that allows to understand the pathogenesis of AD and to achieve an adequate treatment [591].

To date, several genes have been identified as the ones that increase the risk of AD, for example APOEε4 in late-onset Alzheimer's or PSEN1 and PSEN2 in early-onset affection. Currently, the advances in molecular genetics and sequencing technologies have allowed to identify new genes that may be associated with an increased risk of AD. Most of these genes are related to lipid metabolism such as SORL1, ABCA7 and CLU, inflammation and immune response such as CR1, CD33, MS4A, ABCA7, EPHA1, TREM2 and CLU or axonal transport such as NME8, CELF1 and CASS4. Validation and quantification of the clinical impact of these new genetic risk factors will also undoubtedly contribute to the development of prevention and treatment strategies for AD [592].

New neuroimaging techniques such as Aβ positron emission tomography and their appropriate clinical use should help to establish a relationship between pathophysiological alterations and impaired memory, cognition, behavior and other symptoms in the early stages of the disease [593]. Development of blood and CSF based screening tools, in order to identify AD markers, also may contribute to an early diagnosis of the disease [594 - 598].

Advances in diagnosis must be accompanied by new drugs that are capable of modify the molecular mechanisms of the disease. Drugs that are currently in clinical development are directed towards the main known stages of the pathogenesis of the disease or alternatively they are drugs acting simultaneously on several targets involved in AD. Furthermore, the study of innovative applications of drugs already marketed is being carried out and it may provide a faster answer to treat the disease [599]. In addition, the development of clinical trials with an adequate design and quality seems to be necessary [600].

Taking into account the complexity of the disease, its cure seems a milestone still distant, therefore efforts should also focus on preventing the establishment of AD. Different epidemiological studies indicate that lifestyle including diet [601, 602] and exercise [603, 604] may contribute to AD's onset, mainly in those subjects with a higher genetic risk of developing AD. Therefore, in the next years, early interventions should be directed towards these patients.

CONCLUSION

In the search for new molecules for the treatment/prevention of AD, it has been found that numerous substances of natural origin play a key role in the disease. In recent years it has been established that alterations in plasma and CSF levels of certain hormones, vitamins, amino acids and fatty acids, among other markers, are already detectable even in stages prior to the development of symptoms. Therefore, proper nutrition that maintains adequate levels of these in the body may contribute to delay the establishment of the disease or alleviate its symptoms, as observed in different clinical trials.

On the other hand, many plants used in traditional medicine represent a source of compounds with multiple pharmacological activities and that can be used to develop new drugs with multitarget properties which can be effective in the treatment of AD. Often, the plant extract itself may be used showing fewer side effects than drugs currently marketed for AD therapy. However due to the heterogeneity of the extracts and conducted studies, the results of activity in many cases are inconclusive.

CONFLICT OF INEREST

The authors (editor) declares no conflict of interest, financial or otherwise.

ACKOWLEDGMENT

The authors thank Laura Loche for some parts of the translation work.

REFERENCES

[1]   Wimo A, Winblad B, Aguero-Torres H, von Strauss E. The magnitude of dementia occurrence in the world. Alzheimer Dis Assoc Disord 2003; 17(2): 63-7.
       [http://dx.doi.org/10.1097/00002093-200304000-00002] [PMID: 12794381]

[2]   Pillai JA, Cummings JL. Clinical trials in predementia stages of Alzheimer disease. Med Clin North Am 2013; 97(3): 439-57.
       [http://dx.doi.org/10.1016/j.mcna.2013.01.002] [PMID: 23642580]

[3]   Petersen RC, Doody R, Kurz A, *et al.* Current concepts in mild cognitive impairment. Arch Neurol 2001; 58(12): 1985-92.

[http://dx.doi.org/10.1001/archneur.58.12.1985] [PMID: 11735772]

[4]   Copeland MP, Daly E, Hines V, *et al.* Psychiatric symptomatology and prodromal Alzheimer's disease. Alzheimer Dis Assoc Disord 2003; 17(1): 1-8.
[http://dx.doi.org/10.1097/00002093-200301000-00001] [PMID: 12621314]

[5]   Palmer K, Berger AK, Monastero R, Winblad B, Bäckman L, Fratiglioni L. Predictors of progression from mild cognitive impairment to Alzheimer disease. Neurology 2007; 68(19): 1596-602.
[http://dx.doi.org/10.1212/01.wnl.0000260968.92345.3f] [PMID: 17485646]

[6]   European Medicines Agency. Pre-authorisation evaluation of medicines for human use. Guideline on medicinal products for the treatment of alzheimer's disease and other dementias CPMP/EWP/553/95 Rev 1. 2008.

[7]   United States Food and Drug Administration Peripheral and Central Nervous System Drugs Advisory Committee.   Mild   Cognitive   Impairment 2001.http://www.fda.gov/ohrms/dockets/ac/01/minutes/3724m1.html

[8]   Jelic V, Kivipelto M, Winblad B. Clinical trials in mild cognitive impairment: lessons for the future. J Neurol Neurosurg Psychiatry 2006; 77(4): 429-38.
[http://dx.doi.org/10.1136/jnnp.2005.072926] [PMID: 16306154]

[9]   Auld DS, Kornecook TJ, Bastianetto S, Quirion R. Alzheimer's disease and the basal forebrain cholinergic system: relations to β-amyloid peptides, cognition, and treatment strategies. Prog Neurobiol 2002; 68(3): 209-45.
[http://dx.doi.org/10.1016/S0301-0082(02)00079-5] [PMID: 12450488]

[10]   Farlow MR, Miller ML, Pejovic V. Treatment options in Alzheimer's disease: maximizing benefit, managing expectations. Dement Geriatr Cogn Disord 2008; 25(5): 408-22.
[http://dx.doi.org/10.1159/000122962] [PMID: 18391487]

[11]   Scarmeas N, Stern Y, Mayeux R, Manly JJ, Schupf N, Luchsinger JA. Mediterranean diet and mild cognitive impairment. Arch Neurol 2009; 66(2): 216-25.
[http://dx.doi.org/10.1001/archneurol.2008.536] [PMID: 19204158]

[12]   Scarmeas N, Stern Y, Tang MX, Mayeux R, Luchsinger JA. Mediterranean diet and risk for Alzheimer's disease. Ann Neurol 2006; 59(6): 912-21.
[http://dx.doi.org/10.1002/ana.20854] [PMID: 16622828]

[13]   Kowall NW. Alzheimer's disease 1999: a status report. Alzheimer Dis Assoc Disord 1999; 13 (Suppl. 1): S11-6.
[http://dx.doi.org/10.1097/00002093-199904001-00005] [PMID: 10369512]

[14]   Hardy J, Selkoe DJ. The amyloid hypothesis of Alzheimer's disease: progress and problems on the road to therapeutics. Science 2002; 297(5580): 353-6.
[http://dx.doi.org/10.1126/science.1072994] [PMID: 12130773]

[15]   Walsh DM, Selkoe DJ. A β oligomers - a decade of discovery. J Neurochem 2007; 101(5): 1172-84.
[http://dx.doi.org/10.1111/j.1471-4159.2006.04426.x] [PMID: 17286590]

[16]   Kurz A, Perneczky R. Novel insights for the treatment of Alzheimer's disease. Prog Neuropsychopharmacol Biol Psychiatry 2011; 35(2): 373-9.
[http://dx.doi.org/10.1016/j.pnpbp.2010.07.018] [PMID: 20655969]

[17]   Kumar A, Dogra S. Neuropathology and therapeutic management of Alzheimer's disease –an update. Drugs Future 2008; 33: 433-46.
[http://dx.doi.org/10.1358/dof.2008.033.05.1192677]

[18]   McGeer PL, Itagaki S, Tago H, McGeer EG. Reactive microglia in patients with senile dementia of the Alzheimer type are positive for the histocompatibility glycoprotein HLA-DR. Neurosci Lett 1987; 79(1-2): 195-200.
[http://dx.doi.org/10.1016/0304-3940(87)90696-3] [PMID: 3670729]

[19]   Thal DR. The role of astrocytes in amyloid β-protein toxicity and clearance. Exp Neurol 2012; 236(1): 1-5.
[http://dx.doi.org/10.1016/j.expneurol.2012.04.021] [PMID: 22575598]

[20]   Deane R, Du Yan S, Submamaryan RK, *et al.* RAGE mediates amyloid-beta peptide transport across the blood-brain barrier and accumulation in brain. Nat Med 2003; 9(7): 907-13.
[http://dx.doi.org/10.1038/nm890] [PMID: 12808450]

[21]   Thinakaran G, Koo EH. Amyloid precursor protein trafficking, processing, and function. J Biol Chem 2008; 283(44): 29615-9.
[http://dx.doi.org/10.1074/jbc.R800019200] [PMID: 18650430]

[22]   Hardy J. Alzheimer's disease: the amyloid cascade hypothesis: an update and reappraisal. J Alzheimers Dis 2006; 9(3) (Suppl.): 151-3.
[http://dx.doi.org/10.3233/JAD-2006-9S317] [PMID: 16914853]

[23]   Wang B, Wang Z, Sun L, *et al.* The amyloid precursor protein controls adult hippocampal neurogenesis through GABAergic interneurons. J Neurosci 2014; 34(40): 13314-25.
[http://dx.doi.org/10.1523/JNEUROSCI.2848-14.2014] [PMID: 25274811]

[24]   Iqbal K, Liu F, Gong C-X, Alonso AdelC, Grundke-Iqbal I. Mechanisms of tau-induced neurodegeneration. Acta Neuropathol 2009; 118(1): 53-69.
[http://dx.doi.org/10.1007/s00401-009-0486-3] [PMID: 19184068]

[25]   Ballatore C, Lee VM, Trojanowski JQ. Tau-mediated neurodegeneration in Alzheimer's disease and related disorders. Nat Rev Neurosci 2007; 8(9): 663-72.
[http://dx.doi.org/10.1038/nrn2194] [PMID: 17684513]

[26]   Schweers O, Schönbrunn-Hanebeck E, Marx A, Mandelkow E. Structural studies of tau protein and Alzheimer paired helical filaments show no evidence for β-structure. J Biol Chem 1994; 269(39): 24290-7.
[PMID: 7929085]

[27]   Alonso AD, Grundke-Iqbal I, Barra HS, Iqbal K. Abnormal phosphorylation of tau and the mechanism of Alzheimer neurofibrillary degeneration: sequestration of microtubule-associated proteins 1 and 2 and the disassembly of microtubules by the abnormal tau. Proc Natl Acad Sci USA 1997; 94(1): 298-303.
[http://dx.doi.org/10.1073/pnas.94.1.298] [PMID: 8990203]

[28]   Selkoe DJ. Developing preventive therapies for chronic diseases: lessons learned from Alzheimer's disease. Nutr Rev 2007; 65(12 Pt 2): S239-43.
[http://dx.doi.org/10.1301/nr.2007.dec.S239-S243] [PMID: 18240556]

[29]   Bohm C, Chen F, Sevalle J, *et al.* Current and future implications of basic and translational research on amyloid-β peptide production and removal pathways. Mol Cell Neurosci 2015; 66(Pt A): 3-11.
[http://dx.doi.org/10.1016/j.mcn.2015.02.016] [PMID: 25748120]

[30]   Umeda T, Tomiyama T, Kitajima E, *et al.* Hypercholesterolemia accelerates intraneuronal accumulation of Aβ oligomers resulting in memory impairment in Alzheimer's disease model mice. Life Sci 2012; 91(23-24): 1169-76.
[http://dx.doi.org/10.1016/j.lfs.2011.12.022] [PMID: 22273754]

[31]   LeBlanc A, Liu H, Goodyer C, Bergeron C, Hammond J. Caspase-6 role in apoptosis of human neurons, amyloidogenesis, and Alzheimer's disease. J Biol Chem 1999; 274(33): 23426-36.
[http://dx.doi.org/10.1074/jbc.274.33.23426] [PMID: 10438520]

[32]   Guo H, Albrecht S, Bourdeau M, Petzke T, Bergeron C, LeBlanc AC. Active caspase-6 and caspase--cleaved tau in neuropil threads, neuritic plaques, and neurofibrillary tangles of Alzheimer's disease. Am J Pathol 2004; 165(2): 523-31.
[http://dx.doi.org/10.1016/S0002-9440(10)63317-2] [PMID: 15277226]

[33]   LeBlanc AC. Caspase-6 as a novel early target in the treatment of Alzheimer's disease. Eur J Neurosci

2013; 37(12): 2005-18.
[http://dx.doi.org/10.1111/ejn.12250] [PMID: 23773070]

[34]   Gilgun-Sherki Y, Melamed E, Offen D. Oxidative stress induced-neurodegenerative diseases: the need for antioxidants that penetrate the blood brain barrier. Neuropharmacology 2001; 40(8): 959-75.
[http://dx.doi.org/10.1016/S0028-3908(01)00019-3] [PMID: 11406187]

[35]   Velayutham M, Hemann C, Zweier JL. Removal of H_2O_2 and generation of superoxide radical: role of cytochrome c and NADH. Free Radic Biol Med 2011; 51(1): 160-70.
[http://dx.doi.org/10.1016/j.freeradbiomed.2011.04.007] [PMID: 21545835]

[36]   Uttara B, Singh AV, Zamboni P, Mahajan RT. Oxidative stress and neurodegenerative diseases: a review of upstream and downstream antioxidant therapeutic options. Curr Neuropharmacol 2009; 7(1): 65-74.
[http://dx.doi.org/10.2174/157015909787602823] [PMID: 19721819]

[37]   Nemoto S, Takeda K, Yu ZX, Ferrans VJ, Finkel T. Role for mitochondrial oxidants as regulators of cellular metabolism. Mol Cell Biol 2000; 20(19): 7311-8.
[http://dx.doi.org/10.1128/MCB.20.19.7311-7318.2000] [PMID: 10982848]

[38]   Zhao Y, Zhao B. Oxidative stress and the pathogenesis of Alzheimer's disease. Oxid Med Cell Longev 2013; 2013: 316523.
[http://dx.doi.org/10.1155/2013/316523]

[39]   Esteras N, Dinkova-Kostova AT, Abramov AY. Nrf2 activation in the treatment of neurodegenerative diseases: a focus on its role in mitochondrial bioenergetics and function. Biol Chem 2016; 397(5): 383-400.
[http://dx.doi.org/10.1515/hsz-2015-0295] [PMID: 26812787]

[40]   Glass CK, Saijo K, Winner B, Marchetto MC, Gage FH. Mechanisms underlying inflammation in neurodegeneration. Cell 2010; 140(6): 918-34.
[http://dx.doi.org/10.1016/j.cell.2010.02.016] [PMID: 20303880]

[41]   Mizuno T. The biphasic role of microglia in Alzheimer's disease. Int J Alzheimers Dis 2012. Article ID 737846.
[http://dx.doi.org/10.1155/2012/737846] [PMID: 22655214]

[42]   Tuppo EE, Arias HR. The role of inflammation in Alzheimer's disease. Int J Biochem Cell Biol 2005; 37(2): 289-305.
[http://dx.doi.org/10.1016/j.biocel.2004.07.009] [PMID: 15474976]

[43]   Talantova M, Sanz-Blasco S, Zhang X, *et al.* Aβ induces astrocytic glutamate release, extrasynaptic NMDA receptor activation, and synaptic loss. Proc Natl Acad Sci USA 2013; 110(27): E2518-27.
[http://dx.doi.org/10.1073/pnas.1306832110] [PMID: 23776240]

[44]   Lee CY, Landreth GE. The role of microglia in amyloid clearance from the AD brain. J Neural Transm (Vienna) 2010; 117(8): 949-60.
[http://dx.doi.org/10.1007/s00702-010-0433-4] [PMID: 20552234]

[45]   Hickman SE, Allison EK, El Khoury J. Microglial dysfunction and defective β-amyloid clearance pathways in aging Alzheimer's disease mice. J Neurosci 2008; 28(33): 8354-60.
[http://dx.doi.org/10.1523/JNEUROSCI.0616-08.2008] [PMID: 18701698]

[46]   Heneka MT, Nadrigny F, Regen T, *et al.* Locus ceruleus controls Alzheimer's disease pathology by modulating microglial functions through norepinephrine. Proc Natl Acad Sci USA 2010; 107(13): 6058-63.
[http://dx.doi.org/10.1073/pnas.0909586107] [PMID: 20231476]

[47]   Osherovich L. Managing microglia in Alzheirmer's 2010. SciBX 3(14); [doi:10.138/scibx.2010.423].

[48]   Barzilai N, Huffman DM, Muzumdar RH, Bartke A. The critical role of metabolic pathways in aging. Diabetes 2012; 61(6): 1315-22.
[http://dx.doi.org/10.2337/db11-1300] [PMID: 22618766]

[49]   Harris JJ, Jolivet R, Attwell D. Synaptic energy use and supply. Neuron 2012; 75(5): 762-77.
        [http://dx.doi.org/10.1016/j.neuron.2012.08.019] [PMID: 22958818]

[50]   Lourenço CF, Ledo A, Dias C, Barbosa RM, Laranjinha J. Neurovascular and neurometabolic
        derailment in aging and Alzheimer's disease. Front Aging Neurosci 2015; 7: 103.
        [http://dx.doi.org/10.3389/fnagi.2015.00103] [PMID: 26074816]

[51]   Pithadia AS, Lim MH. Metal-associated amyloid-β species in Alzheimer's disease. Curr Opin Chem
        Biol 2012; 16(1-2): 67-73.
        [http://dx.doi.org/10.1016/j.cbpa.2012.01.016] [PMID: 22366383]

[52]   Solomonov I, Korkotian E, Born B, *et al.* Zn^{2+}-Aβ40 complexes form metastable quasi-spherical
        oligomers that are cytotoxic to cultured hippocampal neurons. J Biol Chem 2012; 287(24): 20555-64.
        [http://dx.doi.org/10.1074/jbc.M112.344036] [PMID: 22528492]

[53]   Savelieff MG, Lee S, Liu Y, Lim MH. Untangling amyloid-β, tau, and metals in Alzheimer's disease.
        ACS Chem Biol 2013; 8(5): 856-65.
        [http://dx.doi.org/10.1021/cb400080f] [PMID: 23506614]

[54]   Sayre LM, Perry G, Harris PL, Liu Y, Schubert KA, Smith MA. *In situ* oxidative catalysis by
        neurofibrillary tangles and senile plaques in Alzheimer's disease: a central role for bound transition
        metals. J Neurochem 2000; 74(1): 270-9.
        [http://dx.doi.org/10.1046/j.1471-4159.2000.0740270.x] [PMID: 10617129]

[55]   Hogervorst E, Williams J, Budge M, Barnetson L, Combrinck M, Smith AD. Serum total testosterone
        is lower in men with Alzheimer's disease. Neuroendocrinol Lett 2001; 22(3): 163-8.
        [PMID: 11449190]

[56]   Gouras GK, Xu H, Gross RS, *et al.* Testosterone reduces neuronal secretion of Alzheimer's beta-
        amyloid peptides. Proc Natl Acad Sci USA 2000; 97(3): 1202-5.
        [http://dx.doi.org/10.1073/pnas.97.3.1202] [PMID: 10655508]

[57]   Brinton RD. Impact of estrogen therapy on Alzheimer's disease: a fork in the road? CNS Drugs 2004;
        18(7): 405-22.
        [http://dx.doi.org/10.2165/00023210-200418070-00001] [PMID: 15139797]

[58]   Rosario ER, Chang L, Head EH, Stanczyk FZ, Pike CJ. Brain levels of sex steroid hormones in men
        and women during normal aging and in Alzheimer's disease. Neurobiol Aging 2011; 32(4): 604-13.
        [http://dx.doi.org/10.1016/j.neurobiolaging.2009.04.008] [PMID: 19428144]

[59]   Akhter R, Sanphui P, Biswas SC. The essential role of p53-up-regulated modulator of apoptosis
        (Puma) and its regulation by FoxO3a transcription factor in β-amyloid-induced neuron death. J Biol
        Chem 2014; 289(15): 10812-22.
        [http://dx.doi.org/10.1074/jbc.M113.519355] [PMID: 24567336]

[60]   Yin JC, Tully T. CREB and the formation of long-term memory. Curr Opin Neurobiol 1996; 6(2):
        264-8.
        [http://dx.doi.org/10.1016/S0959-4388(96)80082-1] [PMID: 8725970]

[61]   Tully T. Regulation of gene expression and its role in long-term memory and synaptic plasticity. Proc
        Natl Acad Sci USA 1997; 94(9): 4239-41.
        [http://dx.doi.org/10.1073/pnas.94.9.4239] [PMID: 9113972]

[62]   Bender AT, Beavo JA. Cyclic nucleotide phosphodiesterases: molecular regulation to clinical use.
        Pharmacol Rev 2006; 58(3): 488-520.
        [http://dx.doi.org/10.1124/pr.58.3.5] [PMID: 16968949]

[63]   Nazarali AJ, Reynolds GP. Monoamine neurotransmitters and their metabolites in brain regions in
        Alzheimer's disease: a postmortem study. Cell Mol Neurobiol 1992; 12(6): 581-7.
        [http://dx.doi.org/10.1007/BF00711237] [PMID: 1283363]

[64]   Dringenberg HC. Alzheimer's disease: more than a 'cholinergic disorder' - evidence that cholinergic-

monoaminergic interactions contribute to EEG slowing and dementia. Behav Brain Res 2000; 115(2): 235-49.
[http://dx.doi.org/10.1016/S0166-4328(00)00261-8] [PMID: 11000423]

[65]   Buhot MC. Serotonin receptors in cognitive behaviors. Curr Opin Neurobiol 1997; 7(2): 243-54.
[http://dx.doi.org/10.1016/S0959-4388(97)80013-X] [PMID: 9142756]

[66]   Jeltsch-David H, Koenig J, Cassel JC. Modulation of cholinergic functions by serotonin and possible implications in memory: general data and focus on 5-HT(1A) receptors of the medial septum. Behav Brain Res 2008; 195(1): 86-97.
[http://dx.doi.org/10.1016/j.bbr.2008.02.037] [PMID: 18400315]

[67]   Olvera-Cortés ME, Anguiano-Rodríguez P, López-Vázquez MA, Alfaro JM. Serotonin/dopamine interaction in learning. Prog Brain Res 2008; 172: 567-602.
[http://dx.doi.org/10.1016/S0079-6123(08)00927-8] [PMID: 18772051]

[68]   Kepe V, Barrio JR, Huang SC, *et al.* Serotonin 1A receptors in the living brain of Alzheimer's disease patients. Proc Natl Acad Sci USA 2006; 103(3): 702-7.
[http://dx.doi.org/10.1073/pnas.0510237103] [PMID: 16407119]

[69]   Lyness SA, Zarow C, Chui HC. Neuron loss in key cholinergic and aminergic nuclei in Alzheimer disease: a meta-analysis. Neurobiol Aging 2003; 24(1): 1-23.
[http://dx.doi.org/10.1016/S0197-4580(02)00057-X] [PMID: 12493547]

[70]   Burke WJ, Chung HD, Huang JS, *et al.* Evidence for retrograde degeneration of epinephrine neurons in Alzheimer's disease. Ann Neurol 1988; 24(4): 532-6.
[http://dx.doi.org/10.1002/ana.410240409] [PMID: 3239955]

[71]   Harkany T, Dijkstra IM, Oosterink BJ, *et al.* Increased amyloid precursor protein expression and serotonergic sprouting following excitotoxic lesion of the rat magnocellular nucleus basalis: neuroprotection by Ca^{2+} antagonist nimodipine. Neuroscience 2000; 101(1): 101-14.
[http://dx.doi.org/10.1016/S0306-4522(00)00296-7] [PMID: 11068140]

[72]   Peddie CJ, Davies HA, Colyer FM, Stewart MG, Rodríguez JJ. Dendritic colocalisation of serotonin1B receptors and the glutamate NMDA receptor subunit NR1 within the hippocampal dentate gyrus: an ultrastructural study. J Chem Neuroanat 2008; 36(1): 17-26.
[http://dx.doi.org/10.1016/j.jchemneu.2008.05.001] [PMID: 18572381]

[73]   Rodríguez JJ, Noristani HN, Verkhratsky A. The serotonergic system in ageing and Alzheimer's disease. Prog Neurobiol 2012; 99(1): 15-41.
[http://dx.doi.org/10.1016/j.pneurobio.2012.06.010] [PMID: 22766041]

[74]   Ouchi Y, Yoshikawa E, Futatsubashi M, Yagi S, Ueki T, Nakamura K. Altered brain serotonin transporter and associated glucose metabolism in Alzheimer disease. J Nucl Med 2009; 50(8): 1260-6.
[http://dx.doi.org/10.2967/jnumed.109.063008] [PMID: 19617327]

[75]   Hu M, Retz W, Baader M, *et al.* Promoter polymorphism of the 5-HT transporter and Alzheimer's disease. Neurosci Lett 2000; 294(1): 63-5.
[http://dx.doi.org/10.1016/S0304-3940(00)01544-5] [PMID: 11044587]

[76]   Storga D, Vrecko K, Birkmayer JG, Reibnegger G. Monoaminergic neurotransmitters, their precursors and metabolites in brains of Alzheimer patients. Neurosci Lett 1996; 203(1): 29-32.
[http://dx.doi.org/10.1016/0304-3940(95)12256-7] [PMID: 8742039]

[77]   Tohgi H, Ueno M, Abe T, Takahashi S, Nozaki Y. Concentrations of monoamines and their metabolites in the cerebrospinal fluid from patients with senile dementia of the Alzheimer type and vascular dementia of the Binswanger type. J Neural Transm Park Dis Dement Sect 1992; 4(1): 69-77.
[http://dx.doi.org/10.1007/BF02257623] [PMID: 1540305]

[78]   Liu HC, Yang JC, Chang YF, Liu TY, Chi CW. Analysis of monoamines in the cerebrospinal fluid of Chinese patients with Alzheimer's disease. Ann N Y Acad Sci 1991; 640: 215-8.
[http://dx.doi.org/10.1111/j.1749-6632.1991.tb00220.x] [PMID: 1723257]

[79] Kemppainen N, Ruottinen H, Någren K, Rinne JO. PET shows that striatal dopamine D1 and D2 receptors are differentially affected in AD. Neurology 2000; 55(2): 205-9.
[http://dx.doi.org/10.1212/WNL.55.2.205] [PMID: 10908891]

[80] Allard P, Alafuzoff I, Carlsson A, et al. Loss of dopamine uptake sites labeled with [3H]GBR-12935 in Alzheimer's disease. Eur Neurol 1990; 30(4): 181-5.
[http://dx.doi.org/10.1159/000117341] [PMID: 2209670]

[81] Reinikainen KJ, Paljärvi L, Huuskonen M, Soininen H, Laakso M, Riekkinen PJ. A post-mortem study of noradrenergic, serotonergic and GABAergic neurons in Alzheimer's disease. J Neurol Sci 1988; 84(1): 101-16.
[http://dx.doi.org/10.1016/0022-510X(88)90179-7] [PMID: 2452858]

[82] Martignoni E, Blandini F, Petraglia F, Pacchetti C, Bono G, Nappi G. Cerebrospinal fluid norepinephrine, 3-methoxy-4-hydroxyphenylglycol and neuropeptide Y levels in Parkinson's disease, multiple system atrophy and dementia of the Alzheimer type. J Neural Transm Park Dis Dement Sect 1992; 4(3): 191-205.
[http://dx.doi.org/10.1007/BF02260903] [PMID: 1320891]

[83] Iversen LL, Rossor MN, Reynolds GP, et al. Loss of pigmented dopamine-beta-hydroxylase positive cells from locus coeruleus in senile dementia of Alzheimer's type. Neurosci Lett 1983; 39(1): 95-100.
[http://dx.doi.org/10.1016/0304-3940(83)90171-4] [PMID: 6633940]

[84] Youdim MB, Gross A, Finberg JP. Rasagiline [N-propargyl-1R(+)-aminoindan], a selective and potent inhibitor of mitochondrial monoamine oxidase B. Br J Pharmacol 2001; 132(2): 500-6.
[http://dx.doi.org/10.1038/sj.bjp.0703826] [PMID: 11159700]

[85] Panula P, Yang HY, Costa E. Histamine-containing neurons in the rat hypothalamus. Proc Natl Acad Sci USA 1984; 81(8): 2572-6.
[http://dx.doi.org/10.1073/pnas.81.8.2572] [PMID: 6371818]

[86] Nakamura S, Takemura M, Ohnishi K, et al. Loss of large neurons and occurrence of neurofibrillary tangles in the tuberomammillary nucleus of patients with Alzheimer's disease. Neurosci Lett 1993; 151(2): 196-9.
[http://dx.doi.org/10.1016/0304-3940(93)90019-H] [PMID: 8506080]

[87] Inestrosa NC, Alvarez A, Pérez CA, et al. Acetylcholinesterase accelerates assembly of amyloid-bet--peptides into Alzheimer's fibrils: possible role of the peripheral site of the enzyme. Neuron 1996; 16(4): 881-91.
[http://dx.doi.org/10.1016/S0896-6273(00)80108-7] [PMID: 8608006]

[88] Brioni JD, Esbenshade TA, Garrison TR, Bitner SR, Cowart MD. Discovery of histamine H3 antagonists for the treatment of cognitive disorders and Alzheimer's disease. J Pharmacol Exp Ther 2011; 336(1): 38-46.
[http://dx.doi.org/10.1124/jpet.110.166876] [PMID: 20864505]

[89] Rahman A. The role of adenosine in Alzheimer's disease. Curr Neuropharmacol 2009; 7(3): 207-16.
[http://dx.doi.org/10.2174/157015909789152119] [PMID: 20190962]

[90] Ribeiro JA, Sebastião AM, de Mendonça A. Adenosine receptors in the nervous system: pathophysiological implications. Prog Neurobiol 2002; 68(6): 377-92.
[http://dx.doi.org/10.1016/S0301-0082(02)00155-7] [PMID: 12576292]

[91] Kalaria RN, Sromek S, Wilcox BJ, Unnerstall JR. Hippocampal adenosine A1 receptors are decreased in Alzheimer's disease. Neurosci Lett 1990; 118(2): 257-60.
[http://dx.doi.org/10.1016/0304-3940(90)90641-L] [PMID: 2274280]

[92] Rivera-Oliver M, Díaz-Ríos M. Using caffeine and other adenosine receptor antagonists and agonists as therapeutic tools against neurodegenerative diseases: a review. Life Sci 2014; 101(1-2): 1-9.
[http://dx.doi.org/10.1016/j.lfs.2014.01.083] [PMID: 24530739]

[93] Ansoleaga B, Jové M, Schlüter A, et al. Deregulation of purine metabolism in Alzheimer's disease.

Neurobiol Aging 2015; 36(1): 68-80.
[http://dx.doi.org/10.1016/j.neurobiolaging.2014.08.004] [PMID: 25311278]

[94] Millan MJ. The epigenetic dimension of Alzheimer's disease: causal, consequence, or curiosity? Dialogues Clin Neurosci 2014; 16(3): 373-93.
[PMID: 25364287]

[95] Wood PL. Lipidomics of Alzheimer's disease: current status. Alzheimers Res Ther 2012; 4(1): 5.
[http://dx.doi.org/10.1186/alzrt103] [PMID: 22293144]

[96] El Gaamouch F, Jing P, Xia J, Cai D. Alzheimer's disease risk genes and lipid regulators. J Alzheimers Dis 2016; 53(1): 15-29.
[http://dx.doi.org/10.3233/JAD-160169] [PMID: 27128373]

[97] Kalvodova L, Kahya N, Schwille P, *et al.* Lipids as modulators of proteolytic activity of BACE: involvement of cholesterol, glycosphingolipids, and anionic phospholipids *in vitro.* J Biol Chem 2005; 280(44): 36815-23.
[http://dx.doi.org/10.1074/jbc.M504484200] [PMID: 16115865]

[98] Czubowicz K, Strosznajder R. Ceramide in the molecular mechanisms of neuronal cell death. The role of sphingosine-1-phosphate. Mol Neurobiol 2014; 50(1): 26-37.
[http://dx.doi.org/10.1007/s12035-013-8606-4] [PMID: 24420784]

[99] Cutler RG, Kelly J, Storie K, *et al.* Involvement of oxidative stress-induced abnormalities in ceramide and cholesterol metabolism in brain aging and Alzheimer's disease. Proc Natl Acad Sci USA 2004; 101(7): 2070-5.
[http://dx.doi.org/10.1073/pnas.0305799101] [PMID: 14970312]

[100] Markesbery WR, Carney JM. Oxidative alterations in Alzheimer's disease. Brain Pathol 1999; 9(1): 133-46.
[http://dx.doi.org/10.1111/j.1750-3639.1999.tb00215.x] [PMID: 9989456]

[101] Sharma A, Bemis M, Desilets AR. Role of medium chain triglycerides (Axona(R)) in the treatment of mild to moderate Alzheimer's disease. Am J Alzheimers Dis Other Demen 2014; 29(5): 409-14.
[http://dx.doi.org/10.1177/1533317513518650] [PMID: 24413538]

[102] Cunnane SC, Courchesne-Loyer A, St-Pierre V, *et al.* Can ketones compensate for deteriorating brain glucose uptake during aging? Implications for the risk and treatment of Alzheimer's disease. Ann N Y Acad Sci 2016; 1367(1): 12-20.
[http://dx.doi.org/10.1111/nyas.12999] [PMID: 26766547]

[103] Tully AM, Roche HM, Doyle R, *et al.* Low serum cholesteryl ester-docosahexaenoic acid levels in Alzheimer's disease: a case-control study. Br J Nutr 2003; 89(4): 483-9.
[http://dx.doi.org/10.1079/BJN2002804] [PMID: 12654166]

[104] Freund Levi Y, Vedin I, Cederholm T, *et al.* Transfer of omega-3 fatty acids across the blood-brain barrier after dietary supplementation with a docosahexaenoic acid-rich omega-3 fatty acid preparation in patients with Alzheimer's disease: the OmegAD study. J Intern Med 2014; 275(4): 428-36.
[http://dx.doi.org/10.1111/joim.12166] [PMID: 24410954]

[105] Tanabe C, Ebina M, Asai M, *et al.* 1,3-Capryloyl-2-arachidonoyl glycerol activates alpha-secretase activity and suppresses Abeta40 secretion in A172 cells. Neurosci Lett 2009; 450(3): 324-6.
[http://dx.doi.org/10.1016/j.neulet.2008.11.039] [PMID: 19041370]

[106] Guesnet P, Alessandri J-M. Docosahexaenoic acid (DHA) and the developing central nervous system (CNS) - Implications for dietary recommendations. Biochimie 2011; 93(1): 7-12.
[http://dx.doi.org/10.1016/j.biochi.2010.05.005] [PMID: 20478353]

[107] Hillig F, Porscha N, Junne S, Neubauer P. Growth and docosahexaenoic acid production performance of the heterotrophic marine microalgae Crypthecodinium cohnii in the wave-mixed single-use reactor CELL-tainer. Eng Life Sci 2014; 14: 254-63.
[http://dx.doi.org/10.1002/elsc.201400010]

[108] Ren LJ, Sun XM, Ji XJ, Chen SL, Guo DS, Huang H. Enhancement of docosahexaenoic acid synthesis by manipulation of antioxidant capacity and prevention of oxidative damage in Schizochytrium sp. Bioresour Technol 2017; 223: 141-8.
[http://dx.doi.org/10.1016/j.biortech.2016.10.040] [PMID: 27788427]

[109] Monograph. Docosahexaenoic acid (DHA). Altern Med Rev 2009; 14(4): 391-9.
[PMID: 20030466]

[110] Pan Y, Khalil H, Nicolazzo JA. The impact of docosahexaenoic acid on Alzheimer's disease: Is there a role of the blood-brain barrier? Curr Clin Pharmacol 2015; 10(3): 222-41.
[http://dx.doi.org/10.2174/1574884710031508201515532] [PMID: 26338174]

[111] Hashimoto M, Hossain S. Neuroprotective and ameliorative actions of polyunsaturated fatty acids against neuronal diseases: beneficial effect of docosahexaenoic acid on cognitive decline in Alzheimer's disease. J Pharmacol Sci 2011; 116(2): 150-62.
[http://dx.doi.org/10.1254/jphs.10R33FM] [PMID: 21606627]

[112] 2015.https://www.nia.nih.gov/alzheimers/publication/preventing-alzheimers-disease/sear-h-alzheimers-prevention-strategies

[113] Schaefer EJ, Bongard V, Beiser AS, *et al.* Plasma phosphatidylcholine docosahexaenoic acid content and risk of dementia and Alzheimer disease: the Framingham Heart Study. Arch Neurol 2006; 63(11): 1545-50.
[http://dx.doi.org/10.1001/archneur.63.11.1545] [PMID: 17101822]

[114] Ramprasath VR, Eyal I, Zchut S, Jones PJ. Enhanced increase of omega-3 index in healthy individuals with response to 4-week n-3 fatty acid supplementation from krill oil *versus* fish oil. Lipids Health Dis 2013; 12: 178.
[http://dx.doi.org/10.1186/1476-511X-12-178] [PMID: 24304605]

[115] Polotow TG, Poppe SC, Vardaris CV, *et al.* Redox status and neuro inflammation indexes in cerebellum and motor cortex of Wistar rats supplemented with natural sources of Omega-3 fatty acids and Astaxanthin: fish oil, Krill oil, and algal biomass. Mar Drugs 2015; 13(10): 6117-37.
[http://dx.doi.org/10.3390/md13106117] [PMID: 26426026]

[116] Uauy R, Dangour AD. Nutrition in brain development and aging: role of essential fatty acids. Nutr Rev 2006; 64(5 Pt 2): S24-33.
[http://dx.doi.org/10.1301/nr.2006.may.S24-S33] [PMID: 16770950]

[117] Hoshino T, Namba T, Takehara M, *et al.* Prostaglandin E2 stimulates the production of amyloid-beta peptides through internalization of the EP4 receptor. J Biol Chem 2009; 284(27): 18493-502.
[http://dx.doi.org/10.1074/jbc.M109.003269] [PMID: 19407341]

[118] Montine TJ, Sidell KR, Crews BC, *et al.* Elevated CSF prostaglandin E2 levels in patients with probable AD. Neurology 1999; 53(7): 1495-8.
[http://dx.doi.org/10.1212/WNL.53.7.1495] [PMID: 10534257]

[119] Sanchez-Mejia RO, Newman JW, Toh S, *et al.* Phospholipase A2 reduction ameliorates cognitive deficits in a mouse model of Alzheimer's disease. Nat Neurosci 2008; 11(11): 1311-8.
[http://dx.doi.org/10.1038/nn.2213] [PMID: 18931664]

[120] Kotani S, Nakazawa H, Tokimasa T, *et al.* Synaptic plasticity preserved with arachidonic acid diet in aged rats. Neurosci Res 2003; 46(4): 453-61.
[http://dx.doi.org/10.1016/S0168-0102(03)00123-8] [PMID: 12871767]

[121] Kotani S, Sakaguchi E, Warashina S, *et al.* Dietary supplementation of arachidonic and docosahexaenoic acids improves cognitive dysfunction. Neurosci Res 2006; 56(2): 159-64.
[http://dx.doi.org/10.1016/j.neures.2006.06.010] [PMID: 16905216]

[122] Fukaya T, Gondaira T, Kashiyae Y, *et al.* Arachidonic acid preserves hippocampal neuron membrane fluidity in senescent rats. Neurobiol Aging 2007; 28(8): 1179-86.
[http://dx.doi.org/10.1016/j.neurobiolaging.2006.05.023] [PMID: 16790296]

[123] Ma QL, Yang F, Rosario ER, *et al.* Beta-amyloid oligomers induce phosphorylation of tau and inactivation of insulin receptor substrate *via* c-Jun N-terminal kinase signaling: suppression by omega-3 fatty acids and curcumin. J Neurosci 2009; 29(28): 9078-89.
[http://dx.doi.org/10.1523/JNEUROSCI.1071-09.2009] [PMID: 19605645]

[124] Hosono T, Nishitsuji K, Nakamura T, *et al.* Arachidonic acid diet attenuates brain Aβ deposition in Tg2576 mice. Brain Res 2015; 1613: 92-9.
[http://dx.doi.org/10.1016/j.brainres.2015.04.005] [PMID: 25881896]

[125] Kotani S, Sakaguchi E, Warashina S, *et al.* Dietary supplementation of arachidonic and docosahexaenoic acids improves cognitive dysfunction. Neurosci Res 2006; 56(2): 159-64.
[http://dx.doi.org/10.1016/j.neures.2006.06.010] [PMID: 16905216]

[126] Karlamangla AS, Miller-Martinez D, Lachman ME, Tun PA, Koretz BK, Seeman TE. Biological correlates of adult cognition: midlife in the United States (MIDUS). Neurobiol Aging 2014; 35(2): 387-94.
[http://dx.doi.org/10.1016/j.neurobiolaging.2013.07.028] [PMID: 24011541]

[127] Henderson ST, Vogel JL, Barr LJ, Garvin F, Jones JJ, Costantini LC. Study of the ketogenic agent AC-1202 in mild to moderate Alzheimer's disease: a randomized, double-blind, placebo-controlled, multicenter trial. Nutr Metab (Lond) 2009; 6: 31.
[http://dx.doi.org/10.1186/1743-7075-6-31] [PMID: 19664276]

[128] Henderson ST, Poirier J. Pharmacogenetic analysis of the effects of polymorphisms in APOE, IDE and IL1B on a ketone body based therapeutic on cognition in mild to moderate Alzheimer's disease; a randomized, double-blind, placebo-controlled study. BMC Med Genet 2011; 12: 137.
[http://dx.doi.org/10.1186/1471-2350-12-137] [PMID: 21992747]

[129] http://www.alzforum.org/therapeutics/ac-1204

[130] 2015.https://www.nia.nih.gov/alzheimers/clinical-trials/ac-1204-mild-moderate-alzheimers-disease

[131] Albarracin SL, Stab B, Casas Z, *et al.* Effects of natural antioxidants in neurodegenerative disease. Nutr Neurosci 2012; 15(1): 1-9.
[http://dx.doi.org/10.1179/1476830511Y.0000000028] [PMID: 22305647]

[132] Ngoungoure VL, Schluesener J, Moundipa PF, Schluesener H. Natural polyphenols binding to amyloid: a broad class of compounds to treat different human amyloid diseases. Mol Nutr Food Res 2015; 59(1): 8-20.
[http://dx.doi.org/10.1002/mnfr.201400290] [PMID: 25167846]

[133] Ataie A, Shadifar M, Ataee R. Polyphenolic Antioxidants and Neuronal Regeneration. Basic Clin Neurosci 2016; 7(2): 81-90.
[http://dx.doi.org/10.15412/J.BCN.03070201] [PMID: 27303602]

[134] Ebrahimi A, Schluesener H. Natural polyphenols against neurodegenerative disorders: potentials and pitfalls. Ageing Res Rev 2012; 11(2): 329-45.
[http://dx.doi.org/10.1016/j.arr.2012.01.006] [PMID: 22336470]

[135] Zhao L, Wang JL, Liu R, Li XX, Li JF, Zhang L. Neuroprotective, anti-amyloidogenic and neurotrophic effects of apigenin in an Alzheimer's disease mouse model. Molecules 2013; 18(8): 9949-65.
[http://dx.doi.org/10.3390/molecules18089949] [PMID: 23966081]

[136] Venigalla M, Sonego S, Gyengesi E, Sharman MJ, Münch G. Novel promising therapeutics against chronic neuroinflammation and neurodegeneration in Alzheimer's disease. Neurochem Int 2016; 95: 63-74.
[http://dx.doi.org/10.1016/j.neuint.2015.10.011] [PMID: 26529297]

[137] Choi JS, Islam MN, Ali MY, Kim EJ, Kim YM, Jung HA. Effects of C-glycosylation on anti-diabetic, anti-Alzheimer's disease and anti-inflammatory potential of apigenin. Food Chem Toxicol 2014; 64: 27-33.

[http://dx.doi.org/10.1016/j.fct.2013.11.020] [PMID: 24291393]

[138] Liu R, Zhang T, Yang H, Lan X, Ying J, Du G. The flavonoid apigenin protects brain neurovascular coupling against amyloid-β_{25-35}-induced toxicity in mice. J Alzheimers Dis 2011; 24(1): 85-100.
[http://dx.doi.org/10.3233/JAD-2010-101593] [PMID: 21297270]

[139] Choi AY, Choi JH, Lee JY, *et al.* Apigenin protects HT22 murine hippocampal neuronal cells against endoplasmic reticulum stress-induced apoptosis. Neurochem Int 2010; 57(2): 143-52.
[http://dx.doi.org/10.1016/j.neuint.2010.05.006] [PMID: 20493918]

[140] Braidy N, Grant R, Adams S, Guillemin GJ. Neuroprotective effects of naturally occurring polyphenols on quinolinic acid-induced excitotoxicity in human neurons. FEBS J 2010; 277(2): 368-82.
[http://dx.doi.org/10.1111/j.1742-4658.2009.07487.x] [PMID: 20015232]

[141] Zhao L, Wang JL, Wang YR, Fa XZ. Apigenin attenuates copper-mediated β-amyloid neurotoxicity through antioxidation, mitochondrion protection and MAPK signal inactivation in an AD cell model. Brain Res 2013; 1492: 33-45.
[http://dx.doi.org/10.1016/j.brainres.2012.11.019] [PMID: 23178511]

[142] Zhao L, Hou L, Sun HJ, Yan X, Sun XF, Li JG, *et al.* Apigenin isolated from the medicinal plant Elsholtzia rugulosa prevents β-amyloid 25–35-Induces toxicity in rat cerebral microvascular endothelial cells. Molecules 2011; 16(5): 4005-19.
[http://dx.doi.org/10.3390/molecules16054005]

[143] Ono K, Yoshiike Y, Takashima A, Hasegawa K, Naiki H, Yamada M. Potent anti-amyloidogenic and fibril-destabilizing effects of polyphenols *in vitro*: implications for the prevention and therapeutics of Alzheimer's disease. J Neurochem 2003; 87(1): 172-81.
[http://dx.doi.org/10.1046/j.1471-4159.2003.01976.x] [PMID: 12969264]

[144] Heo HJ, Kim DO, Choi SJ, Shin DH, Lee CY. Potent Inhibitory effect of flavonoids in Scutellaria baicalensis on amyloid beta protein-induced neurotoxicity. J Agric Food Chem 2004; 52(13): 4128-32.
[http://dx.doi.org/10.1021/jf049953x] [PMID: 15212458]

[145] Zhang SQ, Obregon D, Ehrhart J, *et al.* Baicalein reduces β-amyloid and promotes nonamyloidogenic amyloid precursor protein processing in an Alzheimer's disease transgenic mouse model. J Neurosci Res 2013; 91(9): 1239-46.
[http://dx.doi.org/10.1002/jnr.23244] [PMID: 23686791]

[146] Choi JS, Islam MN, Ali MY, *et al.* The effects of C-glycosylation of luteolin on its antioxidant, anti-Alzheimer's disease, anti-diabetic, and anti-inflammatory activities. Arch Pharm Res 2014; 37(10): 1354-63.
[http://dx.doi.org/10.1007/s12272-014-0351-3] [PMID: 24988985]

[147] Wilmsen PK, Spada DS, Salvador M. Antioxidant activity of the flavonoid hesperidin in chemical and biological systems. J Agric Food Chem 2005; 53(12): 4757-61.
[http://dx.doi.org/10.1021/jf0502000] [PMID: 15941311]

[148] Hong Y, An Z. Hesperidin attenuates learning and memory deficits in APP/PS1 mice through activation of Akt/Nrf2 signaling and inhibition of RAGE/NF-κB signaling. Arch Pharm Res 2015. [Epub ahead of print]
[http://dx.doi.org/10.1007/s12272-015-0662-z] [PMID: 26391026]

[149] Huang SM, Tsai SY, Lin JA, Wu CH, Yen GC. Cytoprotective effects of hesperetin and hesperidin against amyloid β-induced impairment of glucose transport through downregulation of neuronal autophagy. Mol Nutr Food Res 2012; 56(4): 601-9.
[http://dx.doi.org/10.1002/mnfr.201100682] [PMID: 22383310]

[150] Wang D, Liu L, Zhu X, Wu W, Wang Y. Hesperidin alleviates cognitive impairment, mitochondrial dysfunction and oxidative stress in a mouse model of Alzheimer's disease. Cell Mol Neurobiol 2014; 34(8): 1209-21.
[http://dx.doi.org/10.1007/s10571-014-0098-x] [PMID: 25135708]

[151] Ghofrani S, Joghataei MT, Mohseni S, *et al.* Naringenin improves learning and memory in an Alzheimer's disease rat model: Insights into the underlying mechanisms. Eur J Pharmacol 2015; 764: 195-201.
[http://dx.doi.org/10.1016/j.ejphar.2015.07.001] [PMID: 26148826]

[152] Mandel SA, Amit T, Kalfon L, Reznichenko L, Weinreb O, Youdim MB. Cell signaling pathways and iron chelation in the neurorestorative activity of green tea polyphenols: special reference to epigallocatechin gallate (EGCG). J Alzheimers Dis 2008; 15(2): 211-22.
[http://dx.doi.org/10.3233/JAD-2008-15207] [PMID: 18953110]

[153] http://www.alzforum.org/therapeutics/epigallocatechin-gallate-egcg

[154] Williams RJ, Spencer JP. Flavonoids, cognition, and dementia: actions, mechanisms, and potential therapeutic utility for Alzheimer disease. Free Radic Biol Med 2012; 52(1): 35-45.
[http://dx.doi.org/10.1016/j.freeradbiomed.2011.09.010] [PMID: 21982844]

[155] Spencer JP, Vafeiadou K, Williams RJ, Vauzour D. Neuroinflammation: modulation by flavonoids and mechanisms of action. Mol Aspects Med 2012; 33(1): 83-97.
[http://dx.doi.org/10.1016/j.mam.2011.10.016] [PMID: 22107709]

[156] Shah ZA, Li RC, Ahmad AS, *et al.* The flavanol (-)-epicatechin prevents stroke damage through the Nrf2/HO1 pathway. J Cereb Blood Flow Metab 2010; 30(12): 1951-61.
[http://dx.doi.org/10.1038/jcbfm.2010.53] [PMID: 20442725]

[157] Cao S, Jiang X, Chen J. Effect of Zinc (II) on the interactions of bovine serum albumin with flavonols bearing different number of hydroxyl substituent on B-ring. J Inorg Biochem 2010; 104(2): 146-52.
[http://dx.doi.org/10.1016/j.jinorgbio.2009.10.014] [PMID: 19932510]

[158] DeToma AS, Choi J-S, Braymer JJ, Lim MH. Myricetin: a naturally occurring regulator of metal-induced amyloid-β aggregation and neurotoxicity. ChemBioChem 2011; 12(8): 1198-201.
[http://dx.doi.org/10.1002/cbic.201000790] [PMID: 21538759]

[159] Shimmyo Y, Kihara T, Akaike A, Niidome T, Sugimoto H. Three distinct neuroprotective functions of myricetin against glutamate-induced neuronal cell death: involvement of direct inhibition of caspase-3. J Neurosci Res 2008; 86(8): 1836-45.
[http://dx.doi.org/10.1002/jnr.21629] [PMID: 18265412]

[160] Fiorani M, Guidarelli A, Blasa M, *et al.* Mitochondria accumulate large amounts of quercetin: prevention of mitochondrial damage and release upon oxidation of the extramitochondrial fraction of the flavonoid. J Nutr Biochem 2010; 21(5): 397-404.
[http://dx.doi.org/10.1016/j.jnutbio.2009.01.014] [PMID: 19278846]

[161] Wang D-M, Li S-Q, Wu W-L, Zhu X-Y, Wang Y, Yuan H-Y. Effects of long-term treatment with quercetin on cognition and mitochondrial function in a mouse model of Alzheimer's disease. Neurochem Res 2014; 39(8): 1533-43.
[http://dx.doi.org/10.1007/s11064-014-1343-x] [PMID: 24893798]

[162] Sabogal-Guáqueta AM, Muñoz-Manco JI, Ramírez-Pineda JR, Lamprea-Rodriguez M, Osorio E, Cardona-Gómez GP. The flavonoid quercetin ameliorates Alzheimer's disease pathology and protects cognitive and emotional function in aged triple transgenic Alzheimer's disease model mice. Neuropharmacology 2015; 93: 134-45.
[http://dx.doi.org/10.1016/j.neuropharm.2015.01.027] [PMID: 25666032]

[163] Shimmyo Y, Kihara T, Akaike A, Niidome T, Sugimoto H. Flavonols and flavones as BACE-1 inhibitors: Structure–activity relationship in cell-free, cell-based and in silico studies reveal novel pharmacophore features. Biochimica et Biophysica Acta 2008; 1780: 819-25.
[http://dx.doi.org/10.1016/j.bbagen.2008.01.017] [PMID: 18295609]

[164] Kim JK, Choi S-J, Cho H-Y, *et al.* Protective effects of kaempferol (3,4',5,7-tetrahydroxyflavone) against amyloid beta peptide (Abeta)-induced neurotoxicity in ICR mice. Biosci Biotechnol Biochem 2010; 74(2): 397-401.

[http://dx.doi.org/10.1271/bbb.90585] [PMID: 20139605]

[165]   Kim JK, Shin EC, Kim CR, *et al.* Effects of brussels sprouts and their phytochemical components on oxidative stress-induced neuronal damages in PC12 cells and ICR mice. J Med Food 2013; 16(11): 1057-61.
[http://dx.doi.org/10.1089/jmf.2012.0280] [PMID: 24175656]

[166]   Kelsey N, Hulick W, Winter A, Ross E, Linseman D. Neuroprotective effects of anthocyanins on apoptosis induced by mitochondrial oxidative stress. Nutr Neurosci 2011; 14(6): 249-59.
[http://dx.doi.org/10.1179/1476830511Y.0000000020] [PMID: 22053756]

[167]   Badshah H, Kim TH, Kim MO. Protective effects of anthocyanins against amyloid beta-induced neurotoxicity *in vivo* and *in vitro*. Neurochem Int 2015; 80: 51-9.
[http://dx.doi.org/10.1016/j.neuint.2014.10.009] [PMID: 25451757]

[168]   Monograph. Curcuma longa (turmeric). Altern Med Rev 2001; 6: S62-6.
[PMID: 11591174]

[169]   Jovanovic SV, Steenken S, Boone CW, Simic MG. H-Atom transfer is a preferred antioxidant mechanism of curcumin. J Am Chem Soc 1999; 121: 9677-81.
[http://dx.doi.org/10.1021/ja991446m]

[170]   Krup V, Prakash LH, Harini A. Pharmacological Activities of Turmeric (*Curcuma longa linn*): A review. J Homeop Ayurv Med 2013; 2: 133.
[http://dx.doi.org/10.4172/2167-1206.1000133]

[171]   Kulkarni SK, Dhir A. An overview of curcumin in neurological disorders. Indian J Pharm Sci 2010; 72(2): 149-54.
[http://dx.doi.org/10.4103/0250-474X.65012] [PMID: 20838516]

[172]   Ng TP, Chiam PC, Lee T, Chua HC, Lim L, Kua EH. Curry consumption and cognitive function in the elderly. Am J Epidemiol 2006; 164(9): 898-906.
[http://dx.doi.org/10.1093/aje/kwj267] [PMID: 16870699]

[173]   Chan S, Kantham S, Rao VM, *et al.* Metal chelation, radical scavenging and inhibition of $A\beta_{42}$ fibrillation by food constituents in relation to Alzheimer's disease. Food Chem 2016; 199: 185-94.
[http://dx.doi.org/10.1016/j.foodchem.2015.11.118] [PMID: 26775960]

[174]   Ishrat T, Hoda MN, Khan MB, *et al.* Amelioration of cognitive deficits and neurodegeneration by curcumin in rat model of sporadic dementia of Alzheimer's type (SDAT). Eur Neuropsychopharmacol 2009; 19(9): 636-47.
[http://dx.doi.org/10.1016/j.euroneuro.2009.02.002] [PMID: 19329286]

[175]   Hamaguchi T, Ono K, Yamada M. REVIEW: Curcumin and Alzheimer's disease. CNS Neurosci Ther 2010; 16(5): 285-97.
[http://dx.doi.org/10.1111/j.1755-5949.2010.00147.x] [PMID: 20406252]

[176]   Frautschy SA, Cole GM. Why pleiotropic interventions are needed for Alzheimer's disease. Mol Neurobiol 2010; 41(2-3): 392-409.
[http://dx.doi.org/10.1007/s12035-010-8137-1] [PMID: 20437209]

[177]   Patil SP, Tran N, Geekiyanage H, Liu L, Chan C. Curcumin-induced upregulation of the anti-tau cochaperone BAG2 in primary rat cortical neurons. Neurosci Lett 2013; 554: 121-5.
[http://dx.doi.org/10.1016/j.neulet.2013.09.008] [PMID: 24035895]

[178]   Wahlström B, Blennow G. A study on the fate of curcumin in the rat. Acta Pharmacol Toxicol (Copenh) 1978; 43(2): 86-92.
[http://dx.doi.org/10.1111/j.1600-0773.1978.tb02240.x] [PMID: 696348]

[179]   Shoba G, Joy D, Joseph T, Majeed M, Rajendran R, Srinivas PS. Influence of piperine on the pharmacokinetics of curcumin in animals and human volunteers. Planta Med 1998; 64(4): 353-6.
[http://dx.doi.org/10.1055/s-2006-957450] [PMID: 9619120]

[180] Yang KY, Lin LC, Tseng TY, Wang SC, Tsai TH. Oral bioavailability of curcumin in rat and the herbal analysis from Curcuma longa by LC-MS/MS. J Chromatogr B Analyt Technol Biomed Life Sci 2007; 853(1-2): 183-9.
[http://dx.doi.org/10.1016/j.jchromb.2007.03.010] [PMID: 17400527]

[181] Baum L, Lam CW, Cheung SK, *et al.* Six-month randomized, placebo-controlled, double-blind, pilot clinical trial of curcumin in patients with Alzheimer disease. J Clin Psychopharmacol 2008; 28(1): 110-3.
[http://dx.doi.org/10.1097/jcp.0b013e318160862c] [PMID: 18204357]

[182] Ringman JM, Frautschy SA, Teng E, *et al.* Oral curcumin for Alzheimer's disease: tolerability and efficacy in a 24-week randomized, double blind, placebo-controlled study. Alzheimers Res Ther 2012; 4(5): 43.
[http://dx.doi.org/10.1186/alzrt146] [PMID: 23107780]

[183] Prasad S, Tyagi AK, Aggarwal BB. Recent developments in delivery, bioavailability, absorption and metabolism of curcumin: the golden pigment from golden spice. Cancer Res Treat 2014; 46(1): 2-18.
[http://dx.doi.org/10.4143/crt.2014.46.1.2] [PMID: 24520218]

[184] Gota VS, Maru GB, Soni TG, Gandhi TR, Kochar N, Agarwal MG. Safety and pharmacokinetics of a solid lipid curcumin particle formulation in osteosarcoma patients and healthy volunteers. J Agric Food Chem 2010; 58(4): 2095-9.
[http://dx.doi.org/10.1021/jf9024807] [PMID: 20092313]

[185] Cox KH, Pipingas A, Scholey AB. Investigation of the effects of solid lipid curcumin on cognition and mood in a healthy older population. J Psychopharmacol (Oxford) 2015; 29(5): 642-51.
[http://dx.doi.org/10.1177/0269881114552744] [PMID: 25277322]

[186] DiSilvestro RA, Joseph E, Zhao S, Bomser J. Diverse effects of a low dose supplement of lipidated curcumin in healthy middle aged people. Nutr J 2012; 11: 79-86.
[http://dx.doi.org/10.1186/1475-2891-11-79] [PMID: 23013352]

[187] Frémont L. Biological effects of resveratrol. Life Sci 2000; 66(8): 663-73.
[http://dx.doi.org/10.1016/S0024-3205(99)00410-5] [PMID: 10680575]

[188] Rege SD, Geetha T, Griffin GD, Broderick TL, Babu JR. Neuroprotective effects of resveratrol in Alzheimer disease pathology. Front Aging Neurosci 2014; 6: 218.
[http://dx.doi.org/10.3389/fnagi.2014.00218] [PMID: 25309423]

[189] Wang H, Yang Y-J, Qian H-Y, Zhang Q, Xu H, Li J-J. Resveratrol in cardiovascular disease: what is known from current research? Heart Fail Rev 2012; 17(3): 437-48.
[http://dx.doi.org/10.1007/s10741-011-9260-4] [PMID: 21688187]

[190] Carrizzo A, Forte M, Damato A, *et al.* Antioxidant effects of resveratrol in cardiovascular, cerebral and metabolic diseases. Food Chem Toxicol 2013; 61: 215-26.
[http://dx.doi.org/10.1016/j.fct.2013.07.021] [PMID: 23872128]

[191] Huang TC, Lu KT, Wo YY, Wu YJ, Yang YL. Resveratrol protects rats from Aβ-induced neurotoxicity by the reduction of iNOS expression and lipid peroxidation. PLoS One 2011; 6(12): e29102.
[http://dx.doi.org/10.1371/journal.pone.0029102] [PMID: 22220203]

[192] Granzotto A, Zatta P. Resveratrol acts not through anti-aggregative pathways but mainly *via* its scavenging properties against Aβ and Aβ-metal complexes toxicity. PLoS One 2011; 6(6): e21565.
[http://dx.doi.org/10.1371/journal.pone.0021565] [PMID: 21738712]

[193] Kelsey NA, Wilkins HM, Linseman DA. Nutraceutical antioxidants as novel neuroprotective agents. Molecules 2010; 15(11): 7792-814.
[http://dx.doi.org/10.3390/molecules15117792] [PMID: 21060289]

[194] Lee MK, Kang SJ, Poncz M, Song K-J, Park KS. Resveratrol protects SH-SY5Y neuroblastoma cells from apoptosis induced by dopamine. Exp Mol Med 2007; 39(3): 376-84.

[http://dx.doi.org/10.1038/emm.2007.42] [PMID: 17603292]

[195] Capiralla H, Vingtdeux V, Zhao H, *et al.* Resveratrol mitigates lipopolysaccharide- and Aβ-mediated microglial inflammation by inhibiting the TLR4/NF-κB/STAT signaling cascade. J Neurochem 2012; 120(3): 461-72.
[http://dx.doi.org/10.1111/j.1471-4159.2011.07594.x] [PMID: 22118570]

[196] Frozza RL, Bernardi A, Hoppe JB, *et al.* Lipid-core nanocapsules improve the effects of resveratrol against Abeta-induced neuroinflammation. J Biomed Nanotechnol 2013; 9(12): 2086-104.
[http://dx.doi.org/10.1166/jbn.2013.1709] [PMID: 24266263]

[197] Das S, Das DK. Anti-inflammatory responses of resveratrol. Inflamm Allergy Drug Targets 2007; 6(3): 168-73.
[http://dx.doi.org/10.2174/187152807781696464] [PMID: 17897053]

[198] Kim YA, Lim S-Y, Rhee S-H, *et al.* Resveratrol inhibits inducible nitric oxide synthase and cyclooxygenase-2 expression in beta-amyloid-treated C6 glioma cells. Int J Mol Med 2006; 17(6): 1069-75.
[PMID: 16685418]

[199] Lin YL, Chang HC, Chen TL, *et al.* Resveratrol protects against oxidized LDL-induced breakage of the blood-brain barrier by lessening disruption of tight junctions and apoptotic insults to mouse cerebrovascular endothelial cells. J Nutr 2010; 140(12): 2187-92.
[http://dx.doi.org/10.3945/jn.110.123505] [PMID: 20980646]

[200] Walle T. Bioavailability of resveratrol. Ann N Y Acad Sci 2011; 1215: 9-15.
[http://dx.doi.org/10.1111/j.1749-6632.2010.05842.x] [PMID: 21261636]

[201] Turner RS, Thomas RG, Craft S, *et al.* Alzheimer's Disease Cooperative Study. A randomized, double-blind, placebo-controlled trial of resveratrol for Alzheimer disease. Neurology 2015; 85(16): 1383-91.
[http://dx.doi.org/10.1212/WNL.0000000000002035] [PMID: 26362286]

[202] Augustin MA, Sanguansri L, Lockett T. Nano- and micro-encapsulated systems for enhancing the delivery of resveratrol. Ann N Y Acad Sci 2013; 1290: 107-12.
[http://dx.doi.org/10.1111/nyas.12130] [PMID: 23855472]

[203] Zhang Y, Song H, Shang Z, *et al.* Amino acid-PEGylated resveratrol and its influence on solubility and the controlled release behavior. Biol Pharm Bull 2014; 37(5): 785-93.
[http://dx.doi.org/10.1248/bpb.b13-00863] [PMID: 24599033]

[204] Zhang Z, Wu Z, Zhu X, Hui X, Pan J, Xu Y. Hydroxy-safflor yellow A inhibits neuroinflammation mediated by Aβ$_{1-42}$ in BV-2 cells. Neurosci Lett 2014; 562: 39-44.
[http://dx.doi.org/10.1016/j.neulet.2014.01.005] [PMID: 24412680]

[205] Zhang ZH, Yu LJ, Hui XC, *et al.* Hydroxy-safflor yellow A attenuates Aβ$_{1-42}$-induced inflammation by modulating the JAK2/STAT3/NF-κB pathway. Brain Res 2014; 1563: 72-80.
[http://dx.doi.org/10.1016/j.brainres.2014.03.036] [PMID: 24690200]

[206] Gottlieb M, Leal-Campanario R, Campos-Esparza MR, *et al.* Neuroprotection by two polyphenols following excitotoxicity and experimental ischemia. Neurobiol Dis 2006; 23(2): 374-86.
[http://dx.doi.org/10.1016/j.nbd.2006.03.017] [PMID: 16806951]

[207] Biradar SM, Joshi H, Chheda TK. Neuropharmacological effect of Mangiferin on brain cholinesterase and brain biogenic amines in the management of Alzheimer's disease. Eur J Pharmacol 2012; 683(1-3): 140-7.
[http://dx.doi.org/10.1016/j.ejphar.2012.02.042] [PMID: 22426032]

[208] Urbain A, Marston A, Queiroz EF, Ndjoko K, Hostettmann K. Xanthones from Gentiana campestris as new acetylcholinesterase inhibitors. Planta Med 2004; 70(10): 1011-4.
[http://dx.doi.org/10.1055/s-2004-832632] [PMID: 15490334]

[209] Konrath EL, Passos CdosS, Klein LC Jr, Henriques AT. Alkaloids as a source of potential

anticholinesterase inhibitors for the treatment of Alzheimer's disease. J Pharm Pharmacol 2013; 65(12): 1701-25.
[http://dx.doi.org/10.1111/jphp.12090] [PMID: 24236981]

[210] Ng YP, Or TC, Ip NY. Plant alkaloids as drug leads for Alzheimer's disease. Neurochem Int 2015; 89: 260-70.
[http://dx.doi.org/10.1016/j.neuint.2015.07.018] [PMID: 26220901]

[211] Parys W. Development of Reminyl (R) (Galantamine), a novel acetylcholinesterase inhibitor, for the treatment of Alzheimer's disease. Alzheimers Rep 1998; S19-20.

[212] Harvey AL. The pharmacology of galanthamine and its analogues. Pharmacol Ther 1995; 68(1): 113-28.
[http://dx.doi.org/10.1016/0163-7258(95)02002-0] [PMID: 8604434]

[213] Bitzinger DI, Gruber M, Tümmler S, *et al.* Species- and concentration-dependent differences of acetyl- and butyrylcholinesterase sensitivity to physostigmine and neostigmine. Neuropharmacology 2016; 109: 1-6.
[http://dx.doi.org/10.1016/j.neuropharm.2016.01.005] [PMID: 26772968]

[214] Woodruff-Pak DS, Vogel RW III, Wenk GL. Galantamine: effect on nicotinic receptor binding, acetylcholinesterase inhibition, and learning. Proc Natl Acad Sci USA 2001; 98(4): 2089-94.
[http://dx.doi.org/10.1073/pnas.98.4.2089] [PMID: 11172080]

[215] Matharu B, Gibson G, Parsons R, *et al.* Galantamine inhibits beta-amyloid aggregation and cytotoxicity. J Neurol Sci 2009; 280(1-2): 49-58.
[http://dx.doi.org/10.1016/j.jns.2009.01.024] [PMID: 19249060]

[216] Egea J, Martín-de-Saavedra MD, Parada E, *et al.* Galantamine elicits neuroprotection by inhibiting iNOS, NADPH oxidase and ROS in hippocampal slices stressed with anoxia/reoxygenation. Neuropharmacology 2012; 62(2): 1082-90.
[http://dx.doi.org/10.1016/j.neuropharm.2011.10.022] [PMID: 22085833]

[217] Kita Y, Ago Y, Higashino K, *et al.* Galantamine promotes adult hippocampal neurogenesis *via* M_1 muscarinic and α7 nicotinic receptors in mice. Int J Neuropsychopharmacol 2014; 17(12): 1957-68.
[http://dx.doi.org/10.1017/S1461145714000613] [PMID: 24818616]

[218] Schneider LS, Mangialasche F, Andreasen N, *et al.* Clinical trials and late-stage drug development for Alzheimer's disease: an appraisal from 1984 to 2014. J Intern Med 2014; 275(3): 251-83.
[http://dx.doi.org/10.1111/joim.12191] [PMID: 24605808]

[219] Miranda LF, Gomes KB, Silveira JN, *et al.* Predictive factors of clinical response to cholinesterase inhibitors in mild and moderate Alzheimer's disease and mixed dementia: a one-year naturalistic study. J Alzheimers Dis 2015; 45(2): 609-20.
[PMID: 25589728]

[220] Richarz U, Gaudig M, Rettig K, Schauble B. Galantamine treatment in outpatients with mild Alzheimer's disease. Acta Neurol Scand 2014; 129(6): 382-92.
[http://dx.doi.org/10.1111/ane.12195] [PMID: 24461047]

[221] Ye M, Fu S, Pi R, He F. Neuropharmacological and pharmacokinetic properties of berberine: a review of recent research. J Pharm Pharmacol 2009; 61(7): 831-7.
[http://dx.doi.org/10.1211/jpp.61.07.0001] [PMID: 19589224]

[222] Hu JP, Nishishita K, Sakai E, *et al.* Berberine inhibits RANKL-induced osteoclast formation and survival through suppressing the NF-kappaB and Akt pathways. Eur J Pharmacol 2008; 580(1-2): 70-9.
[http://dx.doi.org/10.1016/j.ejphar.2007.11.013] [PMID: 18083161]

[223] Asai M, Iwata N, Yoshikawa A, *et al.* Berberine alters the processing of Alzheimer's amyloid precursor protein to decrease Abeta secretion. Biochem Biophys Res Commun 2007; 352(2): 498-502.
[http://dx.doi.org/10.1016/j.bbrc.2006.11.043] [PMID: 17125739]

[224] Jia L, Liu J, Song Z, *et al.* Berberine suppresses amyloid-beta-induced inflammatory response in microglia by inhibiting nuclear factor-kappaB and mitogen-activated protein kinase signalling pathways. J Pharm Pharmacol 2012; 64(10): 1510-21.
[http://dx.doi.org/10.1111/j.2042-7158.2012.01529.x] [PMID: 22943182]

[225] Shirwaikar A, Shirwaikar A, Rajendran K, Punitha IS. *In vitro* antioxidant studies on the benzyl tetra isoquinoline alkaloid berberine. Biol Pharm Bull 2006; 29(9): 1906-10.
[http://dx.doi.org/10.1248/bpb.29.1906] [PMID: 16946507]

[226] Kim MH, Kim SH, Yang WM. Mechanisms of action of phytochemicals from medicinal herbs in the treatment of Alzheimer's disease. Planta Med 2014; 80(15): 1249-58.
[http://dx.doi.org/10.1055/s-0034-1383038] [PMID: 25210998]

[227] Abd El-Wahab AE, Ghareeb DA, Sarhan EE, Abu-Serie MM, El Demellawy MA. *In vitro* biological assessment of Berberis vulgaris and its active constituent, berberine: antioxidants, anti-acetylcholinesterase, anti-diabetic and anticancer effects. BMC Complement Altern Med 2013; 13: 218.
[http://dx.doi.org/10.1186/1472-6882-13-218] [PMID: 24007270]

[228] Shi A, Huang L, Lu C, He F, Li X. Synthesis, biological evaluation and molecular modeling of novel triazole-containing berberine derivatives as acetylcholinesterase and β-amyloid aggregation inhibitors. Bioorg Med Chem 2011; 19(7): 2298-305.
[http://dx.doi.org/10.1016/j.bmc.2011.02.025] [PMID: 21397508]

[229] Chen M, Tan M, Jing M, *et al.* Berberine protects homocysteic acid-induced HT-22 cell death: involvement of Akt pathway. Metab Brain Dis 2015; 30(1): 137-42.
[http://dx.doi.org/10.1007/s11011-014-9580-x] [PMID: 25048007]

[230] Durairajan SS, Liu LF, Lu JH, *et al.* Berberine ameliorates β-amyloid pathology, gliosis, and cognitive impairment in an Alzheimer's disease transgenic mouse model. Neurobiol Aging 2012; 33(12): 2903-19.
[http://dx.doi.org/10.1016/j.neurobiolaging.2012.02.016] [PMID: 22459600]

[231] Durairajan SS, Huang YY, Yuen PY, *et al.* Effects of Huanglian-Jie-Du-Tang and its modified formula on the modulation of amyloid-β precursor protein processing in Alzheimer's disease models. PLoS One 2014; 9(3): e92954.
[http://dx.doi.org/10.1371/journal.pone.0092954] [PMID: 24671102]

[232] Zhu F, Wu F, Ma Y, *et al.* Decrease in the production of β-amyloid by berberine inhibition of the expression of β-secretase in HEK293 cells. BMC Neurosci 2011; 12: 125-32.
[http://dx.doi.org/10.1186/1471-2202-12-125] [PMID: 22152059]

[233] Kysenius K, Brunello CA, Huttunen HJ. Mitochondria and NMDA receptor-dependent toxicity of berberine sensitizes neurons to glutamate and rotenone injury. PLoS One 2014; 9(9): e107129.
[http://dx.doi.org/10.1371/journal.pone.0107129] [PMID: 25192195]

[234] Coelho F, Birks J. Physostigmine for Alzheimer's disease. Cochrane Database Syst Rev 2001; 2(2): CD001499.
[http://dx.doi.org/10.1002/14651858.CD001499] [PMID: 11405996]

[235] Becker RE, Greig NH. Fire in the ashes: can failed Alzheimer's disease drugs succeed with second chances? Alzheimers Dement 2013; 9(1): 50-7.
[http://dx.doi.org/10.1016/j.jalz.2012.01.007] [PMID: 22465172]

[236] Greig NH, Sambamurti K, Yu QS, Brossi A, Bruinsma GB, Lahiri DK. An overview of phenserine tartrate, a novel acetylcholinesterase inhibitor for the treatment of Alzheimer's disease. Curr Alzheimer Res 2005; 2(3): 281-90.
[http://dx.doi.org/10.2174/1567205054367829] [PMID: 15974893]

[237] Kadir A, Andreasen N, Almkvist O, *et al.* Effect of phenserine treatment on brain functional activity and amyloid in Alzheimer's disease. Ann Neurol 2008; 63(5): 621-31.

[http://dx.doi.org/10.1002/ana.21345] [PMID: 18300284]

[238]　Wollen KA. Alzheimer's disease: the pros and cons of pharmaceutical, nutritional, botanical, and stimulatory therapies, with a discussion of treatment strategies from the perspective of patients and practitioners. Altern Med Rev 2010; 15(3): 223-44.
[PMID: 21155625]

[239]　Wang R, Yan H, Tang XC. Progress in studies of huperzine A, a natural cholinesterase inhibitor from Chinese herbal medicine. Acta Pharmacol Sin 2006; 27(1): 1-26.
[http://dx.doi.org/10.1111/j.1745-7254.2006.00255.x] [PMID: 16364207]

[240]　Giacobini E. Cholinesterase inhibitors: new roles and therapeutic alternatives. Pharmacol Res 2004; 50(4): 433-40.
[http://dx.doi.org/10.1016/j.phrs.2003.11.017] [PMID: 15304240]

[241]　Tang XC, De Sarno P, Sugaya K, Giacobini E. Effect of huperzine A, a new cholinesterase inhibitor, on the central cholinergic system of the rat. J Neurosci Res 1989; 24(2): 276-85.
[http://dx.doi.org/10.1002/jnr.490240220] [PMID: 2585551]

[242]　Ye JW, Cai JX, Wang LM, Tang XC. Improving effects of huperzine A on spatial working memory in aged monkeys and young adult monkeys with experimental cognitive impairment. J Pharmacol Exp Ther 1999; 288(2): 814-9.
[PMID: 9918593]

[243]　Wang ZF, Tang LL, Yan H, Wang YJ, Tang XC. Effects of huperzine A on memory deficits and neurotrophic factors production after transient cerebral ischemia and reperfusion in mice. Pharmacol Biochem Behav 2006; 83(4): 603-11.
[http://dx.doi.org/10.1016/j.pbb.2006.03.027] [PMID: 16687166]

[244]　Xu SS, Gao ZX, Weng Z, *et al.* Efficacy of tablet huperzine-A on memory, cognition, and behavior in Alzheimer's disease. Zhongguo Yao Li Xue Bao 1995; 16(5): 391-5.
[PMID: 8701750]

[245]　Zhang HY, Zheng CY, Yan H, *et al.* Potential therapeutic targets of huperzine A for Alzheimer's disease and vascular dementia. Chem Biol Interact 2008; 175(1-3): 396-402.
[http://dx.doi.org/10.1016/j.cbi.2008.04.049] [PMID: 18565502]

[246]　Ma X, Tan C, Zhu D, Gang DR, Xiao P. Huperzine A from Huperzia species--an ethnopharmacolgical review. J Ethnopharmacol 2007; 113(1): 15-34.
[http://dx.doi.org/10.1016/j.jep.2007.05.030] [PMID: 17644292]

[247]　Rafii MS, Walsh S, Little JT, *et al.* Alzheimer's Disease Cooperative Study. A phase II trial of huperzine A in mild to moderate Alzheimer disease. Neurology 2011; 76(16): 1389-94.
[http://dx.doi.org/10.1212/WNL.0b013e318216eb7b] [PMID: 21502597]

[248]　Yang G, Wang Y, Tian J, Liu JP. Huperzine A for Alzheimer's disease: a systematic review and meta-analysis of randomized clinical trials. PLoS One 2013; 8(9): e74916.
[http://dx.doi.org/10.1371/journal.pone.0074916] [PMID: 24086396]

[249]　Shao ZQ. Comparison of the efficacy of four cholinesterase inhibitors in combination with memantine for the treatment of Alzheimer's disease. Int J Clin Exp Med 2015; 8(2): 2944-8.
[PMID: 25932260]

[250]　Nehlig A. Are we dependent upon coffee and caffeine? A review on human and animal data. Neurosci Biobehav Rev 1999; 23(4): 563-76.
[http://dx.doi.org/10.1016/S0149-7634(98)00050-5] [PMID: 10073894]

[251]　Fredholm BB, Bättig K, Holmén J, Nehlig A, Zvartau EE. Actions of caffeine in the brain with special reference to factors that contribute to its widespread use. Pharmacol Rev 1999; 51(1): 83-133.
[PMID: 10049999]

[252]　Marques S, Batalha VL, Lopes LV, Outeiro TF. Modulating Alzheimer's disease through caffeine: a putative link to epigenetics. J Alzheimers Dis 2011; 24 (Suppl. 2): 161-71.

[http://dx.doi.org/10.3233/JAD-2011-110032] [PMID: 21427489]

[253]   Arendash GW, Schleif W, Rezai-Zadeh K, *et al.* Caffeine protects Alzheimer's mice against cognitive impairment and reduces brain beta-amyloid production. Neuroscience 2006; 142(4): 941-52.
[http://dx.doi.org/10.1016/j.neuroscience.2006.07.021] [PMID: 16938404]

[254]   Dall'Igna OP, Fett P, Gomes MW, Souza DO, Cunha RA, Lara DR. Caffeine and adenosine A(2a) receptor antagonists prevent beta-amyloid (25-35)-induced cognitive deficits in mice. Exp Neurol 2007; 203(1): 241-5.
[http://dx.doi.org/10.1016/j.expneurol.2006.08.008] [PMID: 17007839]

[255]   Cao C, Cirrito JR, Lin X, *et al.* Caffeine suppresses amyloid-beta levels in plasma and brain of Alzheimer's disease transgenic mice. J Alzheimers Dis 2009; 17(3): 681-97.
[http://dx.doi.org/10.3233/JAD-2009-1071] [PMID: 19581723]

[256]   Dall'Igna OP, Porciúncula LO, Souza DO, Cunha RA, Lara DR. Neuroprotection by caffeine and adenosine A2A receptor blockade of beta-amyloid neurotoxicity. Br J Pharmacol 2003; 138(7): 1207-9.
[http://dx.doi.org/10.1038/sj.bjp.0705185] [PMID: 12711619]

[257]   Arendash GW, Mori T, Cao C, *et al.* Caffeine reverses cognitive impairment and decreases brain amyloid-beta levels in aged Alzheimer's disease mice. J Alzheimers Dis 2009; 17(3): 661-80.
[http://dx.doi.org/10.3233/JAD-2009-1087] [PMID: 19581722]

[258]   Zeitlin R, Patel S, Burgess S, Arendash GW, Echeverria V. Caffeine induces beneficial changes in PKA signaling and JNK and ERK activities in the striatum and cortex of Alzheimer's transgenic mice. Brain Res 2011; 1417: 127-36.
[http://dx.doi.org/10.1016/j.brainres.2011.08.036] [PMID: 21907331]

[259]   Laurent C, Eddarkaoui S, Derisbourg M, *et al.* Beneficial effects of caffeine in a transgenic model of Alzheimer's disease-like tau pathology. Neurobiol Aging 2014; 35(9): 2079-90.
[http://dx.doi.org/10.1016/j.neurobiolaging.2014.03.027] [PMID: 24780254]

[260]   León-Carmona JR, Galano A. Uric and 1-methyluric acids: metabolic wastes or antiradical protectors? J Phys Chem B 2011; 115(51): 15430-8.
[http://dx.doi.org/10.1021/jp209776x] [PMID: 22097927]

[261]   Prasanthi JR, Dasari B, Marwarha G, *et al.* Caffeine protects against oxidative stress and Alzheimer's disease-like pathology in rabbit hippocampus induced by cholesterol-enriched diet. Free Radic Biol Med 2010; 49(7): 1212-20.
[http://dx.doi.org/10.1016/j.freeradbiomed.2010.07.007] [PMID: 20638472]

[262]   Maia L, de Mendonça A. Does caffeine intake protect from Alzheimer's disease? Eur J Neurol 2002; 9(4): 377-82.
[http://dx.doi.org/10.1046/j.1468-1331.2002.00421.x] [PMID: 12099922]

[263]   Vartiainen E, Puska P, Jousilahti P, Korhonen HJ, Tuomilehto J, Nissinen A. Twenty-year trends in coronary risk factors in north Karelia and in other areas of Finland. Int J Epidemiol 1994; 23(3): 495-504.
[http://dx.doi.org/10.1093/ije/23.3.495] [PMID: 7960373]

[264]   Eskelinen MH, Kivipelto M. Caffeine as a protective factor in dementia and Alzheimer's disease. J Alzheimers Dis 2010; 20 (Suppl. 1): S167-74.
[http://dx.doi.org/10.3233/JAD-2010-1404] [PMID: 20182054]

[265]   Kim YS, Kwak SM, Myung SK. Caffeine intake from coffee or tea and cognitive disorders: a meta-analysis of observational studies. Neuroepidemiology 2015; 44(1): 51-63.
[http://dx.doi.org/10.1159/000371710] [PMID: 25721193]

[266]   Kang TH, Murakami Y, Takayama H, *et al.* Protective effect of rhynchophylline and isorhynchophylline on *in vitro* ischemia-induced neuronal damage in the hippocampus: putative neurotransmitter receptors involved in their action. Life Sci 2004; 76(3): 331-43.

[http://dx.doi.org/10.1016/j.lfs.2004.08.012] [PMID: 15531384]

[267] Xian YF, Lin ZX, Mao QQ, *et al.* Isorhynchophylline protects PC12 cells against beta-amyloid induced apoptosis *via* PI3K/Akt signaling pathway. Evid Based Complement Altern Med 2013; 163057.

[268] Xian YF, Mao QQ, Wu JC, *et al.* Isorhynchophylline treatment improves the amyloid-β-induced cognitive impairment in rats *via* inhibition of neuronal apoptosis and tau protein hyperphosphorylation. J Alzheimers Dis 2014; 39(2): 331-46.
[http://dx.doi.org/10.3233/JAD-131457] [PMID: 24164737]

[269] Xian YF, Lin ZX, Mao QQ, Ip SP, Su ZR, Lai XP. Protective effect of isorhynchophylline against β-amyloid-induced neurotoxicity in PC12 cells. Cell Mol Neurobiol 2012; 32(3): 353-60.
[http://dx.doi.org/10.1007/s10571-011-9763-5] [PMID: 22042506]

[270] Shao H, Mi Z, Ji WG, *et al.* Rhynchophylline Protects Against the Amyloid β-Induced Increase of Spontaneous Discharges in the Hippocampal CA1 Region of Rats. Neurochem Res 2015; 40(11): 2365-73.
[http://dx.doi.org/10.1007/s11064-015-1730-y] [PMID: 26441223]

[271] Fu AK, Hung KW, Huang H, *et al.* Blockade of EphA4 signaling ameliorates hippocampal synaptic dysfunctions in mouse models of Alzheimer's disease. Proc Natl Acad Sci USA 2014; 111(27): 9959-64.
[http://dx.doi.org/10.1073/pnas.1405803111] [PMID: 24958880]

[272] Leclerc S, Garnier M, Hoessel R, *et al.* Indirubins inhibit glycogen synthase kinase-3 beta and CDK5/p25, two protein kinases involved in abnormal tau phosphorylation in Alzheimer's disease. A property common to most cyclin-dependent kinase inhibitors? J Biol Chem 2001; 276(1): 251-60.
[http://dx.doi.org/10.1074/jbc.M002466200] [PMID: 11013232]

[273] Selenica ML, Jensen HS, Larsen AK, *et al.* Efficacy of small-molecule glycogen synthase kinase-3 inhibitors in the postnatal rat model of tau hyperphosphorylation. Br J Pharmacol 2007; 152(6): 959-79.
[http://dx.doi.org/10.1038/sj.bjp.0707471] [PMID: 17906685]

[274] Ding Y, Qiao A, Fan GH. Indirubin-3'-monoxime rescues spatial memory deficits and attenuates beta-amyloid-associated neuropathology in a mouse model of Alzheimer's disease. Neurobiol Dis 2010; 39(2): 156-68.
[http://dx.doi.org/10.1016/j.nbd.2010.03.022] [PMID: 20381617]

[275] Zhao T, Wang CH, Wang ZT. Chemical constituents and pharmacologic actions of genus *Peganum*: research advances. J Internat Pharmaceut Res 2010; 37: 333-45.

[276] Frost D, Meechoovet B, Wang T, *et al.* β-carboline compounds, including harmine, inhibit DYRK1A and tau phosphorylation at multiple Alzheimer's disease-related sites. PLoS One 2011; 6(5): e19264.
[http://dx.doi.org/10.1371/journal.pone.0019264] [PMID: 21573099]

[277] Dos Santos RG, Hallak JE. Effects of the Natural β-Carboline Alkaloid Harmine, a Main Constituent of Ayahuasca, in Memory and in the Hippocampus: A Systematic Literature Review of Preclinical Studies. J Psychoactive Drugs 2017; 49(1): 1-10.
[http://dx.doi.org/10.1080/02791072.2016.1260189] [PMID: 27918874]

[278] Rambhia S, Mantione KJ, Stefano GB, Cadet P. Morphine modulation of the ubiquitin-proteasome complex is neuroprotective. Med Sci Monit 2005; 11(11): BR386-96.
[PMID: 16258387]

[279] Qian L, Tan KS, Wei SJ, *et al.* Microglia-mediated neurotoxicity is inhibited by morphine through an opioid receptor-independent reduction of NADPH oxidase activity. J Immunol 2007; 179(2): 1198-209.
[http://dx.doi.org/10.4049/jimmunol.179.2.1198] [PMID: 17617613]

[280] Wang Y, Wang YX, Liu T, *et al.* μ-Opioid receptor attenuates Aβ oligomers-induced neurotoxicity

through mTOR signaling. CNS Neurosci Ther 2015; 21(1): 8-14.
[http://dx.doi.org/10.1111/cns.12316] [PMID: 25146548]

[281] Wang HY, Lee DH, D'Andrea MR, Peterson PA, Shank RP, Reitz AB. β-Amyloid(1-42) binds to alpha7 nicotinic acetylcholine receptor with high affinity. Implications for Alzheimer's disease pathology. J Biol Chem 2000; 275(8): 5626-32.
[http://dx.doi.org/10.1074/jbc.275.8.5626] [PMID: 10681545]

[282] Salomon AR, Marcinowski KJ, Friedland RP, Zagorski MG. Nicotine inhibits amyloid formation by the beta-peptide. Biochemistry 1996; 35(42): 13568-78.
[http://dx.doi.org/10.1021/bi9617264] [PMID: 8885836]

[283] Shim SB, Lee SH, Chae KR, *et al.* Nicotine leads to improvements in behavioral impairment and an increase in the nicotine acetylcholine receptor in transgenic mice. Neurochem Res 2008; 33(9): 1783-8.
[http://dx.doi.org/10.1007/s11064-008-9629-5] [PMID: 18307030]

[284] White HK, Levin ED. Four-week nicotine skin patch treatment effects on cognitive performance in Alzheimer's disease. Psychopharmacology (Berl) 1999; 143(2): 158-65.
[http://dx.doi.org/10.1007/s002130050931] [PMID: 10326778]

[285] Howe MN, Price IR. Effects of transdermal nicotine on learning, memory, verbal fluency, concentration, and general health in a healthy sample at risk for dementia. Int Psychogeriatr 2001; 13(4): 465-75.
[http://dx.doi.org/10.1017/S1041610201007888] [PMID: 12003253]

[286] Elmadfa I, Meyer A, Nowak V, Hasenegger V, Putz P, Verstraeten R, *et al.* European nutrition and health report. Basel: Karger 2009.

[287] Rock CL. Multivitamin-multimineral supplements: who uses them? Am J Clin Nutr 2007; 85(1): 277S-9S.
[PMID: 17209209]

[288] Bailey RL, Carmel R, Green R, *et al.* Monitoring of vitamin B-12 nutritional status in the United States by using plasma methylmalonic acid and serum vitamin B-12. Am J Clin Nutr 2011; 94(2): 552-61.
[http://dx.doi.org/10.3945/ajcn.111.015222] [PMID: 21677051]

[289] Mohajeri MH, Troesch B, Weber P. Inadequate supply of vitamins and DHA in the elderly: implications for brain aging and Alzheimer-type dementia. Nutrition 2015; 31(2): 261-75.
[http://dx.doi.org/10.1016/j.nut.2014.06.016] [PMID: 25592004]

[290] International Institute of Medicine. Standing Committee on the Scientific Evaluation of Dietary Reference Intake of Vitamin C, Vitamin E, Selenium and Carotenoids. Washington, DC: National Academy Press 2000; pp. 58-72.

[291] Narayanankutty A, Kottekkat A, Mathew SE, Illam SP, Suseela IM, Raghavamenon AC. Vitamin E supplementation modulates the biological effects of omega-3 fatty acids in naturally aged rats. Toxicol Mech Methods 2017; 27(3): 207-14.
[http://dx.doi.org/10.1080/15376516.2016.1273431] [PMID: 27996366]

[292] Bacharach AL. Vitamin-E therapy in neuromuscular disorders. BMJ 1941; 2(4217): 618-9.
[http://dx.doi.org/10.1136/bmj.2.4217.618] [PMID: 20783938]

[293] Mangialasche F, Kivipelto M, Mecocci P, *et al.* High plasma levels of vitamin E forms and reduced Alzheimer's disease risk in advanced age. J Alzheimers Dis 2010; 20(4): 1029-37.
[http://dx.doi.org/10.3233/JAD-2010-091450] [PMID: 20413888]

[294] Li FJ, Shen L, Ji HF. Dietary intakes of vitamin E, vitamin C, and β-carotene and risk of Alzheimer's disease: a meta-analysis. J Alzheimers Dis 2012; 31(2): 253-8.
[http://dx.doi.org/10.3233/JAD-2012-120349] [PMID: 22543848]

[295] Giraldo E, Lloret A, Fuchsberger T, Viña J. Aβ and tau toxicities in Alzheimer's are linked *via*

oxidative stress-induced p38 activation: protective role of vitamin E. Redox Biol 2014; 2: 873-7.
[http://dx.doi.org/10.1016/j.redox.2014.03.002] [PMID: 25061569]

[296] Lloret A, Badía MC, Mora NJ, *et al.* Vitamin E paradox in Alzheimer's disease: it does not prevent loss of cognition and may even be detrimental. J Alzheimers Dis 2009; 17(1): 143-9.
[http://dx.doi.org/10.3233/JAD-2009-1033] [PMID: 19494439]

[297] Dysken MW, Sano M, Asthana S, *et al.* Effect of vitamin E and memantine on functional decline in Alzheimer disease: the TEAM-AD VA cooperative randomized trial. JAMA 2014; 311(1): 33-44.
[http://dx.doi.org/10.1001/jama.2013.282834] [PMID: 24381967]

[298] Selhub J. Folate, vitamin B12 and vitamin B6 and one carbon metabolism. J Nutr Health Aging 2002; 6(1): 39-42.
[PMID: 11813080]

[299] Jalili M, Pati S, Rath B, Bjorklun G, Singh RB. Effect of diet and nutrients on molecular mechanism of gene expression mediated by nuclear receptor and epigenetic modulation. Open Nutraceuticals 2013; 6: 27-34.
[http://dx.doi.org/10.2174/1876396001306010027]

[300] Coşar A, Ipçioğlu OM, Ozcan O, Gültepe M. Folate and homocysteine metabolisms and their roles in the biochemical basis of neuropsychiatry. Turk J Med Sci 2014; 44(1): 1-9.
[http://dx.doi.org/10.3906/sag-1211-39] [PMID: 25558551]

[301] Medina M, Urdiales JL, Amores-Sánchez MI. Roles of homocysteine in cell metabolism: old and new functions. Eur J Biochem 2001; 268(14): 3871-82.
[http://dx.doi.org/10.1046/j.1432-1327.2001.02278.x] [PMID: 11453979]

[302] Jin Y, Amaral A, McCann A, Brennan L. Homocysteine levels impact directly on epigenetic reprogramming in astrocytes. Neurochem Int 2011; 58(7): 833-8.
[http://dx.doi.org/10.1016/j.neuint.2011.03.012] [PMID: 21419186]

[303] Sezgin Z, Dincer Y. Alzheimer's disease and epigenetic diet. Neurochem Int 2014; 78: 105-16.
[http://dx.doi.org/10.1016/j.neuint.2014.09.012] [PMID: 25290336]

[304] Wang SC, Oelze B, Schumacher A. Age-specific epigenetic drift in late-onset Alzheimer's disease. PLoS One 2008; 3(7): e2698.
[http://dx.doi.org/10.1371/journal.pone.0002698] [PMID: 18628954]

[305] Sommer BR, Hoff AL, Costa M. Folic acid supplementation in dementia: a preliminary report. J Geriatr Psychiatry Neurol 2003; 16(3): 156-9.
[http://dx.doi.org/10.1177/0891988703256052] [PMID: 12967058]

[306] Araújo JR, Martel F, Borges N, Araújo JM, Keating E. Folates and aging: Role in mild cognitive impairment, dementia and depression. Ageing Res Rev 2015; 22: 9-19.
[http://dx.doi.org/10.1016/j.arr.2015.04.005] [PMID: 25939915]

[307] Gibson GE, Hirsch JA, Cirio RT, Jordan BD, Fonzetti P, Elder J. Abnormal thiamine-dependent processes in Alzheimer's Disease. Lessons from diabetes. Mol Cell Neurosci 2013; 55: 17-25.
[http://dx.doi.org/10.1016/j.mcn.2012.09.001] [PMID: 22982063]

[308] Mimori Y, Katsuoka H, Nakamura S. Thiamine therapy in Alzheimer's disease. Metab Brain Dis 1996; 11(1): 89-94.
[http://dx.doi.org/10.1007/BF02080934] [PMID: 8815393]

[309] Ollat H, Laurent B, Bakchine S, Michel BF, Touchon J, Dubois B. [Effects of the association of sulbutiamine with an acetylcholinesterase inhibitor in early stage and moderate Alzheimer disease]. Encephale 2007; 33(2): 211-5.
[http://dx.doi.org/10.1016/S0013-7006(07)91552-3] [PMID: 17675917]

[310] Armas LA, Hollis BW, Heaney RP. Vitamin D2 is much less effective than vitamin D3 in humans. J Clin Endocrinol Metab 2004; 89(11): 5387-91.
[http://dx.doi.org/10.1210/jc.2004-0360] [PMID: 15531486]

[311] Dickens AP, Lang IA, Langa KM, Kos K, Llewellyn DJ. Vitamin D, cognitive dysfunction and dementia in older adults. CNS Drugs 2011; 25(8): 629-39.
[http://dx.doi.org/10.2165/11593080-000000000-00000] [PMID: 21790207]

[312] Eyles DW, Burne TH, McGrath JJ. Vitamin D, effects on brain development, adult brain function and the links between low levels of vitamin D and neuropsychiatric disease. Front Neuroendocrinol 2013; 34(1): 47-64.
[http://dx.doi.org/10.1016/j.yfrne.2012.07.001] [PMID: 22796576]

[313] Keeney JT, Förster S, Sultana R, *et al.* Dietary vitamin D deficiency in rats from middle to old age leads to elevated tyrosine nitration and proteomics changes in levels of key proteins in brain: implications for low vitamin D-dependent age-related cognitive decline. Free Radic Biol Med 2013; 65: 324-34.
[http://dx.doi.org/10.1016/j.freeradbiomed.2013.07.019] [PMID: 23872023]

[314] Chagas CE, Borges MC, Martini LA, Rogero MM. Focus on vitamin D, inflammation and type 2 diabetes. Nutrients 2012; 4(1): 52-67.
[http://dx.doi.org/10.3390/nu4010052] [PMID: 22347618]

[315] Durk MR, Han K, Chow EC, *et al.* 1α,25-Dihydroxyvitamin D3 reduces cerebral amyloid-β accumulation and improves cognition in mouse models of Alzheimer's disease. J Neurosci 2014; 34(21): 7091-101.
[http://dx.doi.org/10.1523/JNEUROSCI.2711-13.2014] [PMID: 24849345]

[316] Hooshmand B, Lökk J, Solomon A, *et al.* Vitamin D in relation to cognitive impairment, cerebrospinal fluid biomarkers, and brain volumes. J Gerontol A Biol Sci Med Sci 2014; 69(9): 1132-8.
[http://dx.doi.org/10.1093/gerona/glu022] [PMID: 24568931]

[317] Annweiler C, Rolland Y, Schott AM, *et al.* Higher vitamin D dietary intake is associated with lower risk of alzheimer's disease: a 7-year follow-up. J Gerontol A Biol Sci Med Sci 2012; 67(11): 1205-11.
[http://dx.doi.org/10.1093/gerona/gls107] [PMID: 22503994]

[318] Stein MS, Scherer SC, Ladd KS, Harrison LC. A randomized controlled trial of high-dose vitamin D2 followed by intranasal insulin in Alzheimer's disease. J Alzheimers Dis 2011; 26(3): 477-84.
[http://dx.doi.org/10.3233/JAD-2011-110149] [PMID: 21694461]

[319] Carratù MR, Marasco C, Signorile A, Scuderi C, Steardo L. Are retinoids a promise for Alzheimer's disease management? Curr Med Chem 2012; 19(36): 6119-25.
[PMID: 23092137]

[320] Jarvis CI, Goncalves MB, Clarke E, *et al.* Retinoic acid receptor-α signalling antagonizes both intracellular and extracellular amyloid-β production and prevents neuronal cell death caused by amyloid-β. Eur J Neurosci 2010; 32(8): 1246-55.
[http://dx.doi.org/10.1111/j.1460-9568.2010.07426.x] [PMID: 20950278]

[321] Nalivaeva NN, Beckett C, Belyaev ND, Turner AJ. Are amyloid-degrading enzymes viable therapeutic targets in Alzheimer's disease? J Neurochem 2012; 120 (Suppl. 1): 167-85.
[http://dx.doi.org/10.1111/j.1471-4159.2011.07510.x] [PMID: 22122230]

[322] Lee HP, Casadesus G, Zhu X, *et al.* All-trans retinoic acid as a novel therapeutic strategy for Alzheimer's disease. Expert Rev Neurother 2009; 9(11): 1615-21.
[http://dx.doi.org/10.1586/ern.09.86] [PMID: 19903021]

[323] Ding Y, Qiao A, Wang Z, *et al.* Retinoic acid attenuates beta-amyloid deposition and rescues memory deficits in an Alzheimer's disease transgenic mouse model. J Neurosci 2008; 28(45): 11622-34.
[http://dx.doi.org/10.1523/JNEUROSCI.3153-08.2008] [PMID: 18987198]

[324] Fukasawa H, Nakagomi M, Yamagata N, *et al.* Tamibarotene: a candidate retinoid drug for Alzheimer's disease. Biol Pharm Bull 2012; 35(8): 1206-12.
[http://dx.doi.org/10.1248/bpb.b12-00314] [PMID: 22863914]

[325] Yu C, Youmans KL, LaDu MJ. Proposed mechanism for lipoprotein remodelling in the brain. Biochim

Biophys Acta 2010; 1801: 819-23.
[http://dx.doi.org/10.1016/j.bbalip.2010.05.001] [PMID: 20470897]

[326] Corcoran JP, So PL, Maden M. Disruption of the retinoid signalling pathway causes a deposition of amyloid beta in the adult rat brain. Eur J Neurosci 2004; 20(4): 896-902.
[http://dx.doi.org/10.1111/j.1460-9568.2004.03563.x] [PMID: 15305858]

[327] Samad TA, Krezel W, Chambon P, Borrelli E. Regulation of dopaminergic pathways by retinoids: activation of the D2 receptor promoter by members of the retinoic acid receptor-retinoid X receptor family. Proc Natl Acad Sci USA 1997; 94(26): 14349-54.
[http://dx.doi.org/10.1073/pnas.94.26.14349] [PMID: 9405615]

[328] Harrison FE. A critical review of vitamin C for the prevention of age-related cognitive decline and Alzheimer's disease. J Alzheimers Dis 2012; 29(4): 711-26.
[http://dx.doi.org/10.3233/JAD-2012-111853] [PMID: 22366772]

[329] Wengreen HJ, Munger RG, Corcoran CD, *et al.* Antioxidant intake and cognitive function of elderly men and women: the Cache County Study. J Nutr Health Aging 2007; 11(3): 230-7.
[PMID: 17508099]

[330] Zandi PP, Anthony JC, Khachaturian AS, *et al.* Cache County Study Group. Reduced risk of Alzheimer disease in users of antioxidant vitamin supplements: the Cache County Study. Arch Neurol 2004; 61(1): 82-8.
[http://dx.doi.org/10.1001/archneur.61.1.82] [PMID: 14732624]

[331] Harrison FE, May JM. Vitamin C function in the brain: vital role of the ascorbate transporter SVCT2. Free Radic Biol Med 2009; 46(6): 719-30.
[http://dx.doi.org/10.1016/j.freeradbiomed.2008.12.018] [PMID: 19162177]

[332] Padayatty SJ, Katz A, Wang Y, *et al.* Vitamin C as an antioxidant: evaluation of its role in disease prevention. J Am Coll Nutr 2003; 22(1): 18-35.
[http://dx.doi.org/10.1080/07315724.2003.10719272] [PMID: 12569111]

[333] Hodges RE, Baker EM, Hood J, Sauberlich HE, March SC. Experimental scurvy in man. Am J Clin Nutr 1969; 22(5): 535-48.
[PMID: 4977512]

[334] Kaliś K. [Dual action of vitamin C *versus* degradation and supplementation]. Postepy Hig Med Dosw (Online) 2015; 69: 1239-44.
[http://dx.doi.org/10.5604/17322693.1180642] [PMID: 26671914]

[335] Cheng F, Cappai R, Ciccotosto GD, *et al.* Suppression of amyloid β A11 antibody immunoreactivity by vitamin C: possible role of heparan sulfate oligosaccharides derived from glypican-1 by ascorbate-induced, nitric oxide (NO)-catalyzed degradation. J Biol Chem 2011; 286(31): 27559-72.
[http://dx.doi.org/10.1074/jbc.M111.243345] [PMID: 21642435]

[336] Warner TA, Kang JQ, Kennard JA, Harrison FE. Low brain ascorbic acid increases susceptibility to seizures in mouse models of decreased brain ascorbic acid transport and Alzheimer's disease. Epilepsy Res 2015; 110: 20-5.
[http://dx.doi.org/10.1016/j.eplepsyres.2014.11.017] [PMID: 25616451]

[337] Wang T, Chen K, Zeng X, *et al.* The histone demethylases Jhdm1a/1b enhance somatic cell reprogramming in a vitamin-C-dependent manner. Cell Stem Cell 2011; 9(6): 575-87.
[http://dx.doi.org/10.1016/j.stem.2011.10.005] [PMID: 22100412]

[338] Monograph Coenzime Q10. Coenzyme Q10. Monograph. Altern Med Rev 2007; 12(2): 159-68.
[PMID: 17604461]

[339] Spindler M, Beal MF, Henchcliffe C. Coenzyme Q10 effects in neurodegenerative disease. Neuropsychiatr Dis Treat 2009; 5: 597-610.
[PMID: 19966907]

[340] Dumont M, Kipiani K, Yu F, *et al.* Coenzyme Q10 decreases amyloid pathology and improves

behavior in a transgenic mouse model of Alzheimer's disease. J Alzheimers Dis 2011; 27(1): 211-23.
[http://dx.doi.org/10.3233/JAD-2011-110209] [PMID: 21799249]

[341] Abdul HM, Calabrese V, Calvani M, Butterfield DA. Acetyl-L-carnitine-induced up-regulation of heat shock proteins protects cortical neurons against amyloid-beta peptide 1-42-mediated oxidative stress and neurotoxicity: implications for Alzheimer's disease. J Neurosci Res 2006; 84(2): 398-408.
[http://dx.doi.org/10.1002/jnr.20877] [PMID: 16634066]

[342] Brami C, Bao T, Deng G. Natural products and complementary therapies for chemotherapy-induced peripheral neuropathy: A systematic review. Crit Rev Oncol Hematol 2016; 98: 325-34.
[http://dx.doi.org/10.1016/j.critrevonc.2015.11.014] [PMID: 26652982]

[343] Imperato A, Ramacci MT, Angelucci L. Acetyl-L-carnitine enhances acetylcholine release in the striatum and hippocampus of awake freely moving rats. Neurosci Lett 1989; 107(1-3): 251-5.
[http://dx.doi.org/10.1016/0304-3940(89)90826-4] [PMID: 2616037]

[344] Hagen TM, Ingersoll RT, Wehr CM, *et al.* Acetyl-L-carnitine fed to old rats partially restores mitochondrial function and ambulatory activity. Proc Natl Acad Sci USA 1998; 95(16): 9562-6.
[http://dx.doi.org/10.1073/pnas.95.16.9562] [PMID: 9689120]

[345] Paradies G, Petrosillo G, Gadaleta MN, Ruggiero FM. The effect of aging and acetyl-L-carnitine on the pyruvate transport and oxidation in rat heart mitochondria. FEBS Lett 1999; 454(3): 207-9.
[http://dx.doi.org/10.1016/S0014-5793(99)00809-1] [PMID: 10431808]

[346] Dolezal V, Tucek S. Utilization of citrate, acetylcarnitine, acetate, pyruvate and glucose for the synthesis of acetylcholine in rat brain slices. J Neurochem 1981; 36(4): 1323-30.
[http://dx.doi.org/10.1111/j.1471-4159.1981.tb00569.x] [PMID: 6790669]

[347] Epis R, Marcello E, Gardoni F, *et al.* Modulatory effect of acetyl-L-carnitine on amyloid precursor protein metabolism in hippocampal neurons. Eur J Pharmacol 2008; 597(1-3): 51-6.
[http://dx.doi.org/10.1016/j.ejphar.2008.09.001] [PMID: 18801359]

[348] Spagnoli A, Lucca U, Menasce G, *et al.* Long-term acetyl-L-carnitine treatment in Alzheimer's disease. Neurology 1991; 41(11): 1726-32.
[http://dx.doi.org/10.1212/WNL.41.11.1726] [PMID: 1944900]

[349] Pettegrew JW, Klunk WE, Panchalingam K, Kanfer JN, McClure RJ. Clinical and neurochemical effects of acetyl-L-carnitine in Alzheimer's disease. Neurobiol Aging 1995; 16(1): 1-4.
[http://dx.doi.org/10.1016/0197-4580(95)80001-8] [PMID: 7723928]

[350] Thal LJ, Carta A, Clarke WR, *et al.* A 1-year multicenter placebo-controlled study of acetyl--carnitine in patients with Alzheimer's disease. Neurology 1996; 47(3): 705-11.
[http://dx.doi.org/10.1212/WNL.47.3.705] [PMID: 8797468]

[351] Brasse-Lagnel C, Lavoinne A, Husson A. Control of mammalian gene expression by amino acids, especially glutamine. FEBS J 2009; 276(7): 1826-44.
[http://dx.doi.org/10.1111/j.1742-4658.2009.06920.x] [PMID: 19250320]

[352] Chen J, Herrup K. Glutamine acts as a neuroprotectant against DNA damage, beta-amyloid and H2O2-induced stress. PLoS One 2012; 7(3): e33177.
[http://dx.doi.org/10.1371/journal.pone.0033177] [PMID: 22413000]

[353] Gorovits R, Avidan N, Avisar N, Shaked I, Vardimon L. Glutamine synthetase protects against neuronal degeneration in injured retinal tissue. Proc Natl Acad Sci USA 1997; 94(13): 7024-9.
[http://dx.doi.org/10.1073/pnas.94.13.7024] [PMID: 9192685]

[354] Eid T, Ghosh A, Wang Y, *et al.* Recurrent seizures and brain pathology after inhibition of glutamine synthetase in the hippocampus in rats. Brain 2008; 131(Pt 8): 2061-70.
[http://dx.doi.org/10.1093/brain/awn133] [PMID: 18669513]

[355] Butterfield DA, Poon HF, St Clair D, *et al.* Redox proteomics identification of oxidatively modified hippocampal proteins in mild cognitive impairment: insights into the development of Alzheimer's disease. Neurobiol Dis 2006; 22(2): 223-32.

[http://dx.doi.org/10.1016/j.nbd.2005.11.002] [PMID: 16466929]

[356] Smith CD, Carney JM, Starke-Reed PE, *et al*. Excess brain protein oxidation and enzyme dysfunction in normal aging and in Alzheimer disease. Proc Natl Acad Sci USA 1991; 88(23): 10540-3.
[http://dx.doi.org/10.1073/pnas.88.23.10540] [PMID: 1683703]

[357] Wu G, Morris SM Jr. Arginine metabolism: nitric oxide and beyond. Biochem J 1998; 336(Pt 1): 1-17.
[http://dx.doi.org/10.1042/bj3360001] [PMID: 9806879]

[358] Gu Y, Nieves JW, Stern Y, Luchsinger JA, Scarmeas N. Food combination and Alzheimer disease risk: a protective diet. Arch Neurol 2010; 67(6): 699-706.
[http://dx.doi.org/10.1001/archneurol.2010.84] [PMID: 20385883]

[359] Ohtsuka Y, Nakaya J. Effect of oral administration of L-arginine on senile dementia. Am J Med 2000; 108(5): 439.
[http://dx.doi.org/10.1016/S0002-9343(99)00396-4] [PMID: 10759111]

[360] Ignarro LJ, Cirino G, Casini A, Napoli C. Nitric oxide as a signaling molecule in the vascular system: an overview. J Cardiovasc Pharmacol 1999; 34(6): 879-86.
[http://dx.doi.org/10.1097/00005344-199912000-00016] [PMID: 10598133]

[361] Radomski MW, Palmer RM, Moncada S. An L-arginine/nitric oxide pathway present in human platelets regulates aggregation. Proc Natl Acad Sci USA 1990; 87(13): 5193-7.
[http://dx.doi.org/10.1073/pnas.87.13.5193] [PMID: 1695013]

[362] Kubes P, Suzuki M, Granger DN. Nitric oxide: an endogenous modulator of leukocyte adhesion. Proc Natl Acad Sci USA 1991; 88(11): 4651-5.
[http://dx.doi.org/10.1073/pnas.88.11.4651] [PMID: 1675786]

[363] Siani A, Pagano E, Iacone R, Iacoviello L, Scopacasa F, Strazzullo P. Blood pressure and metabolic changes during dietary L-arginine supplementation in humans. Am J Hypertens 2000; 13(5 Pt 1): 547-51.
[http://dx.doi.org/10.1016/S0895-7061(99)00233-2] [PMID: 10826408]

[364] Xu W, Tan L, Wang HF, *et al*. Meta-analysis of modifiable risk factors for Alzheimer's disease. J Neurol Neurosurg Psychiatry 2015; 86(12): 1299-306.
[http://dx.doi.org/10.1136/jnnp-2015-310548] [PMID: 26294005]

[365] Hamel E, Royea J, Ongali B, Tong XK. Neurovascular and Cognitive failure in Alzheimer's Disease: Benefits of Cardiovascular Therapy. Cell Mol Neurobiol 2016; 36(2): 219-32.
[http://dx.doi.org/10.1007/s10571-015-0285-4] [PMID: 26993506]

[366] Wink DA, Hanbauer I, Laval F, Cook JA, Krishna MC, Mitchell JB. Nitric oxide protects against the cytotoxic effects of reactive oxygen species. Ann N Y Acad Sci 1994; 738: 265-78.
[http://dx.doi.org/10.1111/j.1749-6632.1994.tb21812.x] [PMID: 7832437]

[367] Dawson VL, Brahmbhatt HP, Mong JA, Dawson TM. Expression of inducible nitric oxide synthase causes delayed neurotoxicity in primary mixed neuronal-glial cortical cultures. Neuropharmacology 1994; 33(11): 1425-30.
[http://dx.doi.org/10.1016/0028-3908(94)90045-0] [PMID: 7532825]

[368] Wirtz-Brugger F, Giovanni A. Guanosine 3′,5′-cyclic monophosphate mediated inhibition of cell death induced by nerve growth factor withdrawal and beta-amyloid: protective effects of propentofylline. Neuroscience 2000; 99(4): 737-50.
[http://dx.doi.org/10.1016/S0306-4522(00)00243-8] [PMID: 10974437]

[369] Watson GS, Craft S. Modulation of memory by insulin and glucose: neuropsychological observations in Alzheimer's disease. Eur J Pharmacol 2004; 490(1-3): 97-113.
[http://dx.doi.org/10.1016/j.ejphar.2004.02.048] [PMID: 15094077]

[370] Yi J, Horky LL, Friedlich AL, Shi Y, Rogers JT, Huang X. L-arginine and Alzheimer's disease. Int J Clin Exp Pathol 2009; 2(3): 211-38.
[PMID: 19079617]

[371]   Virarkar M, Alappat L, Bradford PG, Awad AB. L-arginine and nitric oxide in CNS function and neurodegenerative diseases. Crit Rev Food Sci Nutr 2013; 53(11): 1157-67.
[http://dx.doi.org/10.1080/10408398.2011.573885] [PMID: 24007420]

[372]   Scarpa S, Fuso A, D'Anselmi F, Cavallaro RA. Presenilin 1 gene silencing by S-adenosylmethionine: a treatment for Alzheimer disease? FEBS Lett 2003; 541(1-3): 145-8.
[http://dx.doi.org/10.1016/S0014-5793(03)00277-1] [PMID: 12706835]

[373]   Rudolph ML, Rabinoff M, Kagan BL. A prospective, Open-Label, 12 Week Trial of S-adenosylmethionine in the Symptomatic Treatment of Alzheimer's Disease. Neurosci Med 2011; 2(3): 222-5.
[http://dx.doi.org/10.4236/nm.2011.23030]

[374]   Bottiglieri T, Godfrey P, Flynn T, Carney MW, Toone BK, Reynolds EH. Cerebrospinal fluid S-adenosylmethionine in depression and dementia: effects of treatment with parenteral and oral S-adenosylmethionine. J Neurol Neurosurg Psychiatry 1990; 53(12): 1096-8.
[http://dx.doi.org/10.1136/jnnp.53.12.1096] [PMID: 2292704]

[375]   Fuso A, Nicolia V, Ricceri L, *et al.* S-adenosylmethionine reduces the progress of the Alzheimer-like features induced by B-vitamin deficiency in mice. Neurobiol Aging 2012; 33(7): 1482.e1-1482.e16.
[http://dx.doi.org/10.1016/j.neurobiolaging.2011.12.013] [PMID: 22221883]

[376]   Chan A, Shea TB. Effects of dietary supplementation with N-acetyl cysteine, acetyl-L-carnitine and S-adenosyl methionine on cognitive performance and aggression in normal mice and mice expressing human ApoE4. Neuromolecular Med 2007; 9(3): 264-9.
[http://dx.doi.org/10.1007/s12017-007-8005-y] [PMID: 17914184]

[377]   Tchantchou F, Graves M, Falcone D, Shea TB. S-adenosylmethionine mediates glutathione efficacy by increasing glutathione S-transferase activity: implications for S-adenosyl methionine as a neuroprotective dietary supplement. J Alzheimers Dis 2008; 14(3): 323-8.
[http://dx.doi.org/10.3233/JAD-2008-14306] [PMID: 18599958]

[378]   Tchantchou F, Graves M, Ortiz D, Shea TB. S-adenosyl methionine: a connection between nutritional and genetic risk factors in Alzheimer's disease. J Nutr Health Aging 2006; 10(6): 541-4.
[PMID: 17183426]

[379]   Montgomery SE, Sepehry AA, Wangsgaard JD, Koenig JE. The effect of S-adenosylmethionine on cognitive performance in mice: an animal model meta-analysis. PLoS One 2014; 9(10): e107756.
[http://dx.doi.org/10.1371/journal.pone.0107756] [PMID: 25347725]

[380]   Cooper JR, Melcer I. The enzymic oxidation of tryptophan to 5-hydroxytryptophan in the biosynthesis of serotonin. J Pharmacol Exp Ther 1961; 132: 265-8.
[PMID: 13695323]

[381]   Porter RJ, Lunn BS, Walker LL, Gray JM, Ballard CG, O'Brien JT. Cognitive deficit induced by acute tryptophan depletion in patients with Alzheimer's disease. Am J Psychiatry 2000; 157(4): 638-40.
[http://dx.doi.org/10.1176/appi.ajp.157.4.638] [PMID: 10739429]

[382]   Noristani HN, Verkhratsky A, Rodríguez JJ. High tryptophan diet reduces CA1 intraneuronal β-amyloid in the triple transgenic mouse model of Alzheimer's disease. Aging Cell 2012; 11(5): 810-22.
[http://dx.doi.org/10.1111/j.1474-9726.2012.00845.x] [PMID: 22702392]

[383]   Robert SJ, Zugaza JL, Fischmeister R, Gardier AM, Lezoualc'h F. The human serotonin 5-HT4 receptor regulates secretion of non-amyloidogenic precursor protein. J Biol Chem 2001; 276(48): 44881-8.
[http://dx.doi.org/10.1074/jbc.M109008200] [PMID: 11584021]

[384]   Takahashi T, Miyazawa M. Serotonin derivatives as inhibitors of beta-secretase (BACE 1). Pharmazie 2011; 66(4): 301-5.
[PMID: 21612159]

[385]   van der Stelt HM, Broersen LM, Olivier B, Westenberg HG. Effects of dietary tryptophan variations

on extracellular serotonin in the dorsal hippocampus of rats. Psychopharmacology (Berl) 2004; 172(2): 137-44.
[http://dx.doi.org/10.1007/s00213-003-1632-6] [PMID: 14647968]

[386] Lee DR, Semba R, Kondo H, Goto S, Nakano K. Decrease in the levels of NGF and BDNF in brains of mice fed a tryptophan-deficient diet. Biosci Biotechnol Biochem 1999; 63(2): 337-40.
[http://dx.doi.org/10.1271/bbb.63.337] [PMID: 10192916]

[387] Jones MG, Hughes J, Tregova A, Milne J, Tomsett AB, Collin HA. Biosynthesis of the flavour precursors of onion and garlic. J Exp Bot 2004; 55(404): 1903-18.
[http://dx.doi.org/10.1093/jxb/erh138] [PMID: 15234988]

[388] Gupta VB, Rao KS. Anti-amyloidogenic activity of S-allyl-L-cysteine and its activity to destabilize Alzheimer's beta-amyloid fibrils *in vitro*. Neurosci Lett 2007; 429(2-3): 75-80.
[http://dx.doi.org/10.1016/j.neulet.2007.09.042] [PMID: 18023978]

[389] Chauhan NB. Effect of aged garlic extract on APP processing and tau phosphorylation in Alzheimer's transgenic model Tg2576. J Ethnopharmacol 2006; 108(3): 385-94.
[http://dx.doi.org/10.1016/j.jep.2006.05.030] [PMID: 16842945]

[390] Li Y, Liu L, Barger SW, Griffin WS. Interleukin-1 mediates pathological effects of microglia on tau phosphorylation and on synaptophysin synthesis in cortical neurons through a p38-MAPK pathway. J Neurosci 2003; 23(5): 1605-11.
[PMID: 12629164]

[391] Borek C. Antioxidant health effects of aged garlic extract. J Nutr 2001; 131(3s): 1010S-5S.
[PMID: 11238807]

[392] Ray B, Chauhan NB, Lahiri DK. The "aged garlic extract:" (AGE) and one of its active ingredients S-allyl-L-cysteine (SAC) as potential preventive and therapeutic agents for Alzheimer's disease (AD). Curr Med Chem 2011; 18(22): 3306-13.
[http://dx.doi.org/10.2174/092986711796504664] [PMID: 21728972]

[393] Ide N, Lau BH. S-allylcysteine attenuates oxidative stress in endothelial cells. Drug Dev Ind Pharm 1999; 25(5): 619-24.
[http://dx.doi.org/10.1081/DDC-100102217] [PMID: 10219531]

[394] Ray B, Chauhan NB, Lahiri DK. Oxidative insults to neurons and synapse are prevented by aged garlic extract and S-allyl-L-cysteine treatment in the neuronal culture and APP-Tg mouse model. J Neurochem 2011; 117(3): 388-402.
[http://dx.doi.org/10.1111/j.1471-4159.2010.07145.x] [PMID: 21166677]

[395] Geng Z, Rong Y, Lau BH. S-allyl cysteine inhibits activation of nuclear factor kappa B in human T cells. Free Radic Biol Med 1997; 23(2): 345-50.
[http://dx.doi.org/10.1016/S0891-5849(97)00006-3] [PMID: 9199898]

[396] Ide N, Lau BH. Garlic compounds minimize intracellular oxidative stress and inhibit nuclear factor-kappa b activation. J Nutr 2001; 131(3s): 1020S-6S.
[PMID: 11238809]

[397] Schneider JA, Arvanitakis Z, Bang W, Bennett DA. Mixed brain pathologies account for most dementia cases in community-dwelling older persons. Neurology 2007; 69(24): 2197-204.
[http://dx.doi.org/10.1212/01.wnl.0000271090.28148.24] [PMID: 17568013]

[398] Vermeer SE, Prins ND, den Heijer T, Hofman A, Koudstaal PJ, Breteler MM. Silent brain infarcts and the risk of dementia and cognitive decline. N Engl J Med 2003; 348(13): 1215-22.
[http://dx.doi.org/10.1056/NEJMoa022066] [PMID: 12660385]

[399] Orozco-Ibarra M, Muñoz-Sánchez J, Zavala-Medina ME, *et al*. Aged garlic extract and S-allylcysteine prevent apoptotic cell death in a chemical hypoxia model. Biol Res 2016; 49: 7.
[http://dx.doi.org/10.1186/s40659-016-0067-6] [PMID: 26830333]

[400] Chauhan NB. Multiplicity of garlic health effects and Alzheimer's disease. J Nutr Health Aging 2005;

9(6): 421-32.
[PMID: 16395514]

[401] Caltagirone C, Ferrannini L, Marchionni N, Nappi G, Scapagnini G, Trabucchi M. The potential protective effect of tramiprosate (homotaurine) against Alzheimer's disease: a review. Aging Clin Exp Res 2012; 24(6): 580-7.
[http://dx.doi.org/10.3275/8585] [PMID: 22961121]

[402] Krzywkowski P, Sebastiani G, Williams S, *et al.* Tramiprosate prevents amyloid Beta-induced Inhibition of Long-term Potentiation in Rat Hippocampal Slices. 8th International Conference AD/PD, Salzburg, Austria. 14-8.

[403] Azzi M, Morissette C, Fallon L, *et al.* Involvement of both GABA-dependent and -independent pathways in tramiprosate neuroprotective effects against amyloid-beta toxicity. 8th International Conference AD/PD, Salzburg, Austria. 14-8.

[404] Aisen PS, Saumier D, Briand R, *et al.* A Phase II study targeting amyloid-beta with 3APS in mild-t--moderate Alzheimer disease. Neurology 2006; 67(10): 1757-63.
[http://dx.doi.org/10.1212/01.wnl.0000244346.08950.64] [PMID: 17082468]

[405] Aisen PS, Gauthier S, Ferris SH, *et al.* Tramiprosate in mild-to-moderate Alzheimer's disease - a randomized, double-blind, placebo-controlled, multi-centre study (the Alphase Study). Arch Med Sci 2011; 7(1): 102-11.
[http://dx.doi.org/10.5114/aoms.2011.20612] [PMID: 22291741]

[406] http://www.alzforum.org/news/research-news/fda-deems-us-alzhemed-trial-results-inconclusive

[407] http://www.biospace.com/news_story.aspx?StoryID=369587

[408] Shifren JL, Hanfling S. Sexuality in Midlife and Beyond: Special Health Report Harvard Health Publications. Boston, MA: Harvard University 2010.

[409] Barron AM, Pike CJ. Sex hormones, aging, and Alzheimer's disease. Front Biosci (Elite Ed) 2012; 4: 976-97.
[PMID: 22201929]

[410] Hardeland R. Melatonin in aging and disease -multiple consequences of reduced secretion, options and limits of treatment. Aging Dis 2012; 3(2): 194-225.
[PMID: 22724080]

[411] Ryan AS. Insulin resistance with aging: effects of diet and exercise. Sports Med 2000; 30(5): 327-46.
[http://dx.doi.org/10.2165/00007256-200030050-00002] [PMID: 11103847]

[412] Boden G. Role of fatty acids in the pathogenesis of insulin resistance and NIDDM. Diabetes 1997; 46(1): 3-10.
[http://dx.doi.org/10.2337/diab.46.1.3] [PMID: 8971073]

[413] Hoyer S. Is sporadic Alzheimer disease the brain type of non-insulin dependent diabetes mellitus? A challenging hypothesis. J Neural Transm (Vienna) 1998; 105(4-5): 415-22.
[http://dx.doi.org/10.1007/s007020050067] [PMID: 9720971]

[414] Hoyer S, Nitsch R, Oesterreich K. Predominant abnormality in cerebral glucose utilization in late-onset dementia of the Alzheimer type: a cross-sectional comparison against advanced late-onset and incipient early-onset cases. J Neural Transm Park Dis Dement Sect 1991; 3(1): 1-14.
[http://dx.doi.org/10.1007/BF02251132] [PMID: 1905936]

[415] Melatonin. Monograph. Altern Med Rev 2005; 10(4): 326-36.
[PMID: 16366741]

[416] Mishima K, Tozawa T, Satoh K, Matsumoto Y, Hishikawa Y, Okawa M. Melatonin secretion rhythm disorders in patients with senile dementia of Alzheimer's type with disturbed sleep-waking. Biol Psychiatry 1999; 45(4): 417-21.
[http://dx.doi.org/10.1016/S0006-3223(97)00510-6] [PMID: 10071710]

[417] Wade AG, Farmer M, Harari G, *et al.* Add-on prolonged-release melatonin for cognitive function and sleep in mild to moderate Alzheimer's disease: a 6-month, randomized, placebo-controlled, multicenter trial. Clin Interv Aging 2014; 9: 947-61.
[http://dx.doi.org/10.2147/CIA.S65625] [PMID: 24971004]

[418] Hardeland R, Cardinali DP, Brown GM, Pandi-Perumal SR. Melatonin and brain inflammaging. Prog Neurobiol 2015; 127-128: 46-63.
[http://dx.doi.org/10.1016/j.pneurobio.2015.02.001] [PMID: 25697044]

[419] Srinivasan V, Kaur C, Pandi-Perumal S, Brown GM, Cardinali DP. Melatonin and its agonist ramelteon in Alzheimer's disease: possible therapeutic value. Int J Alzheimers Dis 2010; 2011: 741974.
[http://dx.doi.org/10.4061/2011/741974]

[420] Cardinali DP, Furio AM, Brusco LI. Clinical aspects of melatonin intervention in Alzheimer's disease progression. Curr Neuropharmacol 2010; 8(3): 218-27.
[http://dx.doi.org/10.2174/157015910792246209] [PMID: 21358972]

[421] Moffat SD, Zonderman AB, Metter EJ, Blackman MR, Harman SM, Resnick SM. Longitudinal assessment of serum free testosterone concentration predicts memory performance and cognitive status in elderly men. J Clin Endocrinol Metab 2002; 87(11): 5001-7.
[http://dx.doi.org/10.1210/jc.2002-020419] [PMID: 12414864]

[422] Hogervorst E, Combrinck M, Smith AD. Testosterone and gonadotropin levels in men with dementia. Neuroendocrinol Lett 2003; 24(3-4): 203-8.
[PMID: 14523358]

[423] Moffat SD, Zonderman AB, Metter EJ, *et al.* Free testosterone and risk for Alzheimer disease in older men. Neurology 2004; 62(2): 188-93.
[http://dx.doi.org/10.1212/WNL.62.2.188] [PMID: 14745052]

[424] Rosario ER, Chang L, Stanczyk FZ, Pike CJ. Age-related testosterone depletion and the development of Alzheimer disease. JAMA 2004; 292(12): 1431-2.
[http://dx.doi.org/10.1001/jama.292.12.1431-b] [PMID: 15383512]

[425] Lv W, Du N, Liu Y, *et al.* Low Testosterone Level and Risk of Alzheimer's Disease in the Elderly Men: a Systematic Review and Meta-Analysis. Mol Neurobiol 2016; 53(4): 2679-84.
[http://dx.doi.org/10.1007/s12035-015-9315-y] [PMID: 26154489]

[426] Rosario ER, Pike CJ. Androgen regulation of beta-amyloid protein and the risk of Alzheimer's disease. Brain Res Brain Res Rev 2008; 57(2): 444-53.
[http://dx.doi.org/10.1016/j.brainresrev.2007.04.012] [PMID: 17658612]

[427] Nguyen TV, Yao M, Pike CJ. Androgens activate mitogen-activated protein kinase signaling: role in neuroprotection. J Neurochem 2005; 94(6): 1639-51.
[http://dx.doi.org/10.1111/j.1471-4159.2005.03318.x] [PMID: 16011741]

[428] Yao M, Nguyen TV, Rosario ER, Ramsden M, Pike CJ. Androgens regulate neprilysin expression: role in reducing beta-amyloid levels. J Neurochem 2008; 105(6): 2477-88.
[http://dx.doi.org/10.1111/j.1471-4159.2008.05341.x] [PMID: 18346198]

[429] Baker LD, Sambamurti K, Craft S, *et al.* 17beta-estradiol reduces plasma Abeta40 for HRT-naïve postmenopausal women with Alzheimer disease: a preliminary study. Am J Geriatr Psychiatry 2003; 11(2): 239-44.
[PMID: 12611754]

[430] Wahjoepramono EJ, Wijaya LK, Taddei K, *et al.* Distinct effects of testosterone on plasma and cerebrospinal fluid amyloid-beta levels. J Alzheimers Dis 2008; 15(1): 129-37.
[http://dx.doi.org/10.3233/JAD-2008-15111] [PMID: 18780973]

[431] Ramsden M, Nyborg AC, Murphy MP, *et al.* Androgens modulate beta-amyloid levels in male rat brain. J Neurochem 2003; 87(4): 1052-5.

[http://dx.doi.org/10.1046/j.1471-4159.2003.02114.x] [PMID: 14622134]

[432] Rosario ER, Carroll JC, Oddo S, LaFerla FM, Pike CJ. Androgens regulate the development of neuropathology in a triple transgenic mouse model of Alzheimer's disease. J Neurosci 2006; 26(51): 13384-9.
[http://dx.doi.org/10.1523/JNEUROSCI.2514-06.2006] [PMID: 17182789]

[433] McAllister C, Long J, Bowers A, *et al.* Genetic targeting aromatase in male amyloid precursor protein transgenic mice down-regulates β-secretase (BACE-1) and prevents Alzheimer-like pathology and cognitive impairment. J Neurosci 2010; 30(21): 7326-34.
[http://dx.doi.org/10.1523/JNEUROSCI.1180-10.2010] [PMID: 20505099]

[434] Tan RS, Pu SJ. A pilot study on the effects of testosterone in hypogonadal aging male patients with Alzheimer's disease. Aging Male 2003; 6(1): 13-7.
[http://dx.doi.org/10.1080/tam.6.1.13.17] [PMID: 12809076]

[435] Wahjoepramono EJ, Asih PR, Aniwiyanti V, *et al.* The effects of testosterone supplementation on cognitive functioning in older men. CNS Neurol Disord Drug Targets 2016; 15(3): 337-43.
[http://dx.doi.org/10.2174/1871527315666151110125704] [PMID: 26553159]

[436] Maki PM, Ernst M, London ED, *et al.* Intramuscular testosterone treatment in elderly men: evidence of memory decline and altered brain function. J Clin Endocrinol Metab 2007; 92(11): 4107-14.
[http://dx.doi.org/10.1210/jc.2006-1805] [PMID: 17726086]

[437] Vaughan C, Goldstein FC, Tenover JL. Exogenous testosterone alone or with finasteride does not improve measurements of cognition in healthy older men with low serum testosterone. J Androl 2007; 28(6): 875-82.
[http://dx.doi.org/10.2164/jandrol.107.002931] [PMID: 17609296]

[438] Lu PH, Masterman DA, Mulnard R, *et al.* Effects of testosterone on cognition and mood in male patients with mild Alzheimer disease and healthy elderly men. Arch Neurol 2006; 63(2): 177-85.
[http://dx.doi.org/10.1001/archneur.63.2.nct50002] [PMID: 16344336]

[439] Mellon SH. Neurosteroid regulation of central nervous system development. Pharmacol Ther 2007; 116(1): 107-24.
[http://dx.doi.org/10.1016/j.pharmthera.2007.04.011] [PMID: 17651807]

[440] Baulieu EE, Robel P, Schumacher M. Neurosteroids: beginning of the story. Int Rev Neurobiol 2001; 46: 1-32.
[http://dx.doi.org/10.1016/S0074-7742(01)46057-0] [PMID: 11599297]

[441] Orentreich N, Brind JL, Rizer RL, Vogelman JH. Age changes and sex differences in serum dehydroepiandrosterone sulfate concentrations throughout adulthood. J Clin Endocrinol Metab 1984; 59(3): 551-5.
[http://dx.doi.org/10.1210/jcem-59-3-551] [PMID: 6235241]

[442] Orentreich N, Brind JL, Vogelman JH, Andres R, Baldwin H. Long-term longitudinal measurements of plasma dehydroepiandrosterone sulfate in normal men. J Clin Endocrinol Metab 1992; 75(4): 1002-4.
[http://dx.doi.org/10.1210/jcem.75.4.1400863] [PMID: 1400863]

[443] Rapp PR, Amaral DG. Individual differences in the cognitive and neurobiological consequences of normal aging. Trends Neurosci 1992; 15(9): 340-5.
[http://dx.doi.org/10.1016/0166-2236(92)90051-9] [PMID: 1382333]

[444] Näsman B, Olsson T, Bäckström T, *et al.* Serum dehydroepiandrosterone sulfate in Alzheimer's disease and in multi-infarct dementia. Biol Psychiatry 1991; 30(7): 684-90.
[http://dx.doi.org/10.1016/0006-3223(91)90013-C] [PMID: 1835658]

[445] Sunderland T, Merril CR, Harrington MG, *et al.* Reduced plasma dehydroepiandrosterone concentrations in Alzheimer's disease. Lancet 1989; 2(8662): 570.
[http://dx.doi.org/10.1016/S0140-6736(89)90700-9] [PMID: 2570275]

[446]   Majewska MD. Neurosteroids: endogenous bimodal modulators of the GABA$_A$ receptor. Mechanism of action and physiological significance. Prog Neurobiol 1992; 38(4): 379-95.
[http://dx.doi.org/10.1016/0301-0082(92)90025-A] [PMID: 1349441]

[447]   Monnet FP, Mahé V, Robel P, Baulieu EE. Neurosteroids, *via* sigma receptors, modulate the [³H]norepinephrine release evoked by N-methyl-D-aspartate in the rat hippocampus. Proc Natl Acad Sci USA 1995; 92(9): 3774-8.
[http://dx.doi.org/10.1073/pnas.92.9.3774] [PMID: 7731982]

[448]   Akan P, Kizildag S, Ormen M, Genc S, Oktem MA, Fadiloglu M. Pregnenolone protects the PC-12 cell line against amyloid beta peptide toxicity but its sulfate ester does not. Chem Biol Interact 2009; 177(1): 65-70.
[http://dx.doi.org/10.1016/j.cbi.2008.09.016] [PMID: 18926803]

[449]   Flood JF, Roberts E. Dehydroepiandrosterone sulfate improves memory in aging mice. Brain Res 1988; 448(1): 178-81.
[http://dx.doi.org/10.1016/0006-8993(88)91116-X] [PMID: 2968829]

[450]   Vallée M, Mayo W, Le Moal M. Role of pregnenolone, dehydroepiandrosterone and their sulfate esters on learning and memory in cognitive aging. Brain Res Brain Res Rev 2001; 37(1-3): 301-12.
[http://dx.doi.org/10.1016/S0165-0173(01)00135-7] [PMID: 11744095]

[451]   Pike CJ, Carroll JC, Rosario ER, Barron AM. Protective actions of sex steroid hormones in Alzheimer's disease. Front Neuroendocrinol 2009; 30(2): 239-58.
[http://dx.doi.org/10.1016/j.yfrne.2009.04.015] [PMID: 19427328]

[452]   Li R, Cui J, Shen Y. Brain sex matters: estrogen in cognition and Alzheimer's disease. Mol Cell Endocrinol 2014; 389(1-2): 13-21.
[http://dx.doi.org/10.1016/j.mce.2013.12.018] [PMID: 24418360]

[453]   Gillies GE, McArthur S. Estrogen actions in the brain and the basis for differential action in men and women: a case for sex-specific medicines. Pharmacol Rev 2010; 62(2): 155-98.
[http://dx.doi.org/10.1124/pr.109.002071] [PMID: 20392807]

[454]   Yue X, Lu M, Lancaster T, *et al.* Brain estrogen deficiency accelerates Abeta plaque formation in an Alzheimer's disease animal model. Proc Natl Acad Sci USA 2005; 102(52): 19198-203.
[http://dx.doi.org/10.1073/pnas.0505203102] [PMID: 16365303]

[455]   Cervellati C, Bergamini CM. Oxidative damage and the pathogenesis of menopause related disturbances and diseases. Clin Chem Lab Med 2016; 54(5): 739-53.
[http://dx.doi.org/10.1515/cclm-2015-0807] [PMID: 26544103]

[456]   Galea LA, Wainwright SR, Roes MM, Duarte-Guterman P, Chow C, Hamson DK. Sex, hormones and neurogenesis in the hippocampus: hormonal modulation of neurogenesis and potential functional implications. J Neuroendocrinol 2013; 25(11): 1039-61.
[http://dx.doi.org/10.1111/jne.12070] [PMID: 23822747]

[457]   Spencer-Segal JL, Tsuda MC, Mattei L, *et al.* Estradiol acts *via* estrogen receptors alpha and beta on pathways important for synaptic plasticity in the mouse hippocampal formation. Neuroscience 2012; 202: 131-46.
[http://dx.doi.org/10.1016/j.neuroscience.2011.11.035] [PMID: 22133892]

[458]   Anastasio TJ. Exploring the contribution of estrogen to amyloid-Beta regulation: a novel multifactorial computational modeling approach. Front Pharmacol 2013; 4: 16.
[http://dx.doi.org/10.3389/fphar.2013.00016] [PMID: 23459573]

[459]   Singh M, Sétáló G Jr, Guan X, Warren M, Toran-Allerand CD. Estrogen-induced activation of mitogen-activated protein kinase in cerebral cortical explants: convergence of estrogen and neurotrophin signaling pathways. J Neurosci 1999; 19(4): 1179-88.
[PMID: 9952396]

[460]   Leon RL, Huber JD, Rosen CL. Potential age-dependent effects of estrogen on neural injury. Am J

Pathol 2011; 178(6): 2450-60.
[http://dx.doi.org/10.1016/j.ajpath.2011.01.057] [PMID: 21641373]

[461]   Henderson VW. Action of estrogens in the aging brain: Dementia and cognitive aging. Biochim Biophys Acta 2010; 1800(10): 1077-83.
[http://dx.doi.org/10.1016/j.bbagen.2009.11.005] [PMID: 19913598]

[462]   Espeland MA, Rapp SR, Shumaker SA, *et al.* Women's Health Initiative Memory Study. Conjugated equine estrogens and global cognitive function in postmenopausal women: Women's Health Initiative Memory Study. JAMA 2004; 291(24): 2959-68.
[http://dx.doi.org/10.1001/jama.291.24.2959] [PMID: 15213207]

[463]   Wharton W, Baker LD, Gleason CE, *et al.* Short-term hormone therapy with transdermal estradiol improves cognition for postmenopausal women with Alzheimer's disease: results of a randomized controlled trial. J Alzheimers Dis 2011; 26(3): 495-505.
[http://dx.doi.org/10.3233/JAD-2011-110341] [PMID: 21694454]

[464]   Asthana S, Baker LD, Craft S, *et al.* High-dose estradiol improves cognition for women with AD: results of a randomized study. Neurology 2001; 57(4): 605-12.
[http://dx.doi.org/10.1212/WNL.57.4.605] [PMID: 11524467]

[465]   Azcoitia I, Moreno A, Carrero P, Palacios S, Garcia-Segura LM. Neuroprotective effects of soy phytoestrogens in the rat brain. Gynecol Endocrinol 2006; 22(2): 63-9.
[http://dx.doi.org/10.1080/09513590500519161] [PMID: 16603429]

[466]   Zhao L, Chen Q, Diaz Brinton R. Neuroprotective and neurotrophic efficacy of phytoestrogens in cultured hippocampal neurons. Exp Biol Med (Maywood) 2002; 227(7): 509-19.
[http://dx.doi.org/10.1177/153537020222700716] [PMID: 12094016]

[467]   Soni M, Rahardjo TB, Soekardi R, *et al.* Phytoestrogens and cognitive function: a review. Maturitas 2014; 77(3): 209-20.
[http://dx.doi.org/10.1016/j.maturitas.2013.12.010] [PMID: 24486046]

[468]   King TL, Brucker MC. Pharmacology for Women's Health. Jones & Bartlett Publishers 2010; pp. 372-3.

[469]   Baulieu E, Schumacher M. Progesterone as a neuroactive neurosteroid, with special reference to the effect of progesterone on myelination. Steroids 2000; 65(10-11): 605-12.
[http://dx.doi.org/10.1016/S0039-128X(00)00173-2] [PMID: 11108866]

[470]   Moralí G, Letechipía-Vallejo G, López-Loeza E, Montes P, Hernández-Morales L, Cervantes M. Post-ischemic administration of progesterone in rats exerts neuroprotective effects on the hippocampus. Neurosci Lett 2005; 382(3): 286-90.
[http://dx.doi.org/10.1016/j.neulet.2005.03.066] [PMID: 15885907]

[471]   Goodman Y, Bruce AJ, Cheng B, Mattson MP. Estrogens attenuate and corticosterone exacerbates excitotoxicity, oxidative injury, and amyloid beta-peptide toxicity in hippocampal neurons. J Neurochem 1996; 66(5): 1836-44.
[http://dx.doi.org/10.1046/j.1471-4159.1996.66051836.x] [PMID: 8780008]

[472]   Yoon BK, Kim DK, Kang Y, Kim JW, Shin MH, Na DL. Hormone replacement therapy in postmenopausal women with Alzheimer's disease: a randomized, prospective study. Fertil Steril 2003; 79(2): 274-80.
[http://dx.doi.org/10.1016/S0015-0282(02)04666-6] [PMID: 12568834]

[473]   Qin Y, Chen Z, Han X, *et al.* Progesterone attenuates $A\beta_{(25-35)}$-induced neuronal toxicity *via* JNK inactivation and progesterone receptor membrane component 1-dependent inhibition of mitochondrial apoptotic pathway. J Steroid Biochem Mol Biol 2015; 154: 302-11.
[http://dx.doi.org/10.1016/j.jsbmb.2015.01.002] [PMID: 25576906]

[474]   Liu S, Wu H, Xue G, *et al.* Metabolic alteration of neuroactive steroids and protective effect of progesterone in Alzheimer's disease-like rats. Neural Regen Res 2013; 8(30): 2800-10.

[PMID: 25206601]

[475] Ishihara Y, Kawami T, Ishida A, Yamazaki T. Allopregnanolone-mediated protective effects of progesterone on tributyltin-induced neuronal injury in rat hippocampal slices. J Steroid Biochem Mol Biol 2013; 135: 1-6.
[http://dx.doi.org/10.1016/j.jsbmb.2012.12.013] [PMID: 23280249]

[476] Gibbs RB. Fluctuations in relative levels of choline acetyltransferase mRNA in different regions of the rat basal forebrain across the estrous cycle: effects of estrogen and progesterone. J Neurosci 1996; 16(3): 1049-55.
[PMID: 8558233]

[477] Rigaud AS, André G, Vellas B, Touchon J, Pere JJ, Loria-Kanza Y. Oestro-progestagen treatment combined with rivastigmine in menopausal women suffering from Alzheimer's disease. The results of a 28-weeks controlled study. Presse Med 2003; 32(35): 1649-54.
[PMID: 14631268]

[478] de la Monte SM. Contributions of brain insulin resistance and deficiency in amyloid-related neurodegeneration in Alzheimer's disease. Drugs 2012; 72(1): 49-66.
[http://dx.doi.org/10.2165/11597760-000000000-00000] [PMID: 22191795]

[479] Hooper C, Killick R, Lovestone S. The GSK3 hypothesis of Alzheimer's disease. J Neurochem 2008; 104(6): 1433-9.
[http://dx.doi.org/10.1111/j.1471-4159.2007.05194.x] [PMID: 18088381]

[480] Lannert H, Hoyer S. Intracerebroventricular administration of streptozotocin causes long-term diminutions in learning and memory abilities and in cerebral energy metabolism in adult rats. Behav Neurosci 1998; 112(5): 1199-208.
[http://dx.doi.org/10.1037/0735-7044.112.5.1199] [PMID: 9829797]

[481] Wang X, Yu S, Hu JP, *et al.* Streptozotocin-induced diabetes increases amyloid plaque deposition in AD transgenic mice through modulating AGEs/RAGE/NF-κB pathway. Int J Neurosci 2014; 124(8): 601-8.
[http://dx.doi.org/10.3109/00207454.2013.866110] [PMID: 24228859]

[482] Salameh TS, Bullock KM, Hujoel IA, *et al.* Central nervous system delivery of intranasal insulin: mechanisms of uptake and effects on cognition. J Alzheimers Dis 2015; 47(3): 715-28.
[http://dx.doi.org/10.3233/JAD-150307] [PMID: 26401706]

[483] Craft S, Baker LD, Montine TJ, *et al.* Intranasal insulin therapy for Alzheimer disease and amnestic mild cognitive impairment: a pilot clinical trial. Arch Neurol 2012; 69(1): 29-38.
[http://dx.doi.org/10.1001/archneurol.2011.233] [PMID: 21911655]

[484] Reger MA, Watson GS, Green PS, *et al.* Intranasal insulin improves cognition and modulates beta-amyloid in early AD. Neurology 2008; 70(6): 440-8.
[http://dx.doi.org/10.1212/01.WNL.0000265401.62434.36] [PMID: 17942819]

[485] Rosenbloom MH, Barclay TR, Pyle M, *et al.* A single-dose pilot trial of intranasal rapid-acting insulin in apolipoprotein E4 carriers with mild-moderate Alzheimer's disease. CNS Drugs 2014; 28(12): 1185-9.
[http://dx.doi.org/10.1007/s40263-014-0214-y] [PMID: 25373630]

[486] Claxton A, Baker LD, Hanson A, *et al.* Long-acting intranasal insulin detemir improves cognition for adults with mild cognitive impairment or early-stage Alzheimer's disease dementia. J Alzheimers Dis 2015; 44(3): 897-906.
[http://dx.doi.org/10.3233/JAD-141791] [PMID: 25374101]

[487] Greco SJ, Bryan KJ, Sarkar S, *et al.* Leptin reduces pathology and improves memory in a transgenic mouse model of Alzheimer's disease. J Alzheimers Dis 2010; 19(4): 1155-67.
[http://dx.doi.org/10.3233/JAD-2010-1308] [PMID: 20308782]

[488] Zeng Z, Zhu J, Chen L, Wen W, Yu R. Biosynthesis pathways of ginkgolides. Pharmacogn Rev 2013;

7(13): 47-52.
[http://dx.doi.org/10.4103/0973-7847.112848] [PMID: 23922456]

[489] Kehr J, Yoshitake S, Ijiri S, Koch E, Nöldner M, Yoshitake T. Ginkgo biloba leaf extract (EGb 761[R]) and its specific acylated favonol constituents increase dopamine and acetylcholine levels in the rat medial prefrontal cortex: possible implication for cognitive enhancing properties of this ginkgo extract. Intern Psychogeriatics 2012; 1: S25-34.
[http://dx.doi.org/10.1017/S1041610212000567] [PMID: 22784425]

[490] Müller WE, Heiser J, Leuner K. Effects of the standardized Ginko biloba extract EGb761[R] on neuroplasicity. Intern Psychogeriatics 2012; 1: S21-4.
[http://dx.doi.org/10.1017/S1041610212000592] [PMID: 22784424]

[491] Eckert A. Mitochondrial effects of Ginkgo biloba extract. Int Psychogeriatr 2012; 24 (Suppl. 1): S18-20.
[http://dx.doi.org/10.1017/S1041610212000531] [PMID: 22784423]

[492] McKenna DJ, Jones K, Hughes K. Efficacy, safety, and use of ginkgo biloba in clinical and preclinical applications. Altern Ther Health Med 2001; 7(5): 70-86, 88-90.
[PMID: 11565403]

[493] Franke AG, Heinrich I, Lieb K, Fellgiebel A. The use of Ginkgo biloba in healthy elderly. Age (Dordr) 2014; 36(1): 435-44.
[http://dx.doi.org/10.1007/s11357-013-9550-y] [PMID: 23736956]

[494] Wang Y, Liu Y, Wu Q, Yao X, Cheng Z. Rapid and Sensitive Determination of Major Active Ingredients and Toxic Components in GinkgoBiloba Leaves Extract (EGb 761) by a Validated UPLC-MS-MS Method. J Chromatogr Sci 2017; 55(4): 459-64.
[http://dx.doi.org/10.1093/chromsci/bmw206] [PMID: 28069691]

[495] Sarris J, McIntyre E, Camfield DA. Plant-based medicines for anxiety disorders, part 2: a review of clinical studies with supporting preclinical evidence. CNS Drugs 2013; 27(4): 301-19.
[http://dx.doi.org/10.1007/s40263-013-0059-9] [PMID: 23653088]

[496] Russo P, Frustaci A, Del Bufalo A, Fini M, Cesario A. Multitarget drugs of plants origin acting on Alzheimer's disease. Curr Med Chem 2013; 20(13): 1686-93.
[http://dx.doi.org/10.2174/0929867311320130008] [PMID: 23410167]

[497] Montes P, Ruiz-Sanchez E, Rojas C, Rojas P. Ginkgo biloba Extract 761: A Review of Basic Studies and Potential Clinical Use in Psychiatric Disorders. CNS Neurol Disord Drug Targets 2015; 14(1): 132-49.
[http://dx.doi.org/10.2174/1871527314666150202151440] [PMID: 25642989]

[498] Singh SK, Barreto GE, Aliev G, Echeverria V. Ginkgo biloba as an alternative medicine in the treatment of anxiety in dementia and other psychiatric disorders. Curr Drug Metab 2017; 18(2): 112-9.
[http://dx.doi.org/10.2174/1389200217666161201112206] [PMID: 27908257]

[499] Abdel-Kader R, Hauptmann S, Keil U, et al. Stabilization of mitochondrial function by Ginkgo biloba extract (EGb 761). Pharmacol Res 2007; 56(6): 493-502.
[http://dx.doi.org/10.1016/j.phrs.2007.09.011] [PMID: 17977008]

[500] Tchantchou F, Xu Y, Wu Y, Christen Y, Luo Y. EGb 761 enhances adult hippocampal neurogenesis and phosphorylation of CREB in transgenic mouse model of Alzheimer's disease. FASEB J 2007; 21(10): 2400-8.
[http://dx.doi.org/10.1096/fj.06-7649com] [PMID: 17356006]

[501] Wu Y, Wu Z, Butko P, et al. Amyloid-β-induced pathological behaviors are suppressed by Ginkgo biloba extract EGb 761 and ginkgolides in transgenic Caenorhabditis elegans. J Neurosci 2006; 26(50): 13102-13.
[http://dx.doi.org/10.1523/JNEUROSCI.3448-06.2006] [PMID: 17167099]

[502] Költringer P, Langsteger W, Ober O. Dose-dependent hemorheological effects and microcirculatory

modifications following intravenous administration of Ginkgo biloba special extract EGb 761. Clin Hemorheol 1995; 15(4): 649-56.

[503]   Gauthier S, Schlaefke S. Efficacy and tolerability of Ginkgo biloba extract EGb 761® in dementia: a systematic review and meta-analysis of randomized placebo-controlled trials. Clin Interv Aging 2014; 9: 2065-77.
[http://dx.doi.org/10.2147/CIA.S72728] [PMID: 25506211]

[504]   Xiang YZ, Shang HC, Gao XM, Zhang BL. A comparison of the ancient use of ginseng in traditional Chinese medicine with modern pharmacological experiments and clinical trials. Phytother Res 2008; 22(7): 851-8.
[http://dx.doi.org/10.1002/ptr.2384] [PMID: 18567057]

[505]   Ong WY, Farooqui T, Koh HL, Farooqui AA, Ling EA. Protective effects of ginseng on neurological disorders. Front Aging Neurosci 2015; 7: 129.
[http://dx.doi.org/10.3389/fnagi.2015.00129] [PMID: 26236231]

[506]   Loh SH, Park JY, Cho EH, Nah SY, Kang YS. Animal lectins: potential receptors for ginseng polysaccharides. J Ginseng Res 2017; 41(1): 1-9.
[http://dx.doi.org/10.1016/j.jgr.2015.12.006] [PMID: 28123316]

[507]   Attele AS, Wu JA, Yuan CS. Ginseng pharmacology: multiple constituents and multiple actions. Biochem Pharmacol 1999; 58(11): 1685-93.
[http://dx.doi.org/10.1016/S0006-2952(99)00212-9] [PMID: 10571242]

[508]   Wang ZY, Liu JG, Li H, Yang HM. Pharmacological Effects of Active Components of Chinese Herbal Medicine in the Treatment of Alzheimer's Disease: A Review. Am J Chin Med 2016; 44(8): 1525-41.
[http://dx.doi.org/10.1142/S0192415X16500853] [PMID: 27848250]

[509]   Cho IH. Effects of Panax ginseng in neurodegenerative diseases. J Ginseng Res 2012; 36(4): 342-53.
[http://dx.doi.org/10.5142/jgr.2012.36.4.342] [PMID: 23717136]

[510]   Zhao H, Li Q, Zhang Z, Pei X, Wang J, Li Y. Long-term ginsenoside consumption prevents memory loss in aged SAMP8 mice by decreasing oxidative stress and up-regulating the plasticity-related proteins in hippocampus. Brain Res 2009; 1256: 111-22.
[http://dx.doi.org/10.1016/j.brainres.2008.12.031] [PMID: 19133247]

[511]   Tu LH, Ma J, Liu HP, Wang RR, Luo J. The neuroprotective effects of ginsenosides on calcineurin activity and tau phosphorylation in SY5Y cells. Cell Mol Neurobiol 2009; 29(8): 1257-64.
[http://dx.doi.org/10.1007/s10571-009-9421-3] [PMID: 19517226]

[512]   Tan X, Gu J, Zhao B, *et al.* Ginseng improves cognitive deficit *via* the RAGE/NF-κB pathway in advanced glycation end product-induced rats. J Ginseng Res 2015; 39(2): 116-24.
[http://dx.doi.org/10.1016/j.jgr.2014.09.002] [PMID: 26045684]

[513]   Zhang Y, Zhang J, Liu C, Yu M, Li S. Extraction, isolation, and aromatase inhibitory evaluation of low-polar ginsenosides from Panax ginseng leaves. J Chromatogr A 2017; 1483: 20-9.
[http://dx.doi.org/10.1016/j.chroma.2016.12.068] [PMID: 28027838]

[514]   Bao L, Cai X, Wang J, Zhang Y, Sun B, Li Y. Anti-Fatigue Effects of Small Molecule Oligopeptides Isolated from Panax ginseng C. A. Meyer in Mice. Nutrients 2016; 8(12): E807.
[http://dx.doi.org/10.3390/nu8120807] [PMID: 27983571]

[515]   Shin KC, Oh DK. Classification of glycosidases that hydrolyze the specific positions and types of sugar moieties in ginsenosides. Crit Rev Biotechnol 2016; 36(6): 1036-49.
[http://dx.doi.org/10.3109/07388551.2015.1083942] [PMID: 26383974]

[516]   Yang WZ, Hu Y, Wu WY, Ye M, Guo DA. Saponins in the genus Panax L. (Araliaceae): a systematic review of their chemical diversity. Phytochemistry 2014; 106: 7-24.
[http://dx.doi.org/10.1016/j.phytochem.2014.07.012] [PMID: 25108743]

[517]   Wu JG, Wang YY, Zhang ZL, Yu B. Herbal medicine in the treatment of Alzheimer's disease. Chin J Integr Med 2015; 21(2): 102-7.

[http://dx.doi.org/10.1007/s11655-014-1337-y] [PMID: 24752473]

[518] Chen CF, Chiou WF, Zhang JT. Comparison of the pharmacological effects of Panax ginseng and Panax quinquefolium. Acta Pharmacol Sin 2008; 29(9): 1103-8.
[http://dx.doi.org/10.1111/j.1745-7254.2008.00868.x] [PMID: 18718179]

[519] Li N, Zhou L, Li W, Liu Y, Wang J, He P. Protective effects of ginsenosides Rg1 and Rb1 on an Alzheimer's disease mouse model: a metabolomics study. J Chromatogr B Analyt Technol Biomed Life Sci 2015; 985: 54-61.
[http://dx.doi.org/10.1016/j.jchromb.2015.01.016] [PMID: 25660715]

[520] Fang F, Chen X, Huang T, Lue LF, Luddy JS, Yan SS. Multi-faced neuroprotective effects of Ginsenoside Rg1 in an Alzheimer mouse model. Biochim Biophys Acta 2012; 1822(2): 286-92.
[http://dx.doi.org/10.1016/j.bbadis.2011.10.004] [PMID: 22015470]

[521] Shi YQ, Huang TW, Chen LM, *et al.* Ginsenoside Rg1 attenuates amyloid-beta content, regulates PKA/CREB activity, and improves cognitive performance in SAMP8 mice. J Alzheimers Dis 2010; 19(3): 977-89.
[http://dx.doi.org/10.3233/JAD-2010-1296] [PMID: 20157253]

[522] Li W, Chu Y, Zhang L, Yin L, Li L. Ginsenoside Rg1 attenuates tau phosphorylation in SK-N-SH induced by Aβ-stimulated THP-1 supernatant and the involvement of p38 pathway activation. Life Sci 2012; 91(15-16): 809-15.
[http://dx.doi.org/10.1016/j.lfs.2012.08.028] [PMID: 22982182]

[523] Wu J, Yang H, Zhao Q, Zhang X, Lou Y. Ginsenoside Rg1 exerts a protective effect against $A\beta_{25-35}$-induced toxicity in primary cultured rat cortical neurons through the NF-κB/NO pathway. Int J Mol Med 2016; 37(3): 781-8.
[http://dx.doi.org/10.3892/ijmm.2016.2485] [PMID: 26865401]

[524] Wang XY, Zhang JT. Effects of ginsenoside Rg1 on synaptic plasticity of freely moving rats and its mechanism of action. Acta Pharmacol Sin 2001; 22(7): 657-62.
[PMID: 11749833]

[525] Huang L, Liu LF, Liu J, *et al.* Ginsenoside Rg1 protects against neurodegeneration by inducing neurite outgrowth in cultured hippocampal neurons. Neural Regen Res 2016; 11(2): 319-25.
[http://dx.doi.org/10.4103/1673-5374.177741] [PMID: 27073387]

[526] Cao GS, Li SX, Wang Y, *et al.* A combination of four effective components derived from Sheng-mai san attenuates hydrogen peroxide-induced injury in PC12 cells through inhibiting Akt and MAPK signaling pathways. Chin J Nat Med 2016; 14(7): 508-17.
[http://dx.doi.org/10.1016/S1875-5364(16)30060-7] [PMID: 27507201]

[527] Li N, Liu B, Dluzen DE, Jin Y. Protective effects of ginsenoside Rg2 against glutamate-induced neurotoxicity in PC12 cells. J Ethnopharmacol 2007; 111(3): 458-63.
[http://dx.doi.org/10.1016/j.jep.2006.12.015] [PMID: 17257792]

[528] Joo SS, Lee DI. Potential effects of microglial activation induced by ginsenoside Rg3 in rat primary culture: enhancement of type A Macrophage Scavenger Receptor expression. Arch Pharm Res 2005; 28(10): 1164-9.
[http://dx.doi.org/10.1007/BF02972981] [PMID: 16276974]

[529] Yang L, Hao J, Zhang J, *et al.* Ginsenoside Rg3 promotes beta-amyloid peptide degradation by enhancing gene expression of neprilysin. J Pharm Pharmacol 2009; 61(3): 375-80.
[http://dx.doi.org/10.1211/jpp.61.03.0013] [PMID: 19222911]

[530] Shieh PC, Tsao CW, Li JS, *et al.* Role of pituitary adenylate cyclase-activating polypeptide (PACAP) in the action of ginsenoside Rh2 against beta-amyloid-induced inhibition of rat brain astrocytes. Neurosci Lett 2008; 434(1): 1-5.
[http://dx.doi.org/10.1016/j.neulet.2007.12.032] [PMID: 18313848]

[531] Hou J, Xue J, Lee M, *et al.* Ginsenoside Rh2 improves learning and memory in mice. J Med Food

2013; 16(8): 772-6.
[http://dx.doi.org/10.1089/jmf.2012.2564] [PMID: 23957360]

[532] Shi J, Xue W, Zhao WJ, Li KX. Pharmacokinetics and dopamine/acetylcholine releasing effects of ginsenoside Re in hippocampus and mPFC of freely moving rats. Acta Pharmacol Sin 2013; 34(2): 214-20.
[http://dx.doi.org/10.1038/aps.2012.147] [PMID: 23202798]

[533] Hwang SH, Shin TJ, Choi SH, *et al.* Gintonin, newly identified compounds from ginseng, is novel lysophosphatidic acids-protein complexes and activates G protein-coupled lysophosphatidic acid receptors with high affinity. Mol Cells 2012; 33(2): 151-62.
[http://dx.doi.org/10.1007/s10059-012-2216-z] [PMID: 22286231]

[534] Kim HJ, Shin EJ, Lee BH, *et al.* Oral Administration of Gintonin Attenuates Cholinergic Impairments by Scopolamine, Amyloid-β Protein, and Mouse Model of Alzheimer's Disease. Mol Cells 2015; 38(9): 796-805.
[http://dx.doi.org/10.14348/molcells.2015.0116] [PMID: 26255830]

[535] Lee ST, Chu K, Sim JY, Heo JH, Kim M. Panax ginseng enhances cognitive performance in Alzheimer disease. Alzheimer Dis Assoc Disord 2008; 22(3): 222-6.
[http://dx.doi.org/10.1097/WAD.0b013e31816c92e6] [PMID: 18580589]

[536] Scholey A, Ossoukhova A, Owen L, *et al.* Effects of American ginseng (Panax quinquefolius) on neurocognitive function: an acute, randomised, double-blind, placebo-controlled, crossover study. Psychopharmacology (Berl) 2010; 212(3): 345-56.
[http://dx.doi.org/10.1007/s00213-010-1964-y] [PMID: 20676609]

[537] Dar NJ, Hamid A, Ahmad M. Pharmacologic overview of Withania somnifera, the Indian Ginseng. Cell Mol Life Sci 2015; 72(23): 4445-60.
[http://dx.doi.org/10.1007/s00018-015-2012-1] [PMID: 26306935]

[538] Jayaprakasam B, Padmanabhan K, Nair MG. Withanamides in Withania somnifera fruit protect PC-12 cells from beta-amyloid responsible for Alzheimer's disease. Phytother Res 2010; 24(6): 859-63.
[PMID: 19957250]

[539] Schliebs R, Liebmann A, Bhattacharya SK, Kumar A, Ghosal S, Bigl V. Systemic administration of defined extracts from Withania somnifera (Indian Ginseng) and Shilajit differentially affects cholinergic but not glutamatergic and GABAergic markers in rat brain. Neurochem Int 1997; 30(2): 181-90.
[http://dx.doi.org/10.1016/S0197-0186(96)00025-3] [PMID: 9017665]

[540] Yenisetti SC, Manjunath MJ, Muralidhara C. Neuropharmacological Properties of Withania somnifera - Indian Ginseng: An Overview on Experimental Evidence with Emphasis on Clinical Trials and Patents. Recent Patents CNS Drug Discov 2016; 10(2): 204-15.
[http://dx.doi.org/10.2174/1574889810666160615014106] [PMID: 27316579]

[541] Kuboyama T, Tohda C, Komatsu K. Effects of Ashwagandha (roots of Withania somnifera) on neurodegenerative diseases. Biol Pharm Bull 2014; 37(6): 892-7.
[http://dx.doi.org/10.1248/bpb.b14-00022] [PMID: 24882401]

[542] Kulkarni SK, Dhir A. Withania somnifera: an Indian ginseng. Prog Neuropsychopharmacol Biol Psychiatry 2008; 32(5): 1093-105.
[http://dx.doi.org/10.1016/j.pnpbp.2007.09.011] [PMID: 17959291]

[543] Tohda C. [Overcoming several neurodegenerative diseases by traditional medicines: the development of therapeutic medicines and unraveling pathophysiological mechanisms]. Yakugaku Zasshi 2008; 128(8): 1159-67.
[http://dx.doi.org/10.1248/yakushi.128.1159] [PMID: 18670181]

[544] Kuboyama T, Tohda C, Zhao J, Nakamura N, Hattori M, Komatsu K. Axon- or dendrite-predominant outgrowth induced by constituents from Ashwagandha. Neuroreport 2002; 13(14): 1715-20.
[http://dx.doi.org/10.1097/00001756-200210070-00005] [PMID: 12395110]

[545] Wake G, Court J, Pickering A, Lewis R, Wilkins R, Perry E. CNS acetylcholine receptor activity in European medicinal plants traditionally used to improve failing memory. J Ethnopharmacol 2000; 69(2): 105-14.
[http://dx.doi.org/10.1016/S0378-8741(99)00113-0] [PMID: 10687867]

[546] Soodi M, Naghdi N, Hajimehdipoor H, Choopani S, Sahraei E. Memory-improving activity of Melissa officinalis extract in naïve and scopolamine-treated rats. Res Pharm Sci 2014; 9(2): 107-14.
[PMID: 25657779]

[547] Soulimani R, Fleurentin J, Mortier F, Misslin R, Derrieu G, Pelt JM. Neurotropic action of the hydroalcoholic extract of Melissa officinalis in the mouse. Planta Med 1991; 57(2): 105-9.
[http://dx.doi.org/10.1055/s-2006-960042] [PMID: 1891490]

[548] Sepand M, Soodi M, Soleimani M, Hajimehdipoor H. Protective effects of Melissa officinalis extract against beta-amyloid-induced oxidative stress in PC12 cells. Faslnamah-i Giyahan-i Daruyi 2012; 11: 74-85.

[549] Akhondzadeh S, Noroozian M, Mohammadi M, Ohadinia S, Jamshidi AH, Khani M. Melissa officinalis extract in the treatment of patients with mild to moderate Alzheimer's disease: a double blind, randomised, placebo controlled trial. J Neurol Neurosurg Psychiatry 2003; 74(7): 863-6.
[http://dx.doi.org/10.1136/jnnp.74.7.863] [PMID: 12810768]

[550] Khazdair MR, Boskabady MH, Hosseini M, Rezaee R, M Tsatsakis A. The effects of Crocus sativus (saffron) and its constituents on nervous system: A review. Avicenna J Phytomed 2015; 5(5): 376-91.
[PMID: 26468457]

[551] Boskabady MH, Farkhondeh T. Antiinflammatory, Antioxidant, and Immunomodulatory Effects of Crocus sativus L. and its Main Constituents. Phytother Res 2016; 30(7): 1072-94.
[http://dx.doi.org/10.1002/ptr.5622] [PMID: 27098287]

[552] Moshiri M, Vahabzadeh M, Hosseinzadeh H. Clinical applications of saffron (Crocus sativus) and its constituents: A review. Drug Res (Stuttg) 2015; 65(6): 287-95.
[http://dx.doi.org/10.1055/s-0034-1375681] [PMID: 24848002]

[553] Mohajeri SA, Hosseinzadeh H, Abbasi-Ghaeni F. Saffron (Crocus sativus L.) and crocin have memory enhancing effect after chronic cerebral hypoperfusion in rats. J Clinbiochem 2011; 8: 274.

[554] Mehri S, Abnous K, Mousavi SH, Shariaty VM, Hosseinzadeh H. Neuroprotective effect of crocin on acrylamide-induced cytotoxicity in PC12 cells. Cell Mol Neurobiol 2012; 32(2): 227-35.
[http://dx.doi.org/10.1007/s10571-011-9752-8] [PMID: 21901509]

[555] Vakili A, Einali MR, Bandegi AR. Protective effect of crocin against cerebral ischemia in a dose-dependent manner in a rat model of ischemic stroke. J Stroke Cerebrovasc Dis 2014; 23(1): 106-13.
[http://dx.doi.org/10.1016/j.jstrokecerebrovasdis.2012.10.008] [PMID: 23182363]

[556] Geromichalos GD, Lamari FN, Papandreou MA, et al. Saffron as a source of novel acetylcholinesterase inhibitors: molecular docking and in vitro enzymatic studies. J Agric Food Chem 2012; 60(24): 6131-8.
[http://dx.doi.org/10.1021/jf300589c] [PMID: 22655699]

[557] Papandreou MA, Tsachaki M, Efthimiopoulos S, Cordopatis P, Lamari FN, Margarity M. Memory enhancing effects of saffron in aged mice are correlated with antioxidant protection. Behav Brain Res 2011; 219(2): 197-204.
[http://dx.doi.org/10.1016/j.bbr.2011.01.007] [PMID: 21238492]

[558] Khalili M, Hamzeh F. Effects of active constituents of Crocus sativus L., crocin on streptozocin-induced model of sporadic Alzheimer's disease in male rats. Iran Biomed J 2010; 14(1-2): 59-65.
[PMID: 20683499]

[559] Akhondzadeh S, Sabet MS, Harirchian MH, et al. Saffron in the treatment of patients with mild to moderate Alzheimer's disease: a 16-week, randomized and placebo-controlled trial. J Clin Pharm Ther 2010; 35(5): 581-8.

[http://dx.doi.org/10.1111/j.1365-2710.2009.01133.x] [PMID: 20831681]

[560] Akhondzadeh S, Shafiee Sabet M, Harirchian MH, *et al.* A 22-week, multicenter, randomized, double-blind controlled trial of Crocus sativus in the treatment of mild-to-moderate Alzheimer's disease. Psychopharmacology (Berl) 2010; 207(4): 637-43.
[http://dx.doi.org/10.1007/s00213-009-1706-1] [PMID: 19838862]

[561] Lin L, Ni B, Lin H, *et al.* Traditional usages, botany, phytochemistry, pharmacology and toxicology of Polygonum multiflorum Thunb.: a review. J Ethnopharmacol 2015; 159: 158-83.
[http://dx.doi.org/10.1016/j.jep.2014.11.009] [PMID: 25449462]

[562] Pharmacopoeia Commission of the Ministry of Health.. Pharmacopoeia of the People's Republic of China (PPRC) 2010.

[563] Um MY, Choi WH, Aan JY, Kim SR, Ha TY. Protective effect of Polygonum multiflorum Thunb on amyloid beta-peptide 25-35 induced cognitive deficits in mice. J Ethnopharmacol 2006; 104(1-2): 144-8.
[http://dx.doi.org/10.1016/j.jep.2005.08.054] [PMID: 16219438]

[564] Bounda GA, Feng YU. Review of clinical studies of Polygonum multiflorum Thunb. and its isolated bioactive compounds. Pharmacognosy Res 2015; 7(3): 225-36.
[http://dx.doi.org/10.4103/0974-8490.157957] [PMID: 26130933]

[565] Hou DR, Wang Y, Xue L, *et al.* Effect of polygonum multiflorum on the fluidity of the mitochondria membrane and activity of COX in the hippocampus of rats with Abeta 1-40-induced Alzheimer's disease. Zhong Nan Da Xue Xue Bao Yi Xue Ban 2008; 33(11): 987-92.
[PMID: 19060365]

[566] Sheng C, Peng W, Chen Z, *et al.* Impact of 2, 3, 5, 4′-tetrahydroxystilbene-2-O-β-D-glucoside on cognitive deficits in animal models of Alzheimer's disease: a systematic review. BMC Complement Altern Med 2016; 16(1): 320.
[http://dx.doi.org/10.1186/s12906-016-1313-8] [PMID: 27565551]

[567] Wang R, Tang Y, Feng B, *et al.* Changes in hippocampal synapses and learning-memory abilities in age-increasing rats and effects of tetrahydroxystilbene glucoside in aged rats. Neuroscience 2007; 149(4): 739-46.
[http://dx.doi.org/10.1016/j.neuroscience.2007.07.065] [PMID: 17935895]

[568] Zhang L, Xing Y, Ye CF, Ai HX, Wei HF, Li L. Learning-memory deficit with aging in APP transgenic mice of Alzheimer's disease and intervention by using tetrahydroxystilbene glucoside. Behav Brain Res 2006; 173(2): 246-54.
[http://dx.doi.org/10.1016/j.bbr.2006.06.034] [PMID: 16901557]

[569] Chang CC, Chang YC, Hu WL, Hung YC. Oxidative Stress and Salvia miltiorrhiza in Aging-Associated Cardiovascular Diseases. Oxid Med Cell Longev 2016; 2016: 4797102.
[http://dx.doi.org/10.1155/2016/4797102] [PMID: 27807472]

[570] Su CY, Ming QL, Rahman K, Han T, Qin LP. Salvia miltiorrhiza: Traditional medicinal uses, chemistry, and pharmacology. Chin J Nat Med 2015; 13(3): 163-82.
[http://dx.doi.org/10.1016/S1875-5364(15)30002-9] [PMID: 25835361]

[571] Wang X, Morris-Natschke SL, Lee KH. New developments in the chemistry and biology of the bioactive constituents of Tanshen. Med Res Rev 2007; 27(1): 133-48.
[http://dx.doi.org/10.1002/med.20077] [PMID: 16888751]

[572] Zhang XZ, Qian SS, Zhang YJ, Wang RQ. Salvia miltiorrhiza: A source for anti-Alzheimer's disease drugs. Pharm Biol 2016; 54(1): 18-24.
[http://dx.doi.org/10.3109/13880209.2015.1027408] [PMID: 25857808]

[573] Durairajan SS, Yuan Q, Xie L, *et al.* Salvianolic acid B inhibits Abeta fibril formation and disaggregates preformed fibrils and protects against Abeta-induced cytotoxicty. Neurochem Int 2008; 52(4-5): 741-50.

[http://dx.doi.org/10.1016/j.neuint.2007.09.006] [PMID: 17964692]

[574]   Mei Z, Situ B, Tan X, *et al.* Cryptotanshinione upregulates alpha-secretase by activation PI3K pathway in cortical neurons. Brain Res 2010; 1348: 165-73.
[http://dx.doi.org/10.1016/j.brainres.2010.05.083] [PMID: 20595051]

[575]   Liu T, Jin H, Sun QR, Xu JH, Hu HT. The neuroprotective effects of tanshinone IIA on β-amyloi--induced toxicity in rat cortical neurons. Neuropharmacology 2010; 59(7-8): 595-604.
[http://dx.doi.org/10.1016/j.neuropharm.2010.08.013] [PMID: 20800073]

[576]   Zhang HA, Gao M, Zhang L, *et al.* Salvianolic acid A protects human SH-SY5Y neuroblastoma cells against H$_2$O$_2$-induced injury by increasing stress tolerance ability. Biochem Biophys Res Commun 2012; 421(3): 479-83.
[http://dx.doi.org/10.1016/j.bbrc.2012.04.021] [PMID: 22516750]

[577]   Liu CS, Chen NH, Zhang JT. Protection of PC12 cells from hydrogen peroxide-induced cytotoxicity by salvianolic acid B, a new compound isolated from Radix Salviae miltiorrhizae. Phytomedicine 2007; 14(7-8): 492-7.
[http://dx.doi.org/10.1016/j.phymed.2006.11.002] [PMID: 17175150]

[578]   Han M, Liu Y, Zhang B, *et al.* Salvianic borneol ester reduces β-amyloid oligomers and prevents cytotoxicity. Pharm Biol 2011; 49(10): 1008-13.
[http://dx.doi.org/10.3109/13880209.2011.559585] [PMID: 21936627]

[579]   Chen Y, Wu X, Yu S, *et al.* Neuroprotective capabilities of Tanshinone IIA against cerebral ischemia/reperfusion injury *via* anti-apoptotic pathway in rats. Biol Pharm Bull 2012; 35(2): 164-70.
[http://dx.doi.org/10.1248/bpb.35.164] [PMID: 22293345]

[580]   Yu H, Yao L, Zhou H, *et al.* Neuroprotection against Aβ25-35-induced apoptosis by Salvia miltiorrhiza extract in SH-SY5Y cells. Neurochem Int 2014; 75: 89-95.
[http://dx.doi.org/10.1016/j.neuint.2014.06.001] [PMID: 24932696]

[581]   Ren Y, Houghton PJ, Hider RC, Howes MJ. Novel diterpenoid acetylcholinesterase inhibitors from Salvia miltiorhiza. Planta Med 2004; 70(3): 201-4.
[http://dx.doi.org/10.1055/s-2004-815535] [PMID: 15114495]

[582]   Senol FS, Ślusarczyk S, Matkowski A, *et al.* Selective in vitro and in silico butyrylcholinesterase inhibitory activity of diterpenes and rosmarinic acid isolated from Perovskia atriplicifolia Benth. and Salvia glutinosa L. Phytochemistry 2017; 133: 33-44.
[http://dx.doi.org/10.1016/j.phytochem.2016.10.012] [PMID: 27817931]

[583]   Kim DH, Jeon SJ, Jung JW, *et al.* Tanshinone congeners improve memory impairments induced by scopolamine on passive avoidance tasks in mice. Eur J Pharmacol 2007; 574(2-3): 140-7.
[http://dx.doi.org/10.1016/j.ejphar.2007.07.042] [PMID: 17714702]

[584]   Choi HS, Cho DI, Choi HK, Im SY, Ryu SY, Kim KM. Molecular mechanisms of inhibitory activities of tanshinones on lipopolysaccharide-induced nitric oxide generation in RAW 264.7 cells. Arch Pharm Res 2004; 27(12): 1233-7.
[http://dx.doi.org/10.1007/BF02975887] [PMID: 15646797]

[585]   Guo G, Li B, Wang Y, *et al.* Effects of salvianolic acid B on proliferation, neurite outgrowth and differentiation of neural stem cells derived from the cerebral cortex of embryonic mice. Sci China Life Sci 2010; 53(6): 653-62.
[http://dx.doi.org/10.1007/s11427-010-3106-5] [PMID: 20602267]

[586]   Bi XB, Deng YB, Gan DH, Wang YZ. Salvianolic acid B promotes survival of transplanted mesenchymal stem cells in spinal cord-injured rats. Acta Pharmacol Sin 2008; 29(2): 169-76.
[http://dx.doi.org/10.1111/j.1745-7254.2008.00710.x] [PMID: 18215345]

[587]   Lu Y, Liu X, Liang X, Xiang L, Zhang W. Metabolomic strategy to study therapeutic and synergistic effects of tanshinone IIA, salvianolic acid B and ginsenoside Rb1 in myocardial ischemia rats. J Ethnopharmacol 2011; 134(1): 45-9.

[http://dx.doi.org/10.1016/j.jep.2010.11.048] [PMID: 21130150]

[588] Bi X, Liu X, Di L, Zu Q. Improved oral bioavailability using a solid self-microemulsifying drug delivery system containing a multicomponent mixture extracted from salvia miltiorrhiza. Molecules 2016; 21(4): 456.
[http://dx.doi.org/10.3390/molecules21040456] [PMID: 27070565]

[589] Miroddi M, Navarra M, Quattropani MC, Calapai F, Gangemi S, Calapai G. Systematic review of clinical trials assessing pharmacological properties of Salvia species on memory, cognitive impairment and Alzheimer's disease. CNS Neurosci Ther 2014; 20(6): 485-95.
[http://dx.doi.org/10.1111/cns.12270] [PMID: 24836739]

[590] Dorszewska J, Prendecki M, Oczkowska A, Dezor M, Kozubski W. Molecular Basis of Familial and Sporadic Alzheimer's Disease. Curr Alzheimer Res 2016; 13(9): 952-63.
[http://dx.doi.org/10.2174/1567205013666160314150501] [PMID: 26971934]

[591] Devineni A, Tohme S, Kody MT, Cowley RA, Harris BT. Stepping back to move forward: a current review of iPSCs in the fight against Alzheimer's disease. Am J Stem Cells 2016; 5(3): 99-106.
[PMID: 27853631]

[592] Giri M, Zhang M, Lü Y. Genes associated with Alzheimer's disease: an overview and current status. Clin Interv Aging 2016; 11: 665-81.
[http://dx.doi.org/10.2147/CIA.S105769] [PMID: 27274215]

[593] Witte MM, Foster NL, Fleisher AS, *et al.* Clinical use of amyloid-positron emission tomography neuroimaging: Practical and bioethical considerations. Alzheimers Dement (Amst) 2015; 1(3): 358-67.
[http://dx.doi.org/10.1016/j.dadm.2015.06.006] [PMID: 27239516]

[594] Lista S, Garaci FG, Ewers M, *et al.* CSF Aβ1-42 combined with neuroimaging biomarkers in the early detection, diagnosis and prediction of Alzheimer's disease. Alzheimers Dement 2014; 10(3): 381-92.
[http://dx.doi.org/10.1016/j.jalz.2013.04.506] [PMID: 23850330]

[595] Mantzavinosa V, Alexiou A, Greig NH, Kamal MA. Biomarkers for Alzheimer's Disease Diagnosis. Curr Alzheimer Res 2017. Epub ahead of print
[http://dx.doi.org/10.2174/1567205014666170203125942] [PMID: 28164766]

[596] Hampel H, Lista S, Teipel SJ, *et al.* Perspective on future role of biological markers in clinical therapy trials of Alzheimer's disease: a long-range point of view beyond 2020. Biochem Pharmacol 2014; 88(4): 426-49.
[http://dx.doi.org/10.1016/j.bcp.2013.11.009] [PMID: 24275164]

[597] Ruan Q, D'Onofrio G, Sancarlo D, *et al.* Emerging biomarkers and screening for cognitive frailty. Aging Clin Exp Res 2017. Epub ahead of print
[http://dx.doi.org/10.1007/s40520-017-0741-8] [PMID: 28260159]

[598] Kim D, Kim YS, Shin DW, Park CS, Kang JH. Harnessing Cerebrospinal Fluid Biomarkers in Clinical Trials for Treating Alzheimer's and Parkinson's Diseases: Potential and Challenges. J Clin Neurol 2016; 12(4): 381-92.
[http://dx.doi.org/10.3988/jcn.2016.12.4.381] [PMID: 27819412]

[599] Bachurin SO, Bovina EV, Ustyugov AA. Drugs in Clinical Trials for Alzheimer's Disease: The Major Trends. Med Res Rev 2017; 37(5): 1186-225.
[http://dx.doi.org/10.1002/med.21434] [PMID: 28084618]

[600] McGhee DJ, Ritchie CW, Zajicek JP, Counsell CE. A review of clinical trial designs used to detect a disease-modifying effect of drug therapy in Alzheimer's disease and Parkinson's disease. BMC Neurol 2016; 16: 92.
[http://dx.doi.org/10.1186/s12883-016-0606-3] [PMID: 27312378]

[601] Shea TB, Remington R. Nutritional supplementation for Alzheimer's disease? Curr Opin Psychiatry 2015; 28(2): 141-7.
[http://dx.doi.org/10.1097/YCO.0000000000000138] [PMID: 25602242]

[602] Athanasopoulos D, Karagiannis G, Tsolaki M. Recent Findings in Alzheimer Disease and Nutrition Focusing on Epigenetics. Adv Nutr 2016; 7(5): 917-27.
[http://dx.doi.org/10.3945/an.116.012229] [PMID: 27633107]

[603] Stephen R, Hongisto K, Solomon A, Lönnroos E. Physical Activity and Alzheimer's Disease: A Systematic Review. J Gerontol A Biol Sci Med Sci 2017; 72(6): 733-9.
[http://dx.doi.org/10.1093/gerona/glw251] [PMID: 28049634]

[604] Cass SP. Alzheimer's Disease and Exercise: A Literature Review. Curr Sports Med Rep 2017; 16(1): 19-22.
[http://dx.doi.org/10.1249/JSR.0000000000000332] [PMID: 28067736]

CHAPTER 4

Pain Management Strategies: Some Uses of Antidepressant Medications and Non-Pharmacological Approaches

Kathy Sexton-Radek[1,*], Antoine Chami[2] and Alissa Rubinfeld[3]

[1] *Suburban Pulmonary & Sleep Associates/Elmhurst College, North Riverside, IL, USA*

[2] *Chicagoland Advanced Pain Specialists, Westmont, IL, USA*

[3] *The Chicago School of Professional Psychology, Chicago, IL, USA*

Abstract: It is estimated that some 25-30% of the population in the United States have experienced chronic pain. Worldwide, estimates reflect values of 35-40%. Chronic pain prevalence figures intensify the treatment approaches to care. Both pharmacological and non-pharmacological care is provided to address pain discomfort—oftentimes, at an individualized level. Relatively new strategies of prescribing additional pain management prescriptions of anti-depressant medications have enhanced the quality of care of the chronic pain patient. Sensory inputs from ascending pathways to the brain are targeted by pain medications. The common pathway to the cerebrum, at the periaqueductal gray which is largely noradrenergic and serotonergic, provides an additional platform to provide pharmacological treatment to the chronic pain patient. Anti-depressant medications act in many regions of the brain including the periaqueductal area. The resultant modulation of pain serves to also enhance the perception of pain relief from other pain medications. The adjuvant pain management treatment of anti-depressant medications is presented in terms of the types of chronic pain treated, the proposed mechanism of action of the antidepressant medications, and the other medications used in conjunction with anti-depressant medications.

Keywords: Pain Medications, Pain, Antidepressant Medication, Cognitive Behavior Therapy, Medication Management, Pharmacological Treatment, Pain Disorders, Pain Questionnaires, Cognitive Therapy, Complex Pain.

INTRODUCTORY ISSUES

Chronic pain is estimated, partially or completely, to disable approximately 50 million people. Chronic pain affects 34 million people causing an estimated $100 billion in medical expenses for pain each year. Further, global cancer rates are

---

* **Corresponding author Kathy Sexton-Radek:** Suburban Pulmonary & Sleep Associates/Elmhurst College, North Riverside, IL, USA; Tel: (708) 442-8946; Email: ksrsleep@aol.com

expected to double to 20 million in the next 20 years. An estimated 70% of elderly in nursing homes have poorly managed pain. The World Health Organization estimates that some 50-80% of patients fail to receive proper pain management. Pain is the number one cause of adult disability. The prototypical pain patient is 45-64 years and with 30% experiencing consistent pain for greater than 24 hours.

Pain is reduced or amplified based on descending pathways from the brain and is affected by characteristics including pain history, attention to symptoms and emotional state. Factors complicating chronic pain include: substance abuse, history of mental disorder, trauma, chronic illness, family discord, grief, systems issues, legal concerns, financial issues, and multicultural issues.

Pain disorders are not classified by the *Diagnostic and Statistical Manual 5*. Currently, the pain-related disorders are considered in categories of psychological factors affecting physical conditions, somatoform disorders, anxiety disorders and psychological factors affecting medical conditions.

It is estimated that fifty million people are partially or completely disabled because of pain with 30-40 million of these individuals suffering from chronic pain. In this grouping of 100 million people chronic pain patient's diabetes is thought to be present in 25.8 million of the cases, coronary heart disease in 16.3 million, stroke in seven million and cancer cases in 11.9 million. Additionally, chronic conditions of arthritis, hypertension, depression, hyperlipidemia, and asthma co-occur with many diagnoses of chronic pain. Table 1 provides an overview of this relationship. The types of pain are described in Table 2.

Table 1. Chronic Conditions.

Arthritis	44%
High blood pressure	37%
Chronic pain	35%
Depression	31%
High Cholesterol	29%
Diabetes	27%
Heart Disease	16%
Asthma	15%

Source: Goldberg & McGee (2011)

Table 2. Types of Pain.

Nociceptive Pain	**Sprains, bone fracture, burns, bruises, special nerve endings (includes neuromuscular)**
Neuropathic Pain	Shingles, neuralgia, phantom limb pain, carpal tunnel syndrome, peripheral neuropathy
Mixed Category Pain	Migraine headache
Central Pain	Dysfunction of nervous system – Fibromyalgia

Assessment

Assessment of different types of pain is conducted with clinical examination, pain questionnaires and diagnostic tests. With a clinical exam, the individual differences in pain experience can be thoroughly explored. Further, physical exam assessments common to a general physical and neurological exam are administered in addition to the clinical examination. For example, sensory-tactile exam results mapped to sensory maps. Additional details of this concept are presented in Table **3**, as depression and anxiety have significant impact on the pain experience.

Table 3. Chronic pain after accident injury and its relationship to depression and anxiety.

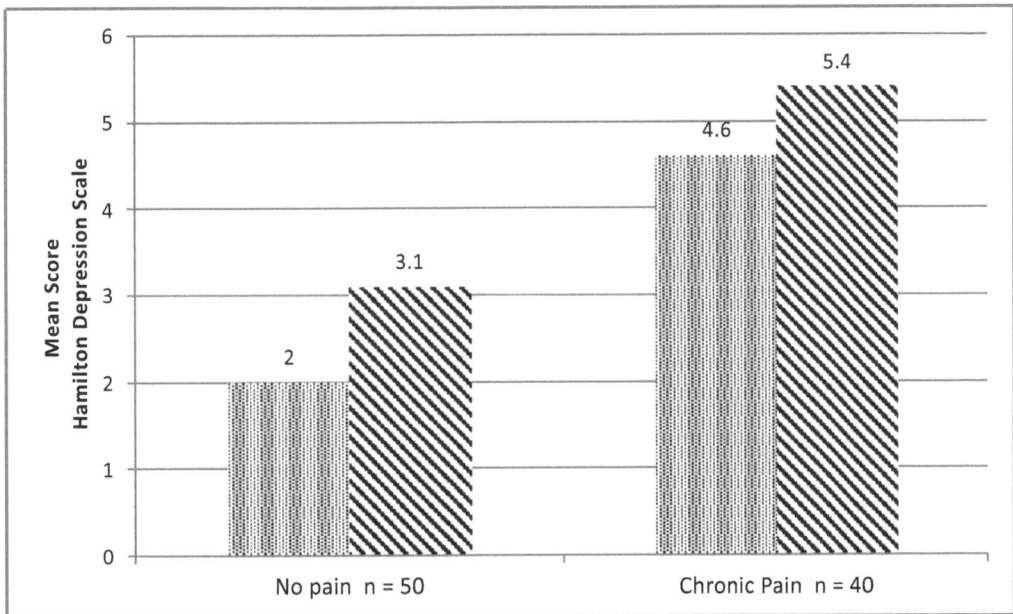

Source: Jenewein, J., *et al.* (2009). *Journal of Psychosomatic Research, 66*: 119-128.

Pain questionnaires provide the assessor with both content and the patient's perception of their pain experience. An example of an interview combined with physical exam is the Leeds Assessment of Neuropathic Symptoms and Signs (*i.e.,* Five question items and two clinical examination items with 82-91 sensitivity and 80-94% specificity). PAINDETECT is another commonly used pain questionnaire and uses a self-report format. Psychological assessments of pain focus on the conduction of a functional analysis of the pain problem where factors that initiated and sustain pain are examined. In addition, at home measures such as visual analogue scales (Rate 1-10, 1 = no pain, 10 = worst pain) with a pain diary are used. In the pain diary, the frequency, duration of the pain and location site of the pain is logged. In cognitively oriented diaries, patients identify precipitants to the pain. In most treatment approaches, the pain diary logging is ongoing and used in the therapeutic process.

Treatment

Health care professionals such as Occupational Therapists and Physical Therapists are involved in the pain management process. De Guchy, Granger and Gorelik [1] described the role of the physical therapist as a primary contact practitioner. In this model, hand fractures, ankle sprains and lumbar pain account for approximately 30% of cases seen. In a sampling of older adults with chronic pain, Song, Graham-Engeland, Mogle and Martire [2] reported that mood and couple interactions on sleep quality substantially facilitated the understanding of the patients' chronic pain. Further, Molton and Terrill [3] report the utility of assessing mood once pain conditions become chronic and persistent for a patient. One in four older adults with persistent pain is at risk for depression and consequences of poorer physical function, disability, social isolation and suicide ideation [3].

Treatment approaches that focus on pain management are typically cognitive or cognitive-behavioral in orientation coupled with medical approaches (*i.e.,* laboratory and medical testing, medical procedures for pain relief, pharmacological management for pain reduction) [4]. Interdisciplinary chronic pain rehabilitation programs have empirical support for their approach to restore function. Therapeutic activities such as education, medical management, physical therapy/occupational therapy and psychological intervention are included [5]. These interdisciplinary programs for the conduction of multi-health care professional assessment and treatment are effective in improvement for those that complete the program [6]. Recent literature has identified the need and utility for a Sleep Education / Sleep Quality focus for patients in interdisciplinary programs for chronic pain (McBeth, Wilkie, Bedson, Chew-Graham and Lacey, 2015). See Table **4** for a presentation of these details.

Table 4. Key Neurobiological Pathways.

Ascending Pathways	
Use of Anti-Depressants	Nor-adrenergic Pathways Serotonergic Pathways
Anterolateral Pathways →	Sensory (pain, temp., crude touch) Spinothalamic Spinomesencophalic Spinoreticular
Periaqueductal Gray	←Central modulation of pain
	←Receives input from hypothalamus, amygdala, cortex and it inhibits pain in dorsal horn via a relay in a region at the pontomedullary junction called rostral ventral medulla (RVM) also, substance P to locus ceruleus modulates
*Opiates * encephalin and dysmorphic containing neurons are concentrated in periaqueductal gray, RVM, spinal cord dorsal horn. * endorphin containing neurons are concentrated in regions of hypothalamus that project to the periaqueductal gray.	

The cognitive / cognitive-behavioral therapy approaches to pain management fit within an interdisciplinary treatment team approach or are conducted in a private practice approach. The approaches begin with assessment and concurrent education about pain and factors that contribute to pain problems. With a cognitive emphasis, beliefs about the pain problem are identified. For example, the Gate Control model of pain is often used to explain, to a general audience of patients, the role of neuro control or the recognition of pain signals. At this point in these approaches, explanations of pain are given in interactional manner to determine which may be personally relevant as an explanation of the pain response. Concurrently now, skills are taught in relaxation and/or mindfulness to provide management for pain response. In addition to the identification of belief about the pain and the subsequent development of alternative assumptions that assist the patient in gaining self-control, a third helpful approach is commonly worked on: Pacing Activities. In therapeutically teaching pacing skills, individuals are provided with the explanation of the impact of over activity and underactivity and understanding of pain [7]. In summary, these therapeutic approaches assist the pain patient to move from the search for a cure to pain management and self-control.

In addition, behavioral therapy approaches with a specific focus to pain management care is affective. The following represents some of these therapies: Pacing, Goal setting, drug use, physical fitness, readiness for change, family issues, work issues, ergonomics, assertiveness, sleep quality, skill training (*e.g.,*

Stress Inoculation Training), problem solving training. In addition, relapse prevention approaches are essential to the success of any treatment approaches.

Relapse prevention entails rehearsal and maintenance, regular practice of techniques to manage pain, prepare for potential flare ups, engage problem solving skills in anticipation, disability does not become extreme.

NEUROPHYSIOLOGICAL THEMES RELATED TO ASSESSMENT AND TREATMENT

Hemington and Coulombe [8] examined the functional connectivity of a key mode of the descending pain modulation pathway, the periaqueductal gray in chronic back pain patients. Altered connectivity was found in patients with back pain in terms of enhanced excitatory signal of norepinephrine as compared to healthy controls.

Muscle trigger points and referred pain areas elicited by muscle pain highlight the pain pathway through the ascending pathway-periventricular region. Alonso-Blanco, de-la-Llave-Rincón and Galan-del-Río [9] identified active trigger points to this pain pathway in patients with myofascial temporomandibular pain and fibromyalgia syndrome. The reader is advised to consider "Neuroanatomy through Clinical Cases" by H. N. Blumenfeld, M.D., Ph.D [10]. for detailed information about pain pathways and clinical cases involving pain diagnoses. The key components are listed in Table **4**.

PSYCHOPHARMACOLOGICAL   COMPONENTS   RELATED   TO ANTIDEPRESSANT ADJUVANT TREATMENT

Complex pain is experienced by the patient as physical, sensory, emotional and cognitive components. Pain is experienced physiologically by both nerve fiber pathways carrying the noxious sensory signal to the brain. Specialty receptor cells such as the mu receptor are activated by glutamate to NMDA receptors in a secondary receptor approach. The mu opioid receptors are involved in analgesia, respiratory depression, relaxed euphoria, sedation, reduced apprehension and concern. The mu receptors are found in high concentration on the ascending pathways. The actions of the mu receptor in the dorsal horn of the peripheral nervous system block the further release of calcium at the cellular level. This action is joined by the central nervous system descending inhibitory systems in the mid brain. Thus, ascending pathways convey the pain sensation to the cortex, and glutamate: NMDA medications (along with GABA neurotransmitter actions, have a role in the cerebral response to pain followed by ascending pathways conveying the sensation of pain. Insufficient sleep impairs the central endogenous opioid-rich regions of pain in the prefrontal cortex and anterior cingulate cortex.

This is depicted in Figs. (**1**, **2**) presents additional brain structures involved in processing pain signals.

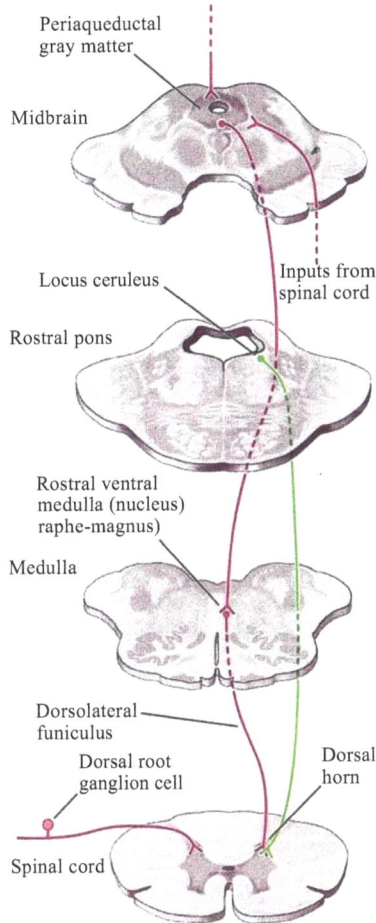

Fig. (1). Central pathways involved in pain modulation. Source: Blumenfeld (2010).

The mesolimbic pathway of serotonin, a brain neurochemical, plays a central role in pain. Richardson and Akil [11] identified the role of serotonin in pain experience with their investigation of the utility of electrical brain stimulation. The periventricular gray matter is amore of the ascending and descending pain modulation system. The interconnection of the periventricular gray matter by projection to spinal cord, cerebrum, cerebellum and subcortex (amygdala, posterior insula) point to its central role in pain. The perfusion of this pathway, by serotonin, an inhibitory neurotransmitter, provides an additional component to understanding Pharmacological Rationale.

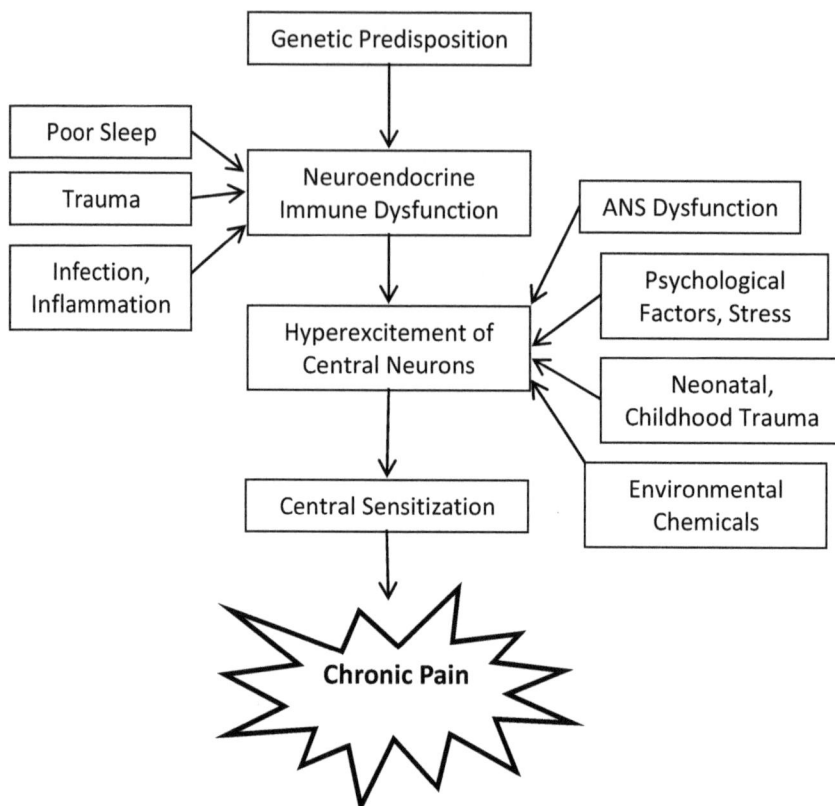

Fig. (2). Neurobiology of chronic pain. Source: Yunus, M.B., Semin Arthritis Rheuma (2007). *38*: 339-358.

TRICYCLIC ANTIDEPRESSANTS AS ADJUVANT PAIN MANAGEMENT

Tricyclic antidepressant (TCA) medication is a functional treatment for neuropathic pain [12]. Tricyclic antidepressant medications such as amitriptyline, nortriptyline and imipramine adherence patterns have been examined. Bordson, Atayee and Best [13] reported a 62% adherence rate from a sampling of 55,296 patients prescribed TCAs for pain. The urinalyses confirmed the adherence and non-adherence. Adherence was found to be highest among older and female subjects with the number/type of other medications not affecting the rate. The urinalysis results provided evidence for the non-adherence that 3% of the general population of prescribers were non-informed of their patients' non-adherence. Table **5** lists the tricyclic antidepressants commonly used for pain. The mechanism of serotonin neurotransmitter release and targeted receptor (5HT 1a) are considered from various levels with each type of antidepressant medication. Antidepressants are classified by targeted receptor, and by targeted pathway. Additionally, the side effect profile is considered when selecting and deselecting an antidepressant as, for example, the sedation side effect may be desired for treatment or not. Further more, about pain and complex pain, the use of Duloxetine (Cymbalta) is considered for the treatment of depression, severe anxiety, fibromyalgia, osteoarthritis, chronic low back pain and diabetic neuropathy. The Federal Drug Administration approved Cymbalta for pharmacological treatment of chronic musculoskeletal pain. While the mechanism of Cymbalta is poorly understood, it is classified as dual action SSRI as it activated receptors for Serotonin and downregulated Norepinephrine neurotransmitter transmission.

Table 5. Tricyclic Antidepressants Commonly Used for Pain.

Amitriptyline
Imipramine (Tofranil) Duloxetine (Cymbalta)
Clomipramine (Anafranil)
Doxepin
Nortriptyline (Pamelor)
Desipramine (Norpramin)

An analysis of the literature in TCA use for pain management revealed similar findings of adherence and utility. Table **5** lists the commonly prescribed tricyclic antidepressants used as adjuvant pain management. In a review of the empirical literature, TCAs have proven to be effective. Table **6** lists a representation of empirical studies of TCA use in pain management. Beakley, Kaye, and Kaye [14] presented the need to carefully monitor TCA usage as serotonin syndrome and side effects dangers are increasing.

CONCLUSIONS / FUTURE DIRECTIONS

Pain conditions are prevalent. The current practices to pharmacologically manage the complicated conditions of pain vary from narcotic use, Nsaid use and tricyclic antidepressants (TCAs) as an adjuvant treatment. Various assessment; methods and treatment approaches to pain conditions by psychologists (additionally physical therapists, occupational therapists). In general, a pain response from the dorsal root pathway to spinal muscles and nerves ascends to the brain stem, mid brain and sensorimotor striation of the cortex. From this central nervous system location, the role of Glutamate: NMDA actions accentuate these actions. Specifically, the neural pathway through the mid brain includes the periaqueductal brain structure, a source of mu receptors. It is believed the inhibitory actions of the TCAs are quite effective in attenuating the pain pathway. Tables were provided that list the types of TCAs and their measured effectiveness as an adjuvant treatment of pain.

Newer approaches in adjuvant treatment include the use of gabapentin (neurotonin). Additionally, Nibbaya, Morinobu and Damen [15] identified the utility of TCA to significantly increase brain-derived neurotropic factor (BDNF) which is believed to promote neuronal survival and protection of neurons from the changing effects of stress such as pain. Zanoveli, Nogueira and Zangrossi [16] identified the role of serotonin, an active component of TCAs, in terms of the serotonin receptor sensitization in the dorsal periaqueductal area resulted in reduced laboratory anxiety responses.

Other future directions include pharmacological investigations of newer narcotic agents and the formulation of non-narcolitic medications.

In summary, given the effectiveness of TCAs as adjuvant treatment of pain, their ongoing use is both measured and recommended.

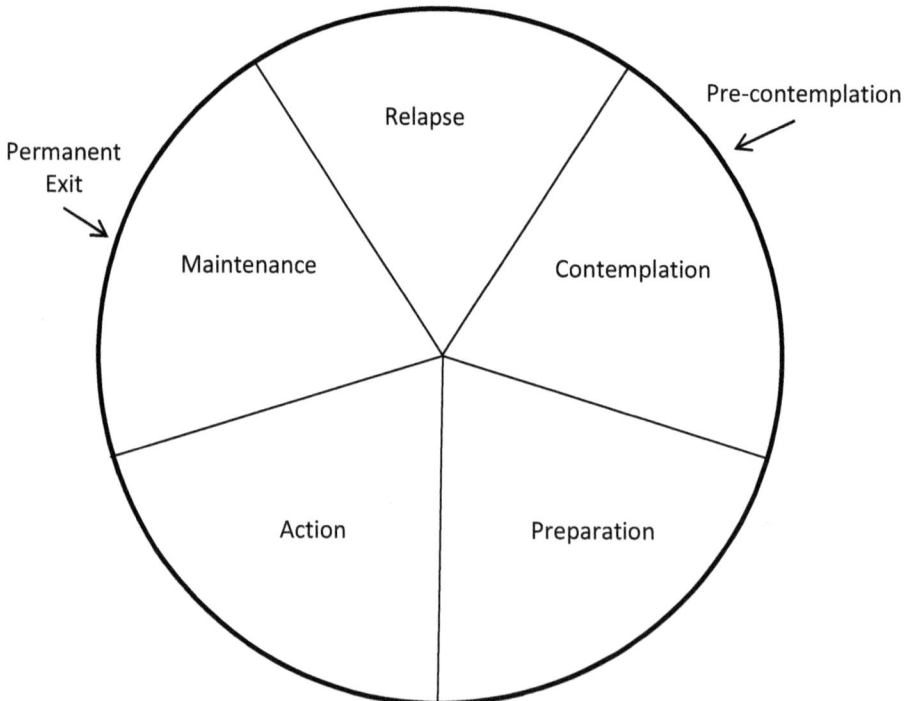

Fig. (3). Cycles of Change.
Source: Prochaska, J.O. & DiClementi, C.C. (1982). *Psychotherapy Theory, Research and Practice, 19*: 276-288.

Table 6. Summary of randomized controlled trial used in systemic reviews of drugs for chronic pain.

Author (year)	Type of study (number of subjects)	Treatment	Control or' reference treatment	Main outcome	Follow-up	Results	Quality	Comments
Antidepressants								
Jenkins et al. (1976)*†	Chronic low back pain (44)	Imipramine	Placebo	Pain	?	U	M	
Goodkin et al. (1990)*†	Chronic low back pain (42)	Trazedone	Placebo	Pain, disability, depression, psychosocial function	?	N	M	Placebo: better physical and psychosocial function
Alcoff et al. (1982)*†	Chronic low back pain (50)	Imipramine	Placebo	Pain, disability, depression	?	P	M	Treatment better in disability, but not pain or depression
Pheasant et al. (1983)*†	Chronic low back pain (18)	Amitriptylene	Placebo	Functional evaluation, number of analgesics	I (6 wk)	P	M	Treatment arm used significantly fewer of analgesics
Ward (1984)*	Community volunteers with depression and low back pain >6 mo (26)	a) Desipramine b) Doxepin		Pain, other symptoms	S (4 wk)	b > a U	?	No difference for depression
Ward (1986)*	Community volunteers with depression and low back pain >6 mo (26)	a) Desipramine b) Doxepin		Pain	S (4 wk)	U	?	Better response in patients with shorter current pain (unclear definition)
Nonsteroidal anti-inflammatory drugs								
Hickey (1992)†	Chronic low back pain (29)	Diflunisal	Paracetamol	Pain	S (4 wk)	P	M	Side effects
Berry (1982)†	Chronic low back pain (37)	Naproxen Diflunisal	Placebo	Pain	S (2 wk)	P	M	Naproxen = reduction of pain
Videman and Osterman (1984)†	Chronic low back pain (28)	Piroxican Indomethacin		Pain	I (6 wk)	U	M	Similar improvement from both drugs, but with side effects
Siegmeth and Sieberer (1978)†	Chronic low back pain (30)	Ibuprofen Diclofenac		Global P	S (4 wk)	U	M	Side effects
Matsumo et al. (1991)†	Chronic low back pain (155)	Ketoprofen Diclofenac			S (2 wk)	N	M	Most patients improved but had side effects
Postacchini et al. (1998)†	Chronic low back pain (369)	Diclofenac	Manipulation Physiotherapy Back school Antidema gel	Global	L (6 mo)	U	L	Side effects
Petri et al. (1987)‡	Painful shoulder	Naproxen Naproxen and lignocaine injection Lignocaine injection	Placebo Triamcinolone and lignocaine injection Saline injection	Pain Range of motion	S (4 wk)	U	M	Improvement in range of motion but not pain
Adebajo et al. (1990)‡	Rotator cuff tendinitis	Diclofenac Triamcinolone and lignocaine injection Saline injection	Placebo Triamcinolone and lignocaine injection Saline injection	Pain Range of motion	S (4 wk)	U	M	Improvement in range of motion but not pain
Muscle relaxants								
Basmajian (1978)§	Unclear (105)	Cyclobenzaprine Diazepam	Placebo	Pain	S (2 wk)	P	H	Not clear which group best
Bercel (1977)§	Unclear (54)	Cyclobenzaprine	Placebo	Pain	S (3 wk)	P	M	
Arbus et al. (1990)†	Chronic low back pain (50)	Tetrazepam	Placebo	Pain	S (4-10 days)	P	M	

S, short term; I, intermediate term (both follow-up); ?, follow-up period not stated; P, positive; U, uncertain; N, negative; H, high quality; M, medium quality; L, low quality; ? (quality), no data on quality of evaluation.
*In systematic review by Turner and Denny (1993). †In systematic review by van Tulder et al. (1997). ‡In systematic review by Green et al. (1998). §In systematic review by Aker et al. (1996).

Reprinted with permission

CONFLICT OF INTEREST

The authors (editor) declares no conflict of interest, financial or otherwise.

ACKNOWLEDGMENT

Declared none.

REFERENCES

[1]   de Gruchy A, Granger C, Gorelik A. Physical therapists as primary practitioners in the emergency department: Six-month prospective practice analysis. Phys Ther 2015; 95(9): 1207-16.
[http://dx.doi.org/10.2522/ptj.20130552] [PMID: 25929528]

[2]   Song S, Graham-Engeland JE, Mogle J, Martire LM. The effects of daily mood and couple interactions on the sleep quality of older adults with chronic pain. J Behav Med 2015; 38(6): 944-55.
[http://dx.doi.org/10.1007/s10865-015-9651-4] [PMID: 26143147]

[3]   Molton IR, Terrill AL. Overview of persistent pain in older adults. Am Psychol 2014; 69(2): 197-207.
[http://dx.doi.org/10.1037/a0035794] [PMID: 24547805]

[4]   Pereira EA, Aziz TZ. Neuropathic pain and deep brain stimulation. Neurotherapeutics 2014; 11(3): 496-507.
[http://dx.doi.org/10.1007/s13311-014-0278-x] [PMID: 24867325]

[5]   Davin S, Wilt J, Covington E, Scheman J. Variability in the relationship between sleep and pain in patients undergoing interdisciplinary rehabilitation for chronic pain. Pain Med 2014; 15(6): 1043-51.
[http://dx.doi.org/10.1111/pme.12438] [PMID: 24716856]

[6]   Barton PM, Schultz GR, Jarrell JF, Becker WJ. A flexible format interdisciplinary treatment and rehabilitation program for chronic daily headache: patient clinical features, resource utilization and outcomes. Headache 2014; 54(8): 1320-36.
[http://dx.doi.org/10.1111/head.12376] [PMID: 24862836]

[7]   McBeth J, Wilkie R, Bedson J, Chew-Graham C, Lacey RJ. Sleep disturbance and chronic widespread pain. Curr Rheumatol Rep 2015; 17(1): 469.
[http://dx.doi.org/10.1007/s11926-014-0469-9] [PMID: 25604572]

[8]   Hemington KS, Coulombe MA. The periaqueductal gray and descending pain modulation: why should we study them and what role do they play in chronic pain? J Neurophysiol 2015; 114(4): 2080-3.
[http://dx.doi.org/10.1152/jn.00998.2014] [PMID: 25673745]

[9]   Alonso-Blanco C, Fernández-de-Las-Peñas C, de-la-Llave-Rincón AI, Zarco-Moreno P, Galán-De--Río F, Svensson P. Characteristics of referred muscle pain to the head from active trigger points in women with myofascial temporomandibular pain and fibromyalgia syndrome. J Headache Pain 2012; 13(8): 625-37.
[http://dx.doi.org/10.1007/s10194-012-0477-y] [PMID: 22935970]

[10]   Blumenfeld H. Neuroanatomy through Clinical Cases. Massachusetts: Sinauer Associates, Inc. 2010.

[11]   Richardson DE, Akil H. Pain reduction by electrical brain stimulation in man. Part 2: Chronic self-administration in the periventricular gray matter. J Neurosurg 1977; 47(2): 184-94.
[http://dx.doi.org/10.3171/jns.1977.47.2.0184] [PMID: 301558]

[12]   Sammons MT. Pharmacological management of chronic pain: I. Fibromyalgia and neuropathic pain. Prof Psychol Res Pr 2004; 35(2): 206-10.
[http://dx.doi.org/10.1037/0735-7028.35.2.206]

[13]   Bordsen SJ, Atayee RS, Ma JD, Bet B. Tricyclic antidepressants: Is your patient taking them? Observations on adherence and unreported use using prescriber-reported medication lists and urine drug testing. DOI: http://dx.doi.org/10.11.11/pmg.12300 355-363. March 1 2014.

[14]   Beakley BD, Kaye AM, Kaye AD. Tramadol, pharmacology, side effects and serotonin syndrome: A review. Pain Physician 2015; 18(4): 395-400.
[PMID: 26218943]

[15]   Nibuya M, Morinobu S, Duman RS. Regulation of BDNF and trkB mRNA in rat brain by chronic electroconvulsive seizure and antidepressant drug treatments. J Neurosci 1995; 15(11): 7539-47.
[PMID: 7472505]

[16]   Zanoveli JM, Nogueira RL, Zangrossi H Jr. Chronic imipramine treatment sensitizes 5-HT1A and 5-HT 2 A receptors in the dorsal periaqueductal gray matter: evidence from the elevated T-maze test of anxiety. Behav Pharmacol 2005; 16(7): 543-52.
[http://dx.doi.org/10.1097/01.fbp.0000179280.05654.5a] [PMID: 16170231]

CHAPTER 5

Biosensors for Detection of Neurotransmitters and Neurodegenerative Related Diseases

Gennady A. Evtugyn[1,2], Tibor Hianik[2,3], Georgia-Paraskevi Nikoleli[4] and **Dimitrios P. Nikolelis[5,*]**

[1] *Analytical Chemistry Department of Kazan Federal University, 18 Kremlevskaya Street, Kazan, 420008, Russian Federation*

[2] *OpenLab "DNA-Sensors" of Kazan Federal University, 18 Kremlevskaya Street, Kazan, 420008, Russian Federation*

[3] *Faculty of Mathematics, Physics and Informatics, Comenius University, Mlynska dolina F1, 84248 Bratislava, Slovakia*

[4] *Laboratory of Inorganic & Analytical Chemistry, School of Chemical Engineering, Dept. 1, Chemical Sciences, National Technical University of Athens, 9 Iroon Polytechniou St., Athens 157 80, Greece*

[5] *Laboratory of Environmental Chemistry, Department of Chemistry, University of Athens, Panepistimiopolis-Kouponia, 15771, Athens, Greece*

Abstract: Biosensors are devices that are composed of recognition element of primary biological nature or mimicking natural receptors, transducer that converts primary chemical signal into the measured physical value (current or voltage) and analyzer that evaluates the physical signal and allows preparing plot of the sensor response *vs.* the concentration of an analyte. Advantage of the biosensor over traditional analytical techniques, such as chromatography, mass spectroscopy and others is its ease of use, fast response and direct detection of analyte without additional labeling. Current trends are focused on using artificial receptors such as calixarenes and DNA/RNA aptamers for recognition of the molecules significant in medical diagnostics, in particularly, related to neurological disorder. This chapter reviews current state of the art in biosensor development toward detection of neurotransmitters using calixarenes and aptamers as well as prion diseases by DNA aptamer-based biosensors. Biosensors for early diagnostics of Alzheimer's disease are considered with particular emphasis to the optical and electrochemical detection. The detection of appropriate biomarkers, *e.g.*, beta- amyloids, tau and ApoE proteins and miRNA in cerebrospinal fluid and blood is characterized from the point of view of biochemical recognition and signal generation principles. Besides, the problem of screening and sensitive detection of Alzheimer's disease drugs based on their anti-cholinesterase effect is discussed with examples of appropriate biosensors.

** **Corresponding author Dimitrios P. Nikolelis:** Laboratory of Environmental Chemistry, Department of Chemistry, University of Athens, Panepistimiopolis-Kouponia, 15771 Athens, Greece; Tel/Fax: +30 210 8957817; E-mail: dnikolel@chem.uoa.gr*

Keywords: Alzheimer's disease, Amyloid-β, Anti-cholinesterase drug, Aptamer, Biosensor, Calixarene, Carbon nanotubes, Dopamine, Neurotransmitters, Prions.

INTRODUCTION

Neurotransmitters are endogenous compounds responsible for regulation of cell signaling in the peripheral and in the central nervous system (CNS). Catecholamines such as dopamine (DA), noradrenaline (also called nor-epinephrine) and adrenaline (also called epinephrine) are important representatives of this class of neurotransmitters. Among catecholamines, DA is of special interest (Fig. **1A**). It is known, that DA is related to Parkinson's disease, Alzheimer's disease, schizophrenia and hyperactivity. DA plays an important role also in the renal, hormonal and cardiovascular systems. It can influence the heart rate and the blood pressure. Low level of DA in CNS is one of important indicators of several neurological diseases. Therefore, there is considerable interest in the measurement of the DA concentration in biological fluids in order to understand possible stage of disease and effectivity of the therapy. The progress in the DA detection by electrochemical biosensors has been recently analyzed by Ribeiro *et al.* [1]. The neurodegenerative diseases (NDD) such as Parkinson's, Alzheimer's, Huntington's diseases (HD) and prion related diseases: Creutzfeldt-Jakob disease (CJD), bovine spongiform encephalopathy (BSE) are accompanied by aggregation of some proteins [2]. The World Health Organization (WHO) estimated that by 2040 NDD will exceed the number of cancer diseases [3]. The success in therapy of these diseases depends on the early diagnosis. The development of fast and effective methods of detection of biomarkers related to these diseases, is therefore in demand. The traditional analytical methods, such as high performance liquid chromatography (HPLC), mass spectroscopy, electrophoresis and others require expensive instrumentation, sample pretreatment, organic solvents and are rather time consuming. Biosensors are an alternative to this approach. Biosensors are devices that are composed of recognition element of primary biological nature or that mimicking natural receptors, transducer that converts primary chemical signal into the measured physical value (current or voltage) and analyzer that evaluates the physical signal and allows preparing plot of the sensor response *vs.* the concentration of an analyte. Advantage of the biosensor over traditional analytical techniques include their easy use, fast response and direct detection of an analyte without additional labeling. Current trends are focused on using artificial receptors such as calixarenes and DNA/RNA aptamers for recognition of the molecules significant in medical diagnostics, in particularly, related to neurological disorder. This chapter reviews current progress in the biosensor development toward detection of DA using calixarenes as well as prion diseases by DNA aptamer based biosensors. Biosensors for early diagnostics of Alzheimer's disease are considered with particular emphasis to the optical and electrochemical detection. The detection of appropriate biomarkers, *e.g.*, beta-amyloids, tau and ApoE proteins and miRNA in cerebrospinal fluid and blood is characterized from the point of view of biochemical recognition and signal generation principles. Besides, the problem of screening and sensitive detection of the Alzheimer's disease drugs based on their anti-cholinesterase effect is discussed with examples of appropriate biosensors.

(A)

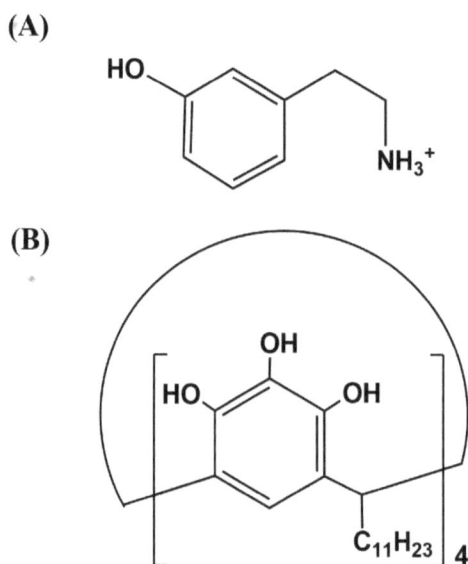

Fig. (1). The structures of (A) dopamine, (B) calix[4]resorcinarene.

The chapter is organized in four subsections. In first one the analysis of current trends in development of biosensors for detection neurotransmitters using calixarenes and DNA/RNA aptamers as receptors is presented. The aptamer-based biosensors for detection prion proteins are described in second subsection. Current trends in treatment of Alzheimer's disease is connected with application of reversible inhibitors of acetylcholinesterase (AChE). The methods of detection these inhibitors by AChE based biosensors are presented in third subsection. The final subsection is focused on biosensors for detection of amyloid- β proteins that are responsible for the formation of senile plaques in brain.

DETERMINATION OF DOPAMINE BY CALIXARENE AND APTAMER-BASED BIOSENSORS

The biosensor technology is rather promising for detection catecholamines. However, so far the focus has been mostly concentrated on the electrochemical detection of these compounds, including DA. This is due to the fact that catecholamines are easily oxidized at various metal interfaces, such are gold, platinum, and on glassy carbon. However, the problem of detection consists in the effect of various interferences. Particularly, ascorbic and uric acid, which are major interferences present in biological liquids show redox properties similar to those of DA and can also be oxidized at the same conditions. Meanwhile. the concentrations of these compounds are much higher than those of DA [1]. Therefore, application of electrochemical DA detection by the sensors that do not include specific receptors is rather difficult. Among possible receptors intended to improve the situation, the calix[n]arenes and DNARNA aptamers are rather effective.

Calix[n]arenes are macrocyclic molecules which originate from the chemical reaction between phenols and aldehydes. The phenolic subunits are connected by methyl groups, which provide their vase-like shape [4]. Other macrocycles containing resorcine or pyrrol subunits have been also synthesized [5]. Hydrophobic cavity of calixarenes is composed by phenolic fragments and serves as a binding site for various low- and macromolecular compounds. The side groups of calixarenes can be modified by various ligands, which enhance their selectivity toward biomolecules. This peculiarity can be advantageously used for preparation of the calixarene-based biosensors. Wang and Liu [6] were among first who reported about amperometric sensor for DA detection based on calix[4]arene. This sensor discriminates DA and epinephrine (EP) signals. Limit of detection (LOD) of DA was 3 μM and has not been affected by ascorbic and uric acids. Further, synthetic calix[4]resorcinarenes (Fig. **1B**) were extensively used in development of amperometric sensors. Several biosensors based on the calix[4]resorcinarenes for the detection of catecholamines have been reported by Nikolelis *et al.* [7 - 13]. The appropriate LODs of these sensors were in micromolar range and the interferences with other compounds such as urea and ascorbic acid was insignificant [7 - 13]. In particular, the DA sensitive calix[4]resorcinarene has been incorporated into the free standing planar bilayer lipid membranes (BLMs) composed of egg phosphatidylcholine (PC) and dipalmitoyl phosphatidic acid (DPPA) [7]. The DA detection was performed by means of transient current measured with electrometric amplifier at a constant potential of 50 mV. The voltage has been applied to BLM through Ag/AgCl electrodes. The scheme of measuring set up is presented on Fig. (**2**).

This sensor rapidly detected DA with response time of about 10 s and sensitivity in μM range. Surface stabilized bilayer lipid membranes (ssBLMs) have also been used for the DA detection [10]. The mixed lipid-calix[4]resorcinarene films were formed on the surface of microporous filter. The stabilization of lipid films was provided by polymerization using methacrylic acid as the functional monomer, ethylene glycol dimethacrylate as a cross-linker and 2,2'-azobis- (2-methyl propionitrile) as initiator (see [10]). The ssBLMs revealed sensitivity similar to that of BLMs. However, their stability was much higher. The interferences with ascorbic acid were minimized due to negative charge of DPPA. The ssBLMs of the same lipid composition with incorporated resorcin [4]arene have been used also for ephedrine (EP) detection [8] with sensitivity similar to that of the DA determination. This sensor was compared with those based on ssBLM with incorporated DNA. Due to its negative charge, DNA accelerated electrostatic interaction with EP. The micromolar range of detection was obtained, but response time was much faster for the ssBLM modified with calix[4]resorcinarene (approx. 20 s) than that of modified by DNA or without any modifications (minutes). A detailed study of the interaction of calix[4]resorcinarene with lipid films has been performed by differential scanning calorimetry [9]. It was confirmed that calix[4]resorcinarene receptors interact with phospholipids by their hydrophobic part, while hydrophilic groups containing in macrocycles with water phase. Stabilized lipid films with incorporated resorcin[4]arene were used also for amperometric detection of DA, EP and adrenaline in flow injection analysis [12]. Injection of the compound studied increased the current. The sensitivity of detection was fond to be several micromoles per liter. Repetitive injections confirmed stable signal. The sensor was validated in urine samples without significant interferences with various compounds like uric acid *etc.*

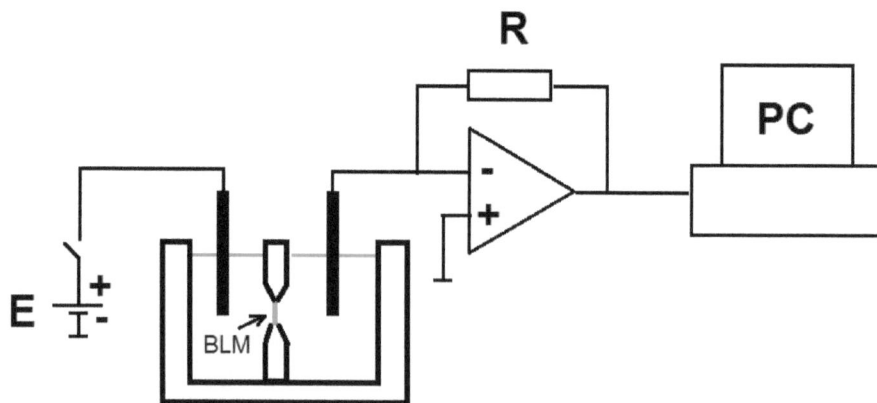

Fig. (2). The set up for measuring the current through BLM modified by calix[4]resorcinarene. The constant voltage E is applied through Ag/AgCl electrodes to the BLM and amplified by current to voltage converting amplifier with high resistance feedback resistor. The output voltage is recorded by PC or by chart recorder.

A novel method of the fluorescence detection of DA in human urine using polymer stabilized lipid films with incorporated calix[4]resorcinarene receptor has been reported [11]. These films provided fluorescence under irradiation with UV lamp. The presence of receptor resulted in fluorescence quenching, but addition of DA switched the fluorescence on. The color of the samples depended on the DA concentration. An increased concentration of DA in the range 10-100 nM resulted in deeper blue color, while samples without DA or at the DA concentration below 10 nM were purple. This could be observed by naked eye or by digital camera. The interference with various compounds such us ascorbic aid, glucose, leucine, glycine, citrate, bicarbonate, caffeine and human albumin was insignificant. It was found that the colors remained stable in the DA containing samples for more than two months. This sensor is suitable for the DA detection in urine of athletes. Similar approach, but utilizing resorcin[4]arene with the OH groups transformed into methoxy groups was reported for the EP detection in human urine [13]. A drop of urine containing EP at the concentration of 10 nM resulted in switching on the fluorescence. Potential interferences like ascorbic and uric acids or DA did not affect the signal at concentrations below 1 µM.

We have also developed the calixarene-based biosensors for detection of DA and EP using electrochemical impedance spectroscopy (EIS) at presence of the redox probe $[Fe(CN)_6]^{3-/4-}$ [14]. In this work, we reported biosensor based on supported lipid films. The sensing layer was formed by liposome fusion at the surface of gold electrode modified by octadecane thiol. The small unilamellar liposomes from soybean phosphatidylcholine (SBPC) were modified with two types of calixarenes: calix[4]arene [14] and calix[6]arene [15]. Addition of DA increased the charge transfer resistance, R_{ct}. This can be due to the fact that DA is incorporated into the calixarene by its positively charged NH_3^+ group, while two hydroxyl groups remain outside in the buffered media. Dissociation of these groups impart a negative charge to the sensor surface, which cause repelling of the anionic redox probe $[Fe(CN)_6]^{3-/4-}$ and hence the increase of the R_{ct} value takes place. Qualitatively similar results were obtained also for the EP. The significant changes of the R_{ct} were obtained already at 1 µM of DA. We also detected DA and EP by means of measurement of the electrical capacitance with no redox probe. The capacitance was determined using frequency response analyzer (FRA) of the AUTOLAB PGSTAT12 (Eco Chemie, The Netherlands) at frequency 1 kHz and ac voltage of an amplitude 5 mV. We have shown that the sensor tends to decrease with increasing concentration of the DA and EP. This can cause from the increase of the thickness of the self-assembled layer (SAM) as a result of the analyte binding to the calixarene. The changes of dielectric permittivity, which also could affect the electrical capacitance, were probably less significant. If the changes of capacitance were due to changes of dielectric permittivity, we should observe increase of the capacitance instead of decrease. However, the sensitivity of the DA and EP detection was much lower in comparison with that based on the EIS method of detection. Significant sensor response was observed at 30 µM of EP and over 100 µM of DA. Much better LOD of DA (0.5 nM) has been achieved in our work using electrochemical sensors based on a composite containing silver nanoparticles and redox active thiacalixarene (Fig. **3A**) [16]. Silver nanoparticles are often used due to their effectivity in catalysis and biomedical applications [17].

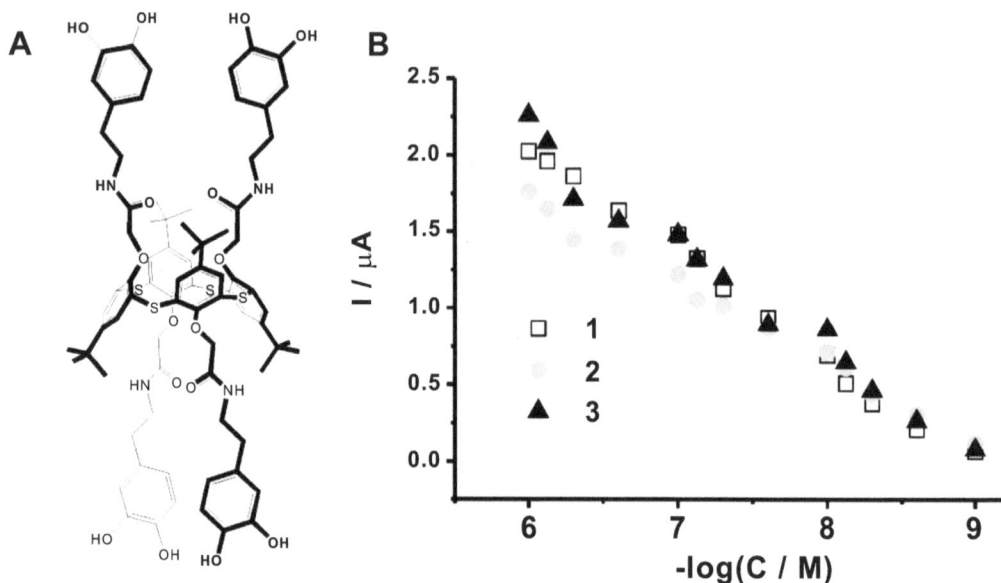

Fig. (3). A. The structure of thiacalixarene. B. The plot of current *vs.* concentration of DA (1) and their mixtures with 10 μM ascorbic (2) and uric (3) acids [16]. With permission of John Wiley and Sons.

Substituted thiacalixarenes can be used for formation of stable composite silver nanoparticles (AgNPs). This has been shown by dynamic light scattering, picrate extraction and potentiometry [18]. We have developed the method of AgNPs synthesis and their stabilization by thiacalix[4]arenes. The DA sensitive sensor was based on glassy carbon electrode modified by above mentioned nanocomposite. In our previous work, we used anodic oxidation of dihydroxyphenolic fragment to the quinone for electrochemical detection of DA. However, the sensor response overlaps those of typical interferences such as ascorbic and uric acids. This problem can be solved by using cathodic signal. The main reason is that the interferences do not exhibit any significant effect on this signal. We used this approach for determination of DA. The current recorded at -700 mV *vs.* Ag/AgCl reference electrode linearly decreased with the logarithm of the DA concentration in the range from 1 nM to 1 μM with the LOD of 0.5 nM (Fig. **3B**). The detection of DA was also performed in the presence of 10 μM of ascorbic or uric acid. Only negligible decrease of the sensitivity of DA detection was observed in this case (Fig. **3B**).

We used also thickness shear mode (TSM) acoustic method for detection of DA using specially synthesized 25,26,27,28-tetrakis (11-sulfanylundecyloxy) calix[4]arene [19]. TSM using quartz crystal transducer with resonant frequency typically in the range of 5 to 10 MHz. In our work, we used 8 MHz transducer. Both side of the traducer are covered by thin gold layers that served as electrodes and sensing surface. However, only one side of transducer is used for preparation of sensing layer. The typical area of this surface is 0.2 cm^2. The high frequency voltage is applied from network analyzer, that measure the complex impedance spectra, *i.e.* impedance as a function of the frequency. Using these spectra, it is possible to determine resonant frequency, f, as well as motional resistance, R_m. The changes in resonant frequency corresponding to the mass or thickness chan-ges, while motional resistance reflects the viscosity contribution related to the fric- tion between sensing layer and surrounding liquid. The simple setup of the TSM is presented on Fig. (**4**). The sensing layer was formed as follows. The calixa- renes or their mixtures with 1-dodecanethiols (DDT) or hexadecanethiols

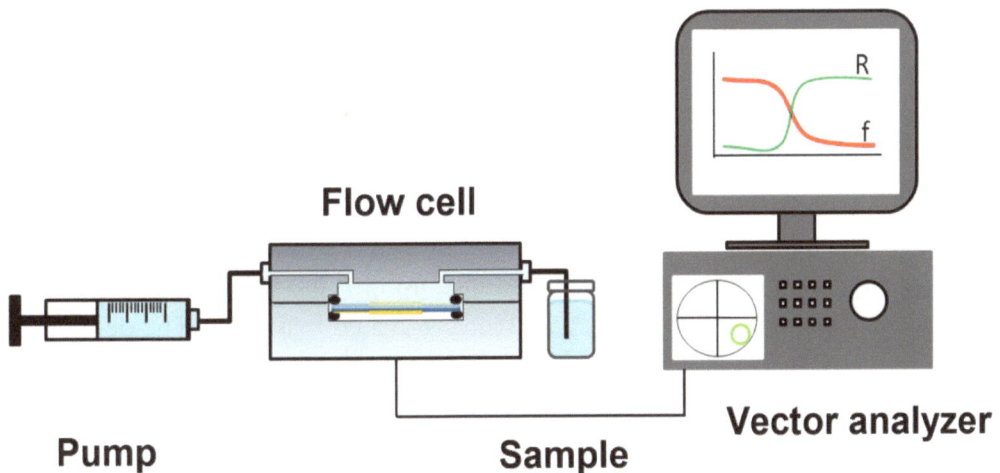

Fig. (4). The experimental set up of thickness shear mode method (TSM). The sample is applied by syringe pump on the surface of TSM transducer. The high frequency voltage is applied from vector analyzer that serves also for signal analysis.

(HDT) formed compact SAM at gold surface of TSM transducer by chemisorption. Addition of DA resulted in the decrease of resonant frequency, f, and in increase of motional resistance, R_m, which is an evidence of the increase of the mass or alternatively in increase of the viscoelastic contribution. This sensor determined DA with sensitivity of 50 pM [19]. Novel approach of DA detection by calixarene based biosensor was proposed recently by Kurzatkowska *et al.* [20] who explored the ability of some calixarenes to form ionic channels. They proposed amperometric ion-channel sensor using the calix[4]arene mercaptoalkyl derivatives chemisorbed at the gold electrode surface. The redox marker $[Ru(NH_3)_6]^{3+}$ has been used for tracking the interactions between calix[4]arene derivative and DA. The advantage of this redox marker consists in the oxidation potential well separated from that of DA. The detection of DA was based on Ostryoung square-wave voltammetry, which allow reducing the capacitive current. Rather high sensitivity was obtained with the LOD of 4.9 pM. Ascorbic and uric acids did not affect the DA detection. Simultaneous electrochemical detection of DA and uric acid though less sensitive than those reported in Ref [20] was published by Wang et al. [21] who used C-undecylcalix[4]resorcinarene (CUCR) film deposited on the glassy carbon electrode by Langmuir-Blodgett method. The sensor allowed detection of the DA in the presence of ascorbic acid with the LOD of 20 nM.

Except calixarenes, the DNA aptamers have been also used for electrochemical DA detection. Aptamers are single-stranded oligonucleotides that are developed in vitro by SELEX (Systematic Evolution of Ligands by EXponencial enrichment) protocol for a variety of low- and macromolecular compounds. Aptamers in a solution of optimal composition fold and create binding site for the target. The affinity of aptamers is in the range from micromoal to sub-nanomolar values. This is comparable with those of antibodies. Aptamers are, however, more stable and flexible in comparison with them. Aptamers mofified by thiol, amino or biotin groups can be easily adsorbed at various surfaces where serve as the receptors. Modification of aptamers by redox, optical labels, quantum dots and nanoparticles can help in signal detection and amplification. Potyrailo et al. [22] reported first aptamer based sensor for thrombin detection. In this work, the fluorescently labeled aptamer was covalently attached to a glass. The detection of thrombin was performed by measuring changes in the evanescent wave-induced fluorescence anisotropy. The aptasensors were applied so far for the detection of a plenty of molecules including catecholamines (see [23, 24] for review).

The DA-sensitive electrochemical aptasensor was proposed by Liu et al. [25]. The 58-mer DA-sensitive aptamer of the composition 5'-GTC TCT GTG TGC GCC AGA GAA CAC TGG GGC AGA TAT GGG CCA GCA CAG AAT GAG GCC C-3' was immobilized at the surface of glassy carbon electrode covered by graphene–polyaniline (GR–PANI) nanocomposite. The determination of DA was based on the measurement of the peaks of square wave voltammetry (SWV) using the redox probe $[Fe(CN)_6]^{4-/3-}$. This aptasensor allowed detection of DA with LOD of approx. 2 pM. The sensor revealed good selectivity at the presence of other interfering compounds like ascorbic acid, tyramine and L-3-hydroxytyrosine as well as low matrix effect during validation of the sensor in human blood serum. The recovery was in the range of 99-105%. The detection of DA has been performed also in a volume using gold nanoparticles modified by the same DA specific 58-mer aptamer. For this purpose, the fluorescence resonance energy transfer (FRET) between rhodamine B (RB) and gold nanoparticles (AuNPs) has been measured [26]. The aptamers protect AuNPs from salt-induced aggregation, resulting in the fluorescence quenching of RB *via* FRET. Binding of DA to the aptamer in the aggregation of AuNPs was indicated by decrease of fluorescence quenching. The sensitivity of the DA detection in this assay was 2 nM, which is much less in comparison with above mentioned electrochemical sensor. Simple electrochemical DA-sensitive aptasensor was reported by Zhou *et al.* [27]. 58-mer thiolated aptamers were immobilized at gold surface by chemisorption. Methylene blue (MB) has been used as a redox probe. MB is known to intercalate between guanine-rich bases. At presence of the DA, the MB was released from the guanine quadruplex of the aptamer which resulted in decrease of the current measured by SWV. The LOD was 1 nM. Similar LOD of 2.1 nM has been achieved by aptasensor based on gold electrodes modified by dispersed AuNPs and oxidized carbon nanotubes (CNTs-COOH) functionalized with polyethyleneimine (PEI). Amine-terminated ssDNA1 ($3'$-NH_2-$(CH_2)_6$-CAG AGA CAC ACG-$5'$) served as a capture probe was immobilized at the surface and DA-sensitive aptamer (ssDNA2) (58-mer aptamer [25]) hybridized with the probe. The interaction between aptamer and DA resulted in release of MB from the sensor surface which decreased the current. Biosensor was validated in blood serum samples [28]. Label-free electrochemical aptasensor based on immobilization of 57-mer thiolated DNA aptamer to a gold support, however, showed much lower sensitivity with LOD of 26.8 μM. The sensitivity was improved using another method of aptamer immobilization on AuNP/reduced graphene oxide-modified glassy carbon electrode (LOD 3.36 μM [29]). Even in this case the sensitivity remained much lower in comparison with the sensors utilizing redox probe. Thus, the sensitivity of the DA detection of most sensitive aptasensor [25] was comparable with that of calixarene-based sensor [20].

APTAMER BIOSENSORS FOR DETECTION OF CELLULAR PRIONS

The review on affinity biosensors for detection prion proteins has been recently published [30]. In this paper, we also compared the sensitivity of antibody and aptamer based biosensors and showed that aptasensors have comparable and in some cases even better sensitivity than antibodies. In this part, we focus on description of aptasensors for detection cellular prions with updating new results that appeared recently. RNA aptamers that selectively bind recombinant hamster [31], human prions [32, 33] bovine PrP fragment [34] and DNA aptamers specific to cellular prions (PrP^C) [35] has been published. DNA aptamers are more advantageous over RNA aptamers due to their higher stability. Additional advantage of DNA aptamers is possibility of their easy molecular engineering. By means of hybridization of an aptamer supporting part it is possible to prepare homo- or heterodimers. So-called aptabodies have been used for the first time in our work for detection of thrombin. The aptamers were immobilized at the surface of multiwalled carbon nanotubes (MWCNTs) and the interaction with thrombin was monitored by electrochemical quartz crystal microbalance (EQCM) [36]. We showed, that aptabodies revealed about twice higher sensitivity for thrombin detection in comparison with conventional single stranded aptamers. We used similar approach for detection of the cellular prions PrP^C. The DNA aptamer reported in [35], but with extended supporting part by 16 thymidines and modified by biotin (5'-CGG TGG GGC AAT TTC TCC TAC TGT TTT TTT TTT TTT TTT-3'-biotin) served as receptor specific to recombinant PrP^C. In solution, these aptamers form loop-like structure (Fig. **5A**). We used EQCM method for detection of aptamer-prion interactions. The sensitivity of aptasensor has been compared with those of monoclonal antibodies specific to PrP^C such as BAR 233 and PRI 308 (SPIbio, Montigny, France) that bind to 141-152 and 106-126 amino acid residues of $recPrP^C$, respectively. 8 MHz QCM transducers covered by thin gold layer (working area 0.2 cm^2) and modified by MWCNTs have been used for immobilization of aptamers and antibodies. The sensor has been inserted into the Teflon EQCM cell (volume 1.5 ml). EQCM method allows simultaneous measurement of cyclic voltammograms and oscillation frequency. Application of cycled voltage was used for improving immobilization of receptors onto the MWCNTs surface. The plot of frequency changes, Δf, *vs.* PrP^C concentration for aptamer and antibody based sensors is on Fig. (**5B**).

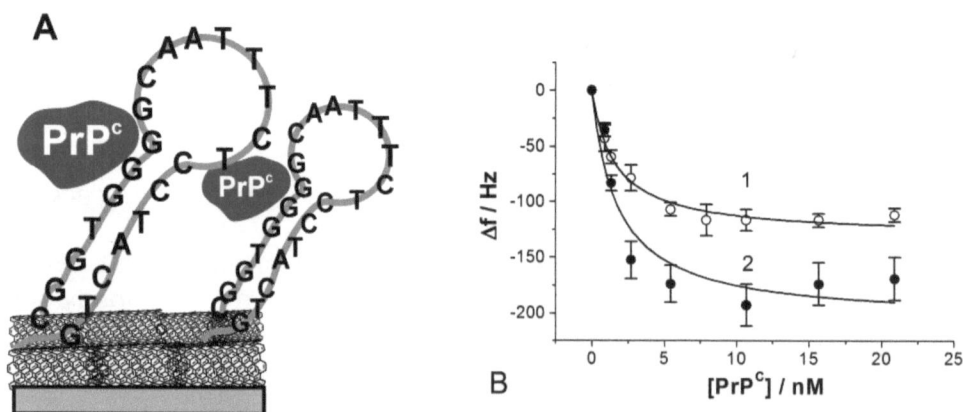

Fig. (5). A. Configuration of aptamers sensitive to PrP^C and scheme of their immobilization at MWCNTs. B. The dependence of frequency changes on PrP^C concentration for (1) aptamer and (2) antibody BAR 223 based sensors. The results represent mean ± SD obtained from 3 independently prepared sensors. Lines are fits according to Langmuir isotherm [37].

The sensor response was higher than those based on BAR 223 antibodies, while similar sensitivity was obtained for PRI 308 and aptasensor. Table **1** summarizes sensitivity of so far published antibody and aptamer based biosensors for prion detection.

Table 1. The comparison of PrP^C detection by antibody and aptamer based sensors of various detection and immobilization methods [30].

Method of detection	Sensing support	LOD, pM	Linear range	Response time	Ref.
Immunosensors					
Resonant mirror, direct	CMD/Plasminogen	2400	N/A	1 min	[46]
SPR, direct	Au/Ab	570	1-200 nM	3-5 min	[47]
TSM, direct	Au/PAMAM-HDT/PrA/Ab	800*	1-25 nM	30 min	[48]
EQCM, direct	Au/MWCNT/Ab	20**	0-5 nM	10 min	[37]
Aptasensors					
EQCM, direct	Au/MWCNT/APT	50	0-5 nM	10 min	[37]
SPR, direct	Au/PPY/Bio/SA	4000	0-400 nM	5 min	[39]
Amperometry, ferrocene as a label	Au/MWCNT/PAMAM/ Fc/APT	0.5	1 pM-1 µM	N/A	[40]
Amperometry, ferrocene as a label	Au/PAMAM/Fc/APT	0.8	1 pM-1 nM	N/A	[41]
Light scattering assay, gold nanoparticles	AuNPs/APT	70	0-50 nM	N/A	[43]
Fluorescence, direct assay	-	2000	0-2 µM	N/A	[44]
Metal-enhanced fluorescence, direct assay	AgNPs-SiO2/APT	50	0-300 pM	N/A	[45]

*Most sensitive biosensor based on PRI 308 Ab, **BAR 223 Ab. Ab – antibody; APT – aptamer; Au – gold; AuNPs – gold nanoparticles; Bio – biotin; CMD – carboxymethyl-dextran; Fc – ferrocene; HDT – hexadecanethiol; PAMAM – polyamidoamine dendrimer; PPY – polypyrrole; PrA – protein A; SA – streptavidin.

The selectivity of the sensor response has been analyzed at presence of 20 µM human serum albumin (HSA). It can be seen from Fig. (**5**) that substantially less changes in the frequency took place. Thus, the sensitivity and selectivity of the developed biosensors is sufficient for detection PrP^C in blood, considering that physiological concentration of this protein is around 20 nM [38]. We reported also the PrP^C aptasensor formed at the polypyrrol polymerized layer (PPY) modified by streptavidin. The biotinylated aptamers of various configuration - single stranded conventional aptamer, aptamer dimers and single stranded aptamer with supporting rigid part formed by hybridization of short nucleotide sequence. The latter aptamer configuration provided best sensitivity (LOD 4 nM) probably due to best orientation of binding site toward target. It has been possible to regenerate this sensor using 0.1 M NaOH [39].

The advantage of mass sensitive acoustic aptasensors is that they need no labels that for detection of PrP^C. Label-free electrochemical aptasensors are also possible, but for this purpose mostly EIS is used. However, even in this case the redox probe such as $[Fe(CN)_6]^{3-/4-}$ is required for monitoring changes in the charge transfer resistance. Amperometric aptasensors, however, require certain labeling, using, for example, methylene blue or ferrocene as a reporter groups. In many cases, such labeling resulted in improved sensitivity of the detection [23]. In our recent work, we modified poly-amidoamine dendrimers of fourth generation (PAMAM G4) by ferrocene derivative contained two phtalamido groups. Dendrimers were covalently attached to a surface of MWCNTs and biotinylated aptamers were linked through streptavidin to the dendrimers, so the ferrocenes were located between aptamers and dendrimers. Addition of PrP^C resulted in decrease of the redox current of ferrocene. This approach allowed very high sensitivity of PrP^C detection (LOD 0.5 pM) in a broad linear range of 1 pM - 10 µM. The sensor was validated in a spiked human blood plasma at the same concentration range with recovery of 85% at 1 nM PrP^C [40]. When the dendrimers were immobilized instead of MWCNTs on the PPY layer, similar sensitivity of detection was obtained (0.8 pM). Sensor validation in a spiked human blood plasma demonstrated good recovery of 90% [41]. Ultra-high sensitivity of detection prion proteins has been recently demonstrated using field effect transistors (FET)-based sensor. The prion proteins were adsorbed at thiamine-modified surface of FET. Chemical amination reaction resulted in transformation of anionic to cationic groups of protein caused by variation of the PrP^C charge. The sensitivity of detection of PrP^C by this method was high: 0.4 fM [42]. However, aptamers were not used in this work.

Optical methods are also suitable for detection of proteins using aptamers as receptors. The approved approach based on AuNPs modified by PrP^C sensitive aptamers has been reported [43]. At presence of PrP^C, the aggregation of nanoparticles occurred which caused increase of the light scattering Using this method, it was possible to detect PrP^C with the LOD of 70 pM in the range from 0.2 to 50 nM. Xiao *et al.* [44] reported fluorescence detection of PrP^C using approved assay in which aptamer is in loop conformation. 5' end was modified by fluorophore and 3' end by quencher. In this work, the tetramethyl-6-carboxyrhodamine (TAMRA) served as fluorophore and guanines as quenchers. In loop conformation (without PrP^C) the fluorophore and quencher were close to each other and fluorescence was minimal. Addition of PrP^C resulted in conformational changes of the aptamer that caused increase of distance between fluorophore and quencher and hence increase of fluorescence took place. This assay allowed detection of PrP^C with LOD of 2 nM. Hu *et al.* [45] proposed a metal enhanced fluorescence (MEF) assay for detection PrP^C. In this method PrP^C sensitive DNA aptamers (APT1) were immobilized on SiO_2 outer layer of silver nanoparticles (AgNPs). Second aptamer (APT2) has been modified by fluorophore cyanine 3 (Cy3). Addition of PrP^C induced formation of the complex NPs-APT1- PrP^C -APT2 and increase of fluorescence due to MEF. This method allowed detection of PrP^C in the range of 0.05–0.30 nM including those in complex sample of mice brain homogenates.

Thus, comparison of the aptamer and antibody based sensors summarized in Table **1** clearly demonstrates advantage of aptasensors. Especially high sensitivity of electrochemical aptasensors with incorporated ferrocenyl probe are promising also for detection of pathogenic prions, PrP^{Sc}, the concentrations of which is much lower (sub pM level) in comparison with those of PrP^C.

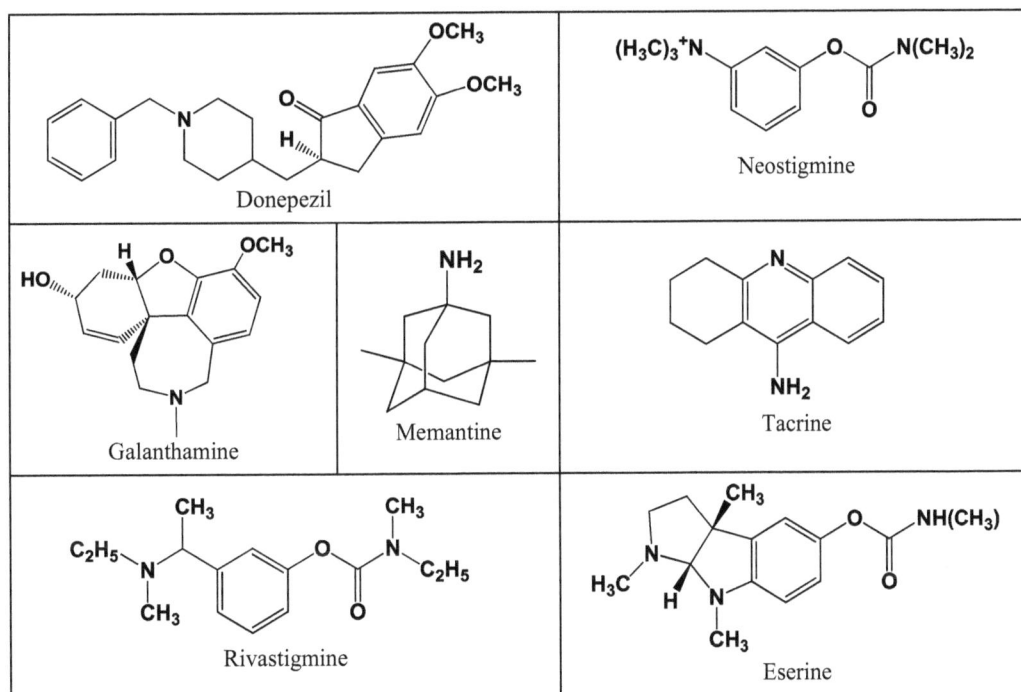

Fig. (6). Anti-cholinesterase drugs used for Alzheimer's disease treatment.

ANTI-CHOLINESTERASE DRUGS

Reversible inhibitors of acetylcholinesterase (AChE) have found increasing attention in the past decades as drugs applied in Alzheimer's disease treatment

[49]. Their use is related to the deterioration of cholinergic neurons of the basal forebrain and neocortex resulted in decreased production of acetylcholine [50]. Accordingly, the AChE inhibitors restore the neurotransmitter level by suppression of its enzymatic hydrolysis [51]. During the past decade, donepezil, neostigmine, galantamine and the *N*-methyl-*D*-aspartate (NMDA) antagonist, memantine, have been approved for treatment of the Alzheimer's disease. The chemical structures of most frequently used formulations are presented in Fig. (**6**).

Considerable interest remains to screening new formulations of known drugs, searching new similar preparations especially taken into account adverse effects often observed for known formulations [52]. Biosensor format assuming compact design, simple and reliable response and application within point-of-care diagnostics concept offers new opportunities for these purposes. To some extent, AChE biosensors can be considered as an alternative to preliminary *in vitro* tests [53] that are commonly conducted on laboratory animal models, which are more expensive, time and labor consuming and rely on behavioral reactions of animals. Conventional method of the AChE activity measurement includes incubation of an enzyme with the synthetic substrate, acetylthiocholine (ATCh), followed by addition of 5,5'-dithiobis-2-nitrobenzoic acid (Ellman reagent [54]) and spectrophotometrical monitoring of yellow product (1) at 412 nm.

$$(CH_3)_3N^+CH_2CH_2SCCH_3 + H_2O \xrightarrow{\text{AChE}} (CH_3)_3N^+CH_2CH_2SH + CH_3COOH$$

Thiocholine

$\xrightarrow{\text{RS-SR Ellman reagent}} RS^-$

$(CH_3)_3N^+CH_2CH_2S\text{-}SR$

$R = O_2N-\text{(benzene ring)}-$, ^-OOC

(1)

Biosensors being much faster and cheaper in mass screening provide a promising alternative to spectrophotometric assay. Immobilization of enzyme offers its multiple use and reduced volume of reagents as well as one step measurement protocol. Many AChE biosensors have been proposed in the last several decades for inhibitor detection [55 - 58]. Among other approaches, oxidation of thiocholine, either direct or mediated, is most frequently used. Fig. (**7**) outlines the measurement scheme with the AChE immobilized on the modified electrode and current of the oxidation referred to the product of enzymatic reaction.

The reaction scheme (2) involves the enzymatic hydrolysis of the ATCh followed by oxidation of thiocholine to disulfide. The use of carbonaceous materials as well as implementation of appropriate mediators accelerate the electron transfer and decrease the working potential from 450 mV to 150-250 mV. The latter one increases the selectivity of the response due to possibility to record oxidation current in the presence of oxidizable components of the samples.

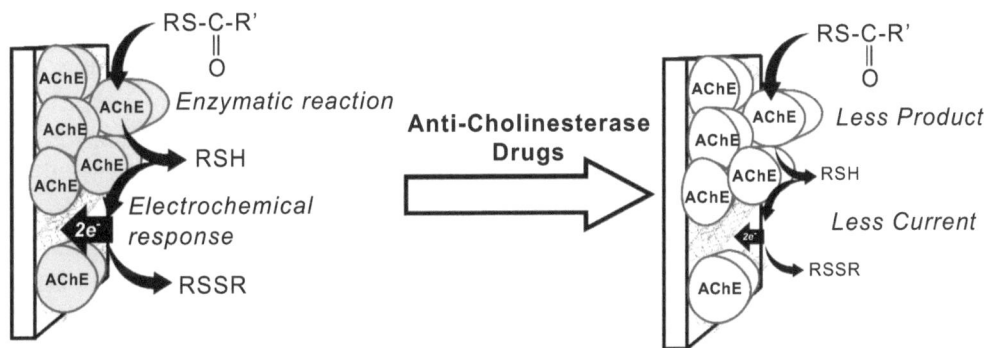

Fig. (7). The AChE biosensor for detection of anti-cholinesterase drugs. The current recorded corresponds to the oxidation of thiol formed in enzymatic reaction. Lesser rate of enzymatic hydrolysis due to inhibition results in smaller currents recorded.

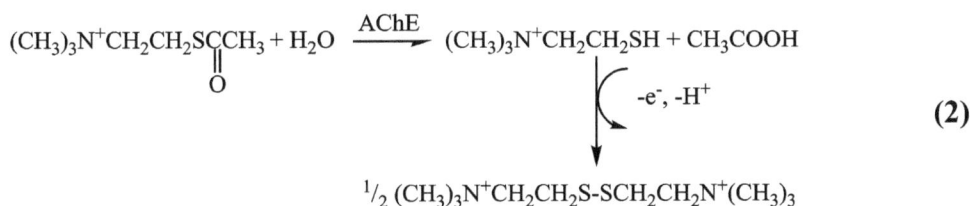

$$(CH_3)_3N^+CH_2CH_2SCCH_3 + H_2O \xrightarrow{\text{AChE}} (CH_3)_3N^+CH_2CH_2SH + CH_3COOH$$
$$\overset{\|}{O}$$

$$\text{-e}^-, \text{-H}^+$$

$$(2)$$

$$^1/_2 (CH_3)_3N^+CH_2CH_2S\text{-}SCH_2CH_2N^+(CH_3)_3$$

Co phtalocyanine [59], Prussian blue [47], tetracyanoquinodimethane (TCNQ) [60], AgNPs [61] are described in the assembly of the AChE biosensors. They are often coupled with carbon nanotubes or carbon black to increase real working surface and simplify immobilization of the enzyme. For acetylcholine (ACh) detection, choline obtained in enzymatic hydrolysis can be oxidized to betaine in the presence of second enzyme, choline oxidase [62]. Electrochemical conditions of signal measurements are mainly selected using direct current voltammetry. However, other protocols, *e.g.*, DPV or SWV, were then used for inhibitor quantification after the selection of working potential. Table **2** summarizes the performance of the AChE biosensors developed especially for Alzheimer's disease drugs.

The use of small volumes of reactants preliminary mixed and then analyzed by electrochemical methods makes it possible to directly compare the inhibition efficiency of free and immobilized enzyme. Meanwhile, entrapment in polymeric materials (sol-gel immobilization) or covalent immobilization by carbodiimide binding are preferable for routine experiments due to reduced cost, faster response and reasonable storage duration. Regarding inhibition measurements, selection of immobilization protocols and especially enzyme matrix can affect the accessibility of enzyme active site and relative ratio substrate/inhibitor molecules competing for enzyme active site [58], but no significant deviation of inhibition efficiency was found at the moment (see Table **2**). Regarding quantification of inhibition, most of the results compare IC_{50} values, *i.e.*, concentrations of an inhibitor resulted in 50% decrease of the signal. For competitive inhibition, this value is equal to the inhibition constant but generally it is used as empirical estimate of relative inhibitor strength. Recently, high potential of application has been shown for biosensors based on ion-selective FETs [69 - 71]. In these devices, a current between two doped areas on Si support is measured. If this distance, Si is covered with oxides or nitrides able to interact with hydrogen ions (SiO_2, Ta_2O_5, Si_3N_4, GaN), pH changes result in the changes of charge and hence the resistance. For this reason, the current recorded becomes sensitive to pH of the solution. Co- immobilization of AChE makes it possible to detect the enzyme activity and hence the quantities of inhibitors. Due to miniature size of such a sensor, ISFETs can be implanted near the neuronal issue and control the ACh release and effect caused by enzyme inhibitor (donepezil [70]). Moreover, similar FET with applied for detection of salivary biomarkers of neurodegenerative diseases (trehalose [71]).

Table 2. Comparison of analytical characteristics of reversible AChE inhibitors determination with appropriate biosensors.

Inhibitor	Immobilization protocol	Signal parameters	Analytical Characteristics	Ref.
Galanthamine, neostigmine	Entrapment in the silicate gel together with chitosan and Au nanoparticles	Oxidation peak current in DC mode, +680-800 mV	K_m 0.14 µM (galanthamine) and 0.19 µM (neostigmine), respectively (AChE from electric eel)	[63]
Donepezil	AChE together with ATCh and inhibitor in phosphate buffer solution	DPV oxidation peak on Au screen-printed electrode at 0.16 V	IC_{50} 28±7 nM, complete inhibition 150 nM (AChE from human erythrocytes)	[64]
Coumarin derivatives, galanthamine	AChE from bovine erythrocytes attached by glutaraldehyde to silica capillary	UV-viz detection after Ellman reagent addition at 412 nm	IC_{50} 12.68±2.40 µM (galanthamine), 17 and 6 µM for potent inhibitors, coumarin derivatives	[65]
Neostigmine, eserine, tacrine, donepezil, rivastigmine, pyridostigmine, galanthamine	AChE from electric eel immobilized on Au disk attached to Ag working electrode	Flow-injection measurements in the range from 0.1 to 10 µM of inhibitor, DC voltammetry at 80 mV	I_{50} 0.50 (donezepril), 0.71 (neostigmine), 0.77 (serine), 1.37 (takrine), 5.45 (berberine), 6.45 (rivastigmine) µM	[66]

(Table 2) contd.....

Inhibitor	Immobilization protocol	Signal parameters	Analytical Characteristics	Ref.
Tacrine	AChE from electric eel, carbodiimide binding to magnetic beads	Screen-printed carbon electrodes with Prussian blue, SWV, current peak square at 1.2-1.4 V	LOD 8.1 μM tacrine, suppression of interferences related to organic solvents	[67]
Donepezil	AChE from electric eel mixed with substrate/inhibitor and moved in the working cell	DPV on carbon paste electrode: peak at 0.5 – 0.75 V	Inhibition up to 80%	[68]

AMYLOID BIOSENSORS

The formation of senile plaques in brain of patients is one of the consequences of Alzheimer's disease that results in memory loss and cognitive declines. Amyloid-β proteins consisting of 39-43 amino acid resides [72] is main constituents of senile plaques. The amyloid cascade theory suggests that such proteins initiate cellular events leading to pathological consequences of Alzheimer's disease [73]. The decrease of amyloid level in cerebrospinal fluid can be caused by its aggregation and plaque deposition assuming its conversion to oligomeric form and finally to amyloid fibril [74]. Hence, increased level of amyloid oligomers especially those containing 50-100 monomers can be regarded as molecular biomarker of Alzheimer's disease and a target for therapeutic treatment [74 - 76]. A simple and reliable electrochemical determination of oligometic amyloid-β protein based on its binding with PrP(95–110) peptide, which is a part of cellular prion protein has been described [77]. The thiolated receptor molecule was attached to golden electrode. After the target binding, the complex was treated with a conjugate of the same PrP(95–110) protein with alkaline phosphatase. The activity of the enzyme was quantified electrochemically by mediated oxidation of ascorbic acid released from its phosphate in the presence of ferrocene methanol. The biosensor makes it possible to determine from 5 pM to 2 nM of oligomeric amyloid-β (LOD 3 pM). The same receptor has been utilized in biosensor with layer-by-layer assembling of recognition layer consisting of electrochemically deposited copolymer of tyramine and 3-(4-hydroxyphenyl)propionic acid [78]. The biotinylated PrP(95–110) peptide was attached to the support by avidin-biotin binding. The target interaction with oligomeric amyloid-β resulted in an increase of the charge transfer resistance measured by EIS. The concentration ranges from 1 pM to 1 μM has been achieved with the LOD of 0.5 pM. What more important is that the biosensor discriminated the response caused by monomeric and oligomeric forms of the amyloid-β protein. Surface electron exchange properties were also monitored with Au electrode modified with monolayer of thioundecanoic acid bearing terminal M13 peptide labeled with ferrocene [79]. The protein M13 specifically binds water soluble oligomer of amyloid-β (1-42). Implementation of the target molecules prevents electron exchange between the ferrocene groups and electrode and this results in significant suppression of the ferrocene peak current measured in SWV regime. Linear dependence of the signal corresponds to the concentration range from 480 pM to 12 nM. Monomeric amyloid-β protein was measured with FET involved carbon nanotubes in gate zone [80]. The reaction of the FET on the target species was recorded using source-drain current shift affected by the charge of the carbon nanotubes walls. For specific response, outer membrane of *Escherichia coli* with autodisplayed Z-domains of protein A from *Staphylococcus aureus* was immobilized on the walls by direct adsorption. The conductance of the gate area increased with the analyte concentration within $1 \times 10^{-12} - 1 \times 10^{-9}$ g/mL with the LOD of 1 pg/mL corresponded to 5% shift of the signal. Sialic acid receptor has been used for selective binding and recognition of amyloid-β (1-40) and (1-42) [81]. Screen-printed carbon or golden strip electrodes were first electrochemically covered with AuNPs. Then, self-assembled monolayer was formed on Au with acetylenyl terminated group involved in the reaction with azide linked sialic acid that was attached to this linker by click chemistry (1.3-dipole cycloaddition). The oxidation peak current attributed to amyloid-β was observed at 0.6 V (tyrosine oxidation) starting from 0.5 µM and with saturation level at 10 µM. No significant difference in the influence of amyloid-β (1-40) and (1-42) was reported. Selectivity of the response was estimated using insulin and BSA which did not alter tyrosine oxidation peaks in the similar concentration interval. Competitive electrochemical immunoassay has been realized with gold electrode covered with specific capture antibodies to amyloid-β [82]. The scheme of competitive electrochemical immunoassay is illustrated in Fig. (**8**).

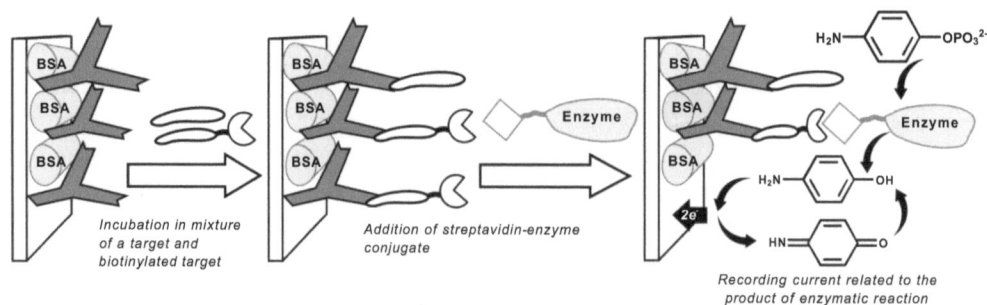

Fig. (8). Electrochemical competitive immunoassay with alkaline phosphatase label.

The electrode was incubated in the sample mixed with biotinylated amyloid-β and then with a conjugate of alkaline phosphatase and streptavidin. The activity of the enzyme was detected with p-aminophenylphosphate as substrate. The redox cycle of *p*-aminophenol and imine quinone was recorded at 0.25 V. To avoid non-specific sorption, the electrode is additionally treated with bovine serum albumin (BSA). The immunosensor can determine either amyloid-β 1-42 protein or total amount of all amyloid oligomers in the sample tested. The LOD was about 5 pM for both targets. Summarizing the results obtained in biosensor format, it should be mentioned that most of the protocols cited make it possible to directly detect amyloid oligomers at their level typical for Alzheimer's disease. The appropriate LOD values are significantly lower or comparable with characteristics of conventional measurement devices, *e.g.*, electrophoresis or SPR techniques. Meanwhile, due to compact design and small working area the biosensors require much lower volumes of biological liquids and are much cheaper even though most of the biosensors are intended for single use. The measurement time remains a weak point of all the biochemical techniques considered. The specific binding requires up to 20 hours so that biosensor format does not provide significant improvement of productivity. Regarding the diagnostics prospects, the discrimination of the signals related to monomeric and oligomeric amyloid peptides is important contribution to the reliable conclusion on neurodegenerative processes and their separation from the influence of other factors like age or other forms of dementia. It is important that many of the biosensors described here were validated on the samples of blood and cerebro-spinal fluid where the content of amyloid-β oligomers correlates with the mini-mental state examination (MMSE) score [83].

CONCLUDING REMARKS

Biosensor technology is rather promising tool in medical diagnostics. Important issue consists in selection of suitable receptors, protocols of their immobilization at transducer surface and the detection of specific interactions. We described several approaches to the development of biosensors for detection neurotransmitters, mostly DA, and for diagnosis of prion-related diseases. Among natural receptors, specially synthesized calixarenes and DNA aptamers are under intensive investigation directed to the development of sensitive and robust biosensors. In some cases, the biosensor recovery is achieved after its contact with biological materials. This increases the probability of their commercialization and medical application. Nevertheless, the distance from laboratory prototypes to conventional diagnostics devices massively produced is still long because of the numerous problems, *e.g.*, insufficient selectivity of direct measurements in body fluids, moderate sensitivity and lack of experience in implantable biosensors. Even in the areas with good laboratory practices, special efforts are required for enhancing the number of specific receptors. The development of pathogen prions (PrP[Sc]) is a good example concerning detection of non-infectious cellular prions

(PrP^C). The required limits of their detection in blood on pM level do not allow using diluted samples necessary to avoid interferences with common components. Most sensitive protocols of biomarker detection require labor consuming stages and specific biochemical equipment and hence are less suitable for real time applications. Integration of sample treatment and signal measurement in lab-on-a-chip sets seems a most probable solution that offers new applications area of such biosensors in the future.

CONFLICT OF INEREST

The authors (editor) declares no conflict of interest, financial or otherwise.

ACKNOWLEDGMENTS

The work is performed according to the Russian Government Program of Competitive Growth of Kazan Federal University. This work was supported by the Slovak Research and Development Agency under the contract No. APVV-14-0267 and by Science Grant Agency VEGA (project No. 1/0152/15).

REFERENCES

[1]   Ribeiro JA, Fernandes PM, Pereira CM, Silva F. Electrochemical sensors and biosensors for determination of catecholamine neurotransmitters: A review. Talanta 2016; 160: 653-79.
[http://dx.doi.org/10.1016/j.talanta.2016.06.066] [PMID: 27591662]

[2]   Chaudhuri TK, Paul S. Protein-misfolding diseases and chaperone-based therapeutic approaches. FEBS J 2006; 273(7): 1331-49.
[http://dx.doi.org/10.1111/j.1742-4658.2006.05181.x] [PMID: 16689923]

[3]   Prince M, Wimo A, Guerchet M, Ali G, Wu Y, Prina M. World Alzheimer's Report 2015, Alzheimer's Disease International. London: ADI 2015.

[4]   Ludwig R, Dzung TK. Calixarene-based molecules for cation recognition. Sensors (Basel) 2002; 2: 397-416.
[http://dx.doi.org/10.3390/s21000397]

[5]   Omar O, Ray AK, Hasan AK, Davis F. Resorcin calixarenes (resorcinarenes): Langmuir–Blodgett films and optical properties. Supramol Sci 1997; 4: 417-21.
[http://dx.doi.org/10.1016/S0968-5677(97)00024-2]

[6]   Wang J, Liu J. Calixarene-coated amperometric detectors. Anal Chim Acta 1994; 294: 201-6.
[http://dx.doi.org/10.1016/0003-2670(94)80195-9]

[7]   Nikolelis DP, Petropoulou SS, Pergel E, Toth K. Biosensors for the rapid detection of dopamine using bilayer lipid membranes (BLMs) with incorporated calix[4]resorcinarene receptor. Electroanalysis 2002; 14: 783-9.
[http://dx.doi.org/10.1002/1521-4109(200206)14:11<783::AID-ELAN783>3.0.CO;2-6]

[8]   Nikolelis DP, Petropoulou SS, Theoharis G. Electrochemical investigation of interactions of bilayer lipid membranes (BLMs) with incorporated resorcin[4]arene receptor with ephedrine for evelopment of a stabilized lipid film biosensor for ephedrine. Electrochim Acta 2002; 47: 3457-67.
[http://dx.doi.org/10.1016/S0013-4686(02)00282-7]

[9]   Nikolelis DP, Petropoulou SS. Investigation of interactions of a resorcin[4]arene receptor with bilayer lipid membranes (BLMs) for the electrochemical biosensing of mixtures of dopamine and ephedrine.

Biochim Biophys Acta 2002; 1558(2): 238-45.
[http://dx.doi.org/10.1016/S0005-2736(01)00438-2] [PMID: 11779572]

[10]   Nikolelis DP, Theoharis G. Biosensor for dopamine based on stabilized lipid films with incorporated resorcin[4]arene receptor. Bioelectrochemistry 2003; 59(1-2): 107-12.
[http://dx.doi.org/10.1016/S1567-5394(03)00009-4] [PMID: 12699826]

[11]   Nikolelis DP, Drivelos DA, Simantiraki MG, Koinis S. An optical spot test for the detection of dopamine in human urine using stabilized in air lipid films. Anal Chem 2004; 76(8): 2174-80.
[http://dx.doi.org/10.1021/ac0499470] [PMID: 15080725]

[12]   Nikolelis DP, Siontorou CG, Theoharis G, Bitter I. Flow injection analysis of mixtures of dopamine, adrenaline and ephedrine in human biofluids using stabilized after storage in air lipid membranes with a novel incorporated resorcin[4]arene receptor. Electroanalysis 2005; 17: 887-94.
[http://dx.doi.org/10.1002/elan.200403168]

[13]   Nikolelis DP, Psaroudakis N, Ferderigos N. Preparation of a selective receptor for ephedrine for the development of an optical spot test for the detection of ephedrine in human urine using stabilized in air lipid films with incorporated receptor. Anal Chem 2005; 77(10): 3217-21.
[http://dx.doi.org/10.1021/ac0484023] [PMID: 15889911]

[14]   Mohsin MA, Hianik T, Banica F-G, Oshima T, Nikolelis DP. The study on the properties of self-assembled bimolecular layers at the gold electrode with incorporated calixarenes for dopamine and epinephrine detection.Sensing in Electroanalysis. Pardubice: University Press Centre 2010; Vol. 5: pp. 185-94.

[15]   Oshima T, Oishi K, Ohto K, Inoue K. Extraction of catecholamines by calixarene carboxylic acid derivatives. J Incl Phenom Macrocycl Chem 2006; 55: 79-85.
[http://dx.doi.org/10.1007/s10847-005-9022-9]

[16]   Evtugyn GA, Shamagsumova RV, Sitdikov RR, *et al.* Dopamine sensor based on a composite of silver nanoparticles implemented in the electroactive matrix of calixarenes. Electroanalysis 2011; 23: 2281-9.
[http://dx.doi.org/10.1002/elan.201100197]

[17]   Willner I, Baron R, Willner B. Integrated nanoparticle-biomolecule systems for biosensing and bioelectronics. Biosens Bioelectron 2007; 22(9-10): 1841-52.
[http://dx.doi.org/10.1016/j.bios.2006.09.018] [PMID: 17071070]

[18]   Evtugyn GA, Stoikov II, Belyakova SV, *et al.* Selectivity of solid-contact Ag potentiometric sensors based on thiacalix[4]arene derivatives. Talanta 2008; 76(2): 441-7.
[http://dx.doi.org/10.1016/j.talanta.2008.03.029] [PMID: 18585303]

[19]   Snejdárková M, Poturnayová A, Rybár P, *et al.* High sensitive calixarene-based sensor for detection of dopamine by electrochemical and acoustic methods. Bioelectrochemistry 2010; 80(1): 55-61.
[http://dx.doi.org/10.1016/j.bioelechem.2010.03.006] [PMID: 20537963]

[20]   Kurzatkowska K, Sayin S, Yilmaz M, Radecka H, Radecki J. Calix[4]arene derivatives as dopamine hosts in electrochemicalsensors. Sens Act B 2015; 218: 111-21.
[http://dx.doi.org/10.1016/j.snb.2015.04.110]

[21]   Wang F, Chia C, Yu B, Ye B. Simultaneous voltammetric determination of dopamine and uric acidbased on Langmuir–Blodgett film of calixarene modified glassycarbon electrode. Sens Act B 2015; 221: 1586-93.
[http://dx.doi.org/10.1016/j.snb.2015.06.155]

[22]   Potyrailo RA, Conrad RC, Ellington AD, Hieftje GM. Adapting selected nucleic acid ligands (aptamers) to biosensors. Anal Chem 1998; 70(16): 3419-25.
[http://dx.doi.org/10.1021/ac9802325] [PMID: 9726167]

[23]   Hianik T, Wang J. Electrochemical aptasensors – recent achievements and perspectives. Electroanalysis 2009; 21: 1223-35.

[http://dx.doi.org/10.1002/elan.200904566]

[24]   Chen A, Yang S. Replacing antibodies with aptamers in lateral flow immunoassay. Biosens Bioelectron 2015; 71: 230-42.
[http://dx.doi.org/10.1016/j.bios.2015.04.041] [PMID: 25912679]

[25]   Liu S, Xing X, Yu J, *et al.* A novel label-free electrochemical aptasensor based on graphene-polyaniline composite film for dopamine determination. Biosens Bioelectron 2012; 36(1): 186-91.
[http://dx.doi.org/10.1016/j.bios.2012.04.011] [PMID: 22560161]

[26]   Xu J, Li Y, Wang L, *et al.* A facile aptamer-based sensing strategy for dopamine through the fluorescence resonance energy transfer between rhodamine B and gold nanoparticles. Dyes Pigments 2015; 123: 55-63.
[http://dx.doi.org/10.1016/j.dyepig.2015.07.019]

[27]   Zhou J, Wang W, Yu P, Xiong E, Zhang X, Chen J. A simple label-free electrochemical aptasensor for dopamine detection. RSC Advances 2014; 4: 52250-5.
[http://dx.doi.org/10.1039/C5RA08090D]

[28]   Azadbakht A, Roushani M, Abbasi AR, Menati S, Derikvand Z. A label-free aptasensor based on polyethyleneimine wrapped carbon nanotubes *in situ* formed gold nanoparticles as signal probe for highly sensitive detection of dopamine. Mater Sci Eng C 2016; 68: 585-93.
[http://dx.doi.org/10.1016/j.msec.2016.05.077] [PMID: 27524058]

[29]   Jarczewska M, Sheelam SR, Ziolkowski R, Gorski L. A label-free electrochemical DNA aptasensor for the detection of dopamine. J Electrochem Soc 2016; 163: B26-31.
[http://dx.doi.org/10.1149/2.0501603jes]

[30]   Hianik T. Affinity biosensors for detection immunoglobulin E and cellular prions. Antibodies *vs.* DNA aptamers. Electroanalysis 2016; 28: 1764-76.
[http://dx.doi.org/10.1002/elan.201600153]

[31]   Weiss S, Proske D, Neumann M, *et al.* RNA aptamers specifically interact with the prion protein PrP. J Virol 1997; 71(11): 8790-7.
[PMID: 9343239]

[32]   Proske D, Gilch S, Wopfner F, Schätzl HM, Winnacker EL, Famulok M. Prion-protein-specific aptamer reduces PrPSc formation. ChemBioChem 2002; 3(8): 717-25.
[http://dx.doi.org/10.1002/1439-7633(20020802)3:8<717::AID-CBIC717>3.0.CO;2-C]   [PMID: 12203970]

[33]   Zeiler B, Adler V, Kryukov V, Grossman A. Concentration and removal of prion proteins from biological solutions. Biotechnol Appl Biochem 2003; 37(Pt 2): 173-82.
[http://dx.doi.org/10.1042/BA20020087] [PMID: 12630906]

[34]   Mashima T, Nishikawa F, Kamatari YO, *et al.* Anti-prion activity of an RNA aptamer and its structural basis. Nucleic Acids Res 2013; 41(2): 1355-62.
[http://dx.doi.org/10.1093/nar/gks1132] [PMID: 23180780]

[35]   Takemura K, Wang P, Vorberg I, *et al.* DNA aptamers that bind to PrP(C) and not PrP(Sc) show sequence and structure specificity. Exp Biol Med (Maywood) 2006; 231(2): 204-14.
[http://dx.doi.org/10.1177/153537020623100211] [PMID: 16446497]

[36]   Hianik T, Porfireva A, Grman I, Evtugyn G. Aptabodies - new type of artificial receptors for detection proteins. Protein Pept Lett 2008; 15(8): 799-805.
[http://dx.doi.org/10.2174/092986608785203656] [PMID: 18855752]

[37]   Hianik T, Porfireva A, Grman I, Evtugyn G. EQCM biosensors based on DNA aptamers and antibodies for rapid detection of prions. Protein Pept Lett 2009; 16(4): 363-7.
[http://dx.doi.org/10.2174/092986609787848090] [PMID: 19356132]

[38]   Prusiner SB. Molecular biology of prion diseases. Science 1991; 252(5012): 1515-22.
[http://dx.doi.org/10.1126/science.1675487] [PMID: 1675487]

[39]   Miodek A, Poturnayová A, Snejdárková M, Hianik T, Korri-Youssoufi H. Binding kinetics of human cellular prion detection by DNA aptamers immobilized on a conducting polypyrrole. Anal Bioanal Chem 2013; 405(8): 2505-14.
[http://dx.doi.org/10.1007/s00216-012-6665-4] [PMID: 23318762]

[40]   Miodek A, Castillo G, Hianik T, Korri-Youssoufi H. Electrochemical aptasensor of human cellular prion based on multiwalled carbon nanotubes modified with dendrimers: a platform for connecting redox markers and aptamers. Anal Chem 2013; 85(16): 7704-12.
[http://dx.doi.org/10.1021/ac500605p] [PMID: 23822753]

[41]   Miodek A, Castillo G, Hianik T, Korri-Youssoufi H. Electrochemical aptasensor of cellular prion protein based on modified polypyrrole with redox dendrimers. Biosens Bioelectron 2014; 56: 104-11.
[http://dx.doi.org/10.1016/j.bios.2013.12.051] [PMID: 24480126]

[42]   Wustoni S, Hideshima S, Kuroiwa S, Nakanishi T, Mori Y, Osaka T. Conversion of protein net charge *via* chemical modification for highly sensitive prion detection using field effect transistor (FET) biosensor. Sens Act B 2016; 230: 374-9.
[http://dx.doi.org/10.1016/j.snb.2016.02.078]

[43]   Zhang HJ, Lu YH, Long YJ, *et al.* An aptamer-functionalized gold nanoparticle biosensor for the detection of prion protein. Anal Methods 2014; 6: 2982-7.
[http://dx.doi.org/10.1039/C3AY42207K]

[44]   Xiao SJ, Hu PP, Li YF, Huang CZ, Huang T, Xiao GF. Aptamer-mediated turn-on fluorescence assay for prion protein based on guanine quenched fluophor. Talanta 2009; 79(5): 1283-6.
[http://dx.doi.org/10.1016/j.talanta.2009.05.040] [PMID: 19635360]

[45]   Hu PP, Zheng LL, Zhan L, *et al.* Metal-enhanced fluorescence of nano-core-shell structure used for sensitive detection of prion protein with a dual-aptamer strategy. Anal Chim Acta 2013; 787: 239-45.
[http://dx.doi.org/10.1016/j.aca.2013.05.061] [PMID: 23830445]

[46]   Cuccioloni M, Amici M, Eleuteri AM, Biagetti M, Barocci S, Angeletti M. Binding of recombinant PrP^c to human plasminogen: kinetic and thermodynamic study using a resonant mirror biosensor. Proteins 2005; 58(3): 728-34.
[http://dx.doi.org/10.1002/prot.20346] [PMID: 15609351]

[47]   Jiayu W, Xiong W, Jiping L, *et al.* A rapid method for detection of PrP by surface plasmon resonance (SPR). Arch Virol 2009; 154(12): 1901-8.
[http://dx.doi.org/10.1007/s00705-009-0532-4] [PMID: 19862471]

[48]   Poturnayova A, Snejdarkova M, Babelova L, Korri-Youssoufi H, Hianik T. Comparative analysis of cellular prion detection by mass-sensitive immunosensors. Electroanalysis 2014; 26: 1312-9.
[http://dx.doi.org/10.1002/elan.201400049]

[49]   Ganesh HV, Chow AM, Kerman K. Recent advances in biosensors for neurodegenerative disease detection. TrAC –. Trends Analyt Chem 2016; 79: 363-70.
[http://dx.doi.org/10.1016/j.trac.2016.02.012]

[50]   Schliebs R, Arendt T. The significance of the cholinergic system in the brain during aging and in Alzheimer's disease. J Neural Trans 2006; 113: 1625-44.

[51]   Ibach B, Haen E. Acetylcholinesterase inhibition in Alzheimer's Disease. Curr Pharm Des 2004; 10(3): 231-51.
[http://dx.doi.org/10.2174/1381612043386509] [PMID: 14754384]

[52]   Giacobini E. Cholinesterases: new roles in brain function and in Alzheimer's disease. Neurochem Res 2003; 28(3-4): 515-22.
[http://dx.doi.org/10.1023/A:1022869222652] [PMID: 12675140]

[53]   Park J, Galligan JJ, Fink GD, Swain GM. *In vitro* continuous amperometry with a diamond microelectrode coupled with video microscopy for simultaneously monitoring endogenous norepinephrine and its effect on the contractile response of a rat mesenteric artery. Anal Chem 2006;

78(19): 6756-64.
[http://dx.doi.org/10.1021/ac060440u] [PMID: 17007494]

[54]   Ellman GL, Courtney KD, Andres V Jr, Feather-Stone RM. A new and rapid colorimetric determination of acetylcholinesterase activity. Biochem Pharmacol 1961; 7: 88-95.
[http://dx.doi.org/10.1016/0006-2952(61)90145-9] [PMID: 13726518]

[55]   Kok FN, Bozoglu F, Hasirci V. Construction of an acetylcholinesterase-choline oxidase biosensor for aldicarb determination. Biosens Bioelectron 2002; 17(6-7): 531-9.
[http://dx.doi.org/10.1016/S0956-5663(02)00009-X] [PMID: 11959475]

[56]   Liu G, Lin Y. Biosensor based on self-assembling acetylcholinesterase on carbon nanotubes for flow injection/amperometric detection of organophosphate pesticides and nerve agents. Anal Chem 2006; 78(3): 835-43.
[http://dx.doi.org/10.1021/ac051559q] [PMID: 16448058]

[57]   Sun X, Wang X. Acetylcholinesterase biosensor based on Prussian blue-modified electrode for detecting organophosphorous pesticides. Biosens Bioelectron 2010; 25(12): 2611-4.
[http://dx.doi.org/10.1016/j.bios.2010.04.028] [PMID: 20466535]

[58]   Amine A, Arduini F, Moscone D, Palleschi G. Recent advances in biosensors based on enzyme inhibition. Biosens Bioelectron 2016; 76: 180-94.
[http://dx.doi.org/10.1016/j.bios.2015.07.010] [PMID: 26227311]

[59]   Ivanov AN, Younusov RR, Evtugyn GA, Arduini F, Moscone D, Palleschi G. Acetylcholinesterase biosensor based on single-walled carbon nanotubes--Co phtalocyanine for organophosphorus pesticides detection. Talanta 2011; 85(1): 216-21.
[http://dx.doi.org/10.1016/j.talanta.2011.03.045] [PMID: 21645691]

[60]   Pauliukaite R, Malinauskas A, Zhylyak G, Spichiger-Keller UE. Conductive organic complex salt TTF-TCNQ as a mediator for biosensors. An overview. Electroanalysis 2007; 19: 2491-8.
[http://dx.doi.org/10.1002/elan.200704035]

[61]   Evtugyn GA, Shamagsumova RV, Padnya PV, Stoikov II, Antipin IS. Cholinesterase sensor based on glassy carbon electrode modified with Ag nanoparticles decorated with macrocyclic ligands. Talanta 2014; 127: 9-17.
[http://dx.doi.org/10.1016/j.talanta.2014.03.048] [PMID: 24913851]

[62]   Lenigk R, Lam E, Lai A, et al. Enzyme biosensor for studying therapeutics of Alzheimer's disease. Biosens Bioelectron 2000; 15(9-10): 541-7.
[http://dx.doi.org/10.1016/S0956-5663(00)00078-6] [PMID: 11419651]

[63]   Du D, Chen S, Cai J, Song D. Comparison of drug sensitivity using acetylcholinesterase biosensor based on nanoparticles-chitosan sol-gel composite. J Electroanal Chem 2007; 611: 60-6.
[http://dx.doi.org/10.1016/j.jelechem.2007.08.007]

[64]   Veloso AJ, Nagy PM, Zhang B, et al. Miniaturized electrochemical system for cholinesterase inhibitor detection. Anal Chim Acta 2013; 774: 73-8.
[http://dx.doi.org/10.1016/j.aca.2013.02.033] [PMID: 23567119]

[65]   da Silva JI, de Moraes MC, Vieira LC, Corrêa AG, Cass QB, Cardoso CL. Acetylcholinesterase capillary enzyme reactor for screening and characterization of selective inhibitors. J Pharm Biomed Anal 2013; 73: 44-52.
[http://dx.doi.org/10.1016/j.jpba.2012.01.026] [PMID: 22391555]

[66]   Vandeput M, Parsajoo C, Vanheuverzwijn J, et al. Flow-through enzyme immobilized amperometric detector for the rapid screening of acetylcholinesterase inhibitors by flow injection analysis. J Pharm Biomed Anal 2015; 102: 267-75.
[http://dx.doi.org/10.1016/j.jpba.2014.09.012] [PMID: 25459923]

[67]   Kostelnik A, Cegan A, Pohanka M. Electrochemical determination of activity of acetylcholinesterase immobilized on magnetic particles. Int J Electrochem Sci 2016; 11: 4840-9.

[http://dx.doi.org/10.20964/2016.06.39]

[68]   Çevik S, Timur S, Anik Ü. Biocentri-voltammetry for the enzyme assay: a model study. RSC Advances 2012; 2: 4299-303.
[http://dx.doi.org/10.1039/c2ra01282k]

[69]   Hai A, Ben-Haim D, Korbakov N, *et al.* Acetylcholinesterase-ISFET based system for the detection of acetylcholine and acetylcholinesterase inhibitors. Biosens Bioelectron 2006; 22(5): 605-12.
[http://dx.doi.org/10.1016/j.bios.2006.01.028] [PMID: 16529923]

[70]   Müntze GM, Pouokam E, Steidle J, *et al. In situ* monitoring of myenteric neuron activity using acetylcholinesterase-modified AlGaN/GaN solution-gate field-effect transistors. Biosens Bioelectron 2016; 77: 1048-54.
[http://dx.doi.org/10.1016/j.bios.2015.10.076] [PMID: 26547432]

[71]   Lau H-C, Lee I-K, Ko P-W, *et al.* Non-invasive screening for Alzheimer's disease by sensing salivary sugar using *Drosophila* cells expressing gustatory receptor (Gr5a) immobilized on an extended gate ion-sensitive field-effect transistor (EG-ISFET) biosensor. PLoS One 2015; 10(2): e0117810.
[http://dx.doi.org/10.1371/journal.pone.0117810] [PMID: 25714733]

[72]   Rauk A. The chemistry of Alzheimer's disease. Chem Soc Rev 2009; 38(9): 2698-715.
[http://dx.doi.org/10.1039/b807980n] [PMID: 19690748]

[73]   Hardy J, Selkoe DJ. The amyloid hypothesis of Alzheimer's disease: progress and problems on the road to therapeutics. Science 2002; 297(5580): 353-6.
[http://dx.doi.org/10.1126/science.1072994] [PMID: 12130773]

[74]   Humpel C. Identifying and validating biomarkers for Alzheimer's disease. Trends Biotechnol 2011; 29(1): 26-32.
[http://dx.doi.org/10.1016/j.tibtech.2010.09.007] [PMID: 20971518]

[75]   Reinke AA, Ung PM, Quintero JJ, Carlson HA, Gestwicki JE. Chemical probes that selectively recognize the earliest Aβ oligomers in complex mixtures. J Am Chem Soc 2010; 132(50): 17655-7.
[http://dx.doi.org/10.1021/ja106291e] [PMID: 21105683]

[76]   Haes AJ, Chang L, Klein WL, Van Duyne RP. Detection of a biomarker for Alzheimer's disease from synthetic and clinical samples using a nanoscale optical biosensor. J Am Chem Soc 2005; 127(7): 2264-71.
[http://dx.doi.org/10.1021/ja044087q] [PMID: 15713105]

[77]   Liu L, Xia N, Jiang M, *et al.* Electrochemical detection of amyloid-β oligomer with the signal amplification of alkaline phosphatase plus electrochemical–chemical–chemical redox cycling. J Electroanal Chem 2015; 754: 40-5.
[http://dx.doi.org/10.1016/j.jelechem.2015.06.017]

[78]   Rushworth JV, Ahmed A, Griffiths HH, Pollock NM, Hooper NM, Millner PA. A label-free electrical impedimetric biosensor for the specific detection of Alzheimer's amyloid-beta oligomers. Biosens Bioelectron 2014; 56: 83-90.
[http://dx.doi.org/10.1016/j.bios.2013.12.036] [PMID: 24480125]

[79]   Li H, Cao Y, Wu X, Ye Z, Li G. Peptide-based electrochemical biosensor for amyloid β 1-42 soluble oligomer assay. Talanta 2012; 93: 358-63.
[http://dx.doi.org/10.1016/j.talanta.2012.02.055] [PMID: 22483923]

[80]   Oh J, Yoo G, Chang YW, *et al.* A carbon nanotube metal semiconductor field effect transistor-based biosensor for detection of amyloid-beta in human serum. Biosens Bioelectron 2013; 50: 345-50.
[http://dx.doi.org/10.1016/j.bios.2013.07.004] [PMID: 23891796]

[81]   Chikae M, Fukuda T, Kerman K, Idegami K, Miura Y, Tamiya E. Amyloid-β detection with saccharide immobilized gold nanoparticle on carbon electrode. Bioelectrochemistry 2008; 74(1): 118-23.
[http://dx.doi.org/10.1016/j.bioelechem.2008.06.005] [PMID: 18676183]

[82]   Liu L, He Q, Zhao F, *et al.* Competitive electrochemical immunoassay for detection of β-amyloid (1-42) and total β-amyloid peptides using p-aminophenol redox cycling. Biosens Bioelectron 2014; 51: 208-12.
[http://dx.doi.org/10.1016/j.bios.2013.07.047] [PMID: 23962708]

[83]   Wang-Dietrich L, Funke SA, Kühbach K, *et al.* The amyloid-β oligomer count in cerebrospinal fluid is a biomarker for Alzheimer's disease. J Alzheimer'ss Dis 2013; 34(4): 985-94.
[PMID: 23313925]

CHAPTER 6

NMDA Receptor as a Molecular Target for Central Nervous System Disorders: The Advances and Contributions of Molecular Modeling

Marcos Vinicius Santana[1], **Helena Carla Castro**[1] and **Paula Alvarez Abreu**[2,*]

[1] *LabiEMol, Laboratório de Antibióticos, Bioquímica, Ensino e Modelagem molecular, Instituto de Biologia, Universidade Federal Fluminense, Niterói, RJ, Brazil*

[2] *LAMCIFAR, Laboratório de Modelagem Molecular e Pesquisa em Ciências Farmacêuticas, Universidade Federal do Rio de Janeiro, Campus UFRJ-Macaé, RJ, Brazil*

Abstract: The N-methyl-D-aspartate receptor (NMDAR) is a glutamate receptor that mediates important physiological functions in the central nervous system (CNS). However, the overstimulation of this receptor is associated with neurodegenerative disorders, including Parkinson, Huntington and Alzheimer diseases. In this new millennium, diseases causing progressive neuronal loss and death have become more frequent and the current therapy still presents several adverse effects and does not block disease progression. In this chapter, we discuss the role of NMDAR in neurodegenerative disorders and its potential as a therapeutic target, the advances in the development of NMDAR antagonists and the contributions of molecular modeling in this field. NMDAR structure is already known allowing the use of molecular modeling tools for the development of new NMDAR antagonists. Studies involving the use of structure based drug-design methods as molecular docking and virtual screening for discovering new NMDAR antagonists were reviewed here; as well as the *in silico* evaluation of pharmacokinetic and toxicological properties. CNS drugs should be capable of effectively cross the blood brain barrier to be active and the early evaluation of the safety profile of these compounds is extremely important to reduce the time and costs to develop new drugs for neurodegenerative disease therapy.

Keywords: ADMET prediction, Central nervous system disease, Docking, Glutamate excitotoxicity, GluN2B antagonists, Molecular modeling, Neurode-generative diseases, NMDA receptor, Structure-based drug design, Virtual screening.

---

* **Corresponding author Paula A. Abreu:** Universidade Federal do Rio de Janeiro, Campus UFRJ-Macaé, RJ, Brazil; Tel/Fax: +55 (22) 2141-3976; Email: abreu_pa@yahoo.com.br

INTRODUCTION

NMDA Receptor Signaling in the CNS

The L-glutamic acid (**1**) (Scheme 1) is the major excitatory neurotransmitter in the mammalian central nervous system (CNS) [1]. This neurotransmitter is stored in high concentrations inside presynaptic vesicles and is released after the stimulus to interact with postsynaptic receptors [1, 2]. Glutamate-mediated excitatory currents are associated with two classes of glutamate receptors (GluR): metabotropic (mGluR) and ionotropic (iGluR). Both classes of GluR are extremely important to the CNS functions, such as synaptic plasticity, memory consolidation, neuronal development and learning [3].

The mGluR are G protein-coupled receptors (GPCR), distributed at perisynaptic and extrasynaptic locations and modulate ion channel activity [4]. As other GPCR, the mGluR has seven transmembrane domains, besides an extracellular N-terminal and an intracellular C-terminal domain. To date, eight mGluR were identified, and classified into three groups (I, II and III) based on agonist binding, signal transduction pathways and homology [5]. Depending on the cell type or neuronal population the signaling cascades activated by the receptors vary [6, 7].

The iGluR mediate fast excitatory currents in the CNS that are the result of Na^+ and Ca^{2+} influx. Three classes of iGluR have been identified and they were named according to its synthetic agonist: N-methyl-D-aspartic acid (NMDA, **4**), kainic acid (KA, **5**) and α-amino-3-hydroxy-5-methyl-4-isoxazolepropionic acid (AMPA, **6**) [6 - 12] (Scheme 1). These receptors are homo or heterotetramers of different subunits which are assembled in four structurally diverse domains: a) the extracellular N-terminal domain (NTD), b) ligand-binding domain (LBD), c) three transmembrane segments (M1, M3 and M4) and a reentrant pore loop (M2) and, d) an intracellular C-terminal domain (CTD) [13, 14].

Both iGluR and mGluR interact with a variety of postsynaptic proteins by their intracellular C-terminal that are responsible for synapse stabilization and postsynaptic effects such as long-term potentiation (LTP) and long-term depression (LTD) [5, 15 - 18]. The postsynaptic density protein of 95 kDa (PSD-95) is a scaffold protein, responsible for the mechanical stabilization of the synapse *via* interaction with neuroligin and β-neurexin, and links glutamate receptors to the actin cytoskeleton [3, 19]. In addition to PSD-95, other proteins are important for signal transduction and stabilization, such as Calcium/Calmodulin-dependent protein kinase II (CaMKII), protein kinase C and A (PKC and PKA, respectively) and the nuclear transcriptional factor cAMP response element-binding protein (CREB).

Among iGluR, the N-methyl-D-aspartate receptor (NMDAR) is essential for normal CNS function, since it is involved in synaptic plasticity and higher cognitive functions [11]. These receptors are unique in requiring both a membrane depolarization and agonist/co-agonist binding for activation, while non-NMDA iGluR requires only L-glutamate for activation [20]. *In vivo*, L-glutamate (**1**) acts as NMDAR agonist, binding to GluN2 subunits, while glycine (**2**) and D-serine (**3**) are co-agonists that can bind to GluN1 or GluN3 subunits (Scheme 1) [10, 18, 21]. However, agonist and co-agonists binding are not enough for receptor activation, because at resting membrane potential magnesium ions block the channel pore. This voltage-dependent blockage is relieved with membrane depolarization, allowing sodium (Na^+) and calcium (Ca^{2+}) influx and potassium (K^+) efflux. NMDAR and AMPAR colocalize in all CNS synapses, leading to coactivation after glutamate release [20]. Also, backpropagating action potentials can facilitate the removal of Mg^{2+}, enhancing the activation of NMDAR by glutamate [22, 23].

NMDA receptors associate with different intracellular proteins responsible for signal transduction and gene expression. It has been shown that synaptic NMDAR stimulates extracellular signal-regulated kinase ½ (ERK½), CREB phosphorylation and is involved in neuroprotection. On the other hand, activation of extrasynaptic NMDAR is associated with cell death *via* interactions with different proteins [24 - 26].

NMDA receptors show high variability in subunit composition, pharmacological properties, and kinetics, and the involvement of these receptors in pathological processes makes them an interesting target for the development of modulators of the receptor function. In this review, we cover the recent advances in NMDAR structure, function and the contributions of computational studies in drug discovery targeting NMDAR

CENTRAL NERVOUS SYSTEM DISORDERS RELATED TO NMDA RECEPTOR

NMDAR function is associated with numerous CNS pathologies, such as Alzheimer (AD), Parkinson (PD), Huntington disease, schizophrenia, epilepsy, neuroinflammation and neuronal death in acute events (*i.e.*, stroke) [27 - 30]. Several studies have linked NMDAR subunit expression, receptor trafficking and localization with these pathological conditions (Table **1**). The drug discovery of NMDAR modulators has focused on subunit-selective antagonists and exploration of allosteric sites, since competitive antagonists have been implicated with unacceptable adverse effects, such as drowsiness and hallucinations [31].

Table 1. NMDA receptor alterations involved in neurodegenerative diseases.

Disease	NMDA receptor involvement	Citation
Alzheimer	Hyperactivation of extrasynaptic NMDAR. Glial relase of L-glutamate and reduced uptake. Extrasynaptic NMDAR-mediated positive feedback of Aβ formation. Promotes tau protein hyperphosphorylation.	[29, 33, 36]
Huntington	Redistribution and hyperactivation of extrasynaptic NMDAR	[27, 31]
Parkinson	Increased GluN2A and redistribution of GluN2B from synaptic to extrasynaptic sites.	[26, 32, 40]

There are two prevailing models for NMDAR involvement in neurological conditions, which are based on receptor localization on neurons and subunit composition [11, 18, 24, 32]. According to the localization model, while synaptic NMDAR triggers cellular cascades associated with cell survival, extrasynaptic NMDAR is linked to cell death pathways [24]. Synaptic NMDAR activation increases cAMP levels, ERK1/2-dependent activation of Jacob protein and its translocation to the nucleus, CREB phosphorylation, and antioxidant defense; whereas extrasynaptic NMDAR activation leads to the transport of non-phosphorylated Jacob to the nucleus and reduces levels of phosphorylated CREB and synapse complexity [32].

In addition to the localization model, several studies have focused on subunit composition and the role of different GluN2 subunits in neurological conditions and the prevailing model is that GluN2A and GluN2B play major roles in neuroprotection and cell death, respectively. A simplified model established that GluN2A and GluN2B subunits are found within and outside the synapse respectively, where extrasynaptic GluN2B-NMDAR triggers cell death and synaptic GluN2A-NMDAR cell survival [18, 33, 34]. However, it is important to point that this model is not absolute and under some experimental conditions synaptic GluN2A activation leads to excitotoxicity [24].

Alzheimer Disease

Alzheimer disease (AD) is one of the most prevalent neurodegenerative diseases on elderly. Thus, many studies have been done to understand its molecular biology and to develop new drugs. Evidence supports the involvement of soluble amyloid-beta (Aβ) oligomers and hyperphosphorylated tau protein as one of the main causes of neurodegeneration associated with AD [32]. NMDAR involvement in AD neurodegeneration is well documented, showing that Aβ oligomers regulate synaptic endocytosis of NMDAR, altering the balance between synaptic and extrasynaptic receptors [35 - 37]. Aβ oligomers also increase presynaptic glutamate and glial release of glutamate in the extracellular space and reduce the expression of glutamate transporters and glial uptake [28]. Furthermore, Aβ expression is dependent on NMDAR localization [36, 37]. Whereas synaptic NMDAR activation has no effect on Aβ expression, extrasynaptic NMDAR activation upregulates amyloid precursor proteins and the production of Aβ oligomers, which can lead to a toxic positive feedback [32].

Hyperphosphorylation of tau is essential for AD neurodegeneration since Aβ cannot elicit toxic effects without tau. Two kinases largely regulate tau phosphorylation and evidence supports that NMDAR regulates this process. Glycogen synthase kinase 3β (GSK-3β) phosphorilates tau, which is in the hyperphosphorylated state, the main component of plaques and neurofibrillary tangles. In addition, it has been shown that the blockade of GluN2B-containing NMDAR leads to inhibition of Ab-induced GSK-3b activation and tau phosphorylation [32, 38, 39]. The other kinase is cyclin-dependent kinase 5 (CDK5) which is also implicated in tau toxicity and extrasynaptic NMDAR increases its activation through S-nitrosylation, contributing to synapse loss [32, 40].

Parkinson Disease

Parkinson disease (PD) is a common neurological condition in elderly affecting primarily people over 60-year-old. The primary cause of PD is the loss of dopaminergic neurons from the substantia nigra that innervate the striatum, with more than 70% loss at the onset [29, 41]. This alteration results in motor symptoms, including slowness, tremors, and loss of balance. In addition, PD is extremely debilitating and imposes a great burden on patients' quality of life.

Current drug therapy includes replacing the dopaminergic deficit with dopamine agonists (ropinirole and pramipexole), or dopamine precursor L-dihydroxyphenylalanine (levodopa or L-DOPA) and monoamine oxidase inhibitors (safinamide, selegiline, and rasagiline) [41]. However, some of these drugs show serious adverse effects as dyskinesia which are the involuntary movements caused by the prolonged use of these drugs, and they result from adaptive changes in brain function [6].

Since dopaminergic and glutamatergic pathways regulate motor functions, it is not surprising that NMDAR is involved in PD symptoms. Studies using models of Parkinson disease as lesion induced by 66-hydroxydopamine- in rat and 1-methy--4-phenyl-1,2,3,6-tetrahydropyridine (MPTP) in primate showed an increase in striatal GluN1 and GluN2B at the membrane, but no changes in the GluN2A levels. In addition, treatment with L-DOPA restored the GluN1 and GluN2B levels [29].

NMDAR antagonists have a potential use as PD drugs, however, they have shown only modest benefits, with only amantadine, a NMDAR channel blocker, being used in clinics to treat PD [31, 41].

Huntington Disease

Huntington disease (HD) is a genetic disorder characterized by extensive CAG repeats on the huntingtin gene, which codes the huntingtin protein. This mutation leads to dementia and motor dysfunction as a result of striatum neurodegeneration. HD patients are also anxious, paranoid and irritable and the outcome of the disease is death as a result of mobility complications [41]. Using mice models, Milnerwood and coworkers (2010) showed that the increase of extrasynaptic NMDAR activity is involved in HD phenotype. In this work, the authors reported increased extrasynaptic activity, expression and reduced striatal CREB phosphorylation, whereas NMDAR synaptic activity remained unaltered. In addition, extrasynaptic activity was linked to GluN2B-containing receptors, and memantine treatment was able to reverse the motor deficits and the reduction of striatal CREB activity [30]. Since HD has no cure and treatments usually have severe adverse effects, the development of other channel blockers or selective GluN2B antagonists could represent a good strategy for new drugs.

Schizophrenia

Schizophrenia is a syndrome characterized by signs of psychosis in adolescence or early adult life. The symptoms of schizophrenia can be classified into positive (hallucinations and paranoia), negative (withdrawn, apathy, lack of pleasure and flattening) and cognitive deficits, such as lack of attention, memory and reduced processing speed [42, 43]. The classical hypothesis of schizophrenia is based on the dysfunction of dopamine signaling. The psychotic episodes of schizophrenia are associated with increased activity of dopamine D2 receptors in the striatum and the nucleus accumbens. Evidence supports the correlation between the blockade of D2 receptors by antipsychotic drugs and the clinical response in schizophrenic patients [42 - 45]. On the other hand, brain imaging showed that the negative and cognitive symptoms are associated with the decrease of D1 receptors activity in the prefrontal cortex (PFC). Although the dopamine hypothesis of schizophrenia has dominated the field for decades, evidence supports that the hypofunction of NMDA receptors might also be involved, even earlier than the dopamine dysfunction [46 - 48].

The glutamatergic hypothesis of schizophrenia states that psychosis and deficits in cognitive and executive functions are a result of an unbalanced excitation and long-term alterations of neural circuitry, and it is due to hypofunction of glutamate transmission, *via* NMDA receptors on GABAergic interneurons [11,

43, 46, 47]. NMDA receptors are crucial for the function of GABAergic interneurons where they are responsible for generating large oscillations (gamma oscillations) for local synchrony of cortical circuits. Furthermore, administration of NMDAR antagonists, such as phencyclidine and ketamine, induced schizophrenia-like symptoms in healthy humans and exacerbate the symptoms in patients with schizophrenia [43, 45, 49, 50]. Ketamine also increased functional connectivity, which is similar to the hyperconnectivity observed in schizophrenia [44].

The schizophrenia phenotype was also reproduced in mice with the deletion of an essential GluN1 subunit of NMDAR from corticolimbic GABAergic interneurons [51]. Furthermore, postmortem analysis of subpopulations of GABAergic interneurons containing the calcium-binding protein parvalbumin (PV) and glutamic acid decarboxylase 67 (GAD67), showed reduced levels of these proteins in patients with schizophrenia [11, 43, 51]. Reduction of NMDAR, GAD67 and PV levels in intact mouse cortex and hippocampus resulted in schizophrenia-like symptoms, such as hyperlocomotion [51, 52]. A similar situation happens after ketamine administration to PFC slices, resulting in the reduction of inhibitory transmission [43]. It seems that NMDAR antagonists decrease GAD67 and PV expression in cortical GABAergic neurons, linking NMDAR hypofunction to dysfunction of GABAergic neurons and the development of schizophrenia, but the mechanisms by which NMDA hypofunction underlies the pathological phenotype of schizophrenia needs future research [51].

Thus, the development of selective NMDAR agonists or compounds that increase agonist/co-agonist concentration is a possible strategy to improve the negative and cognitive deficits in schizophrenia. However, it is important to understand if selective agonists could be more beneficial than non-selective ones. And also, we need to know which NMDAR subunits are the best candidates for modulation. Answering this and other questions would be helpful to develop positive modulators of NMDAR in schizophrenia without affecting its physiological functions.

Epilepsy

Epilepsy is a collection of syndromes characterized by unprovoked epileptic seizures due to excessive or synchronous neuronal activity in the brain. Epilepsy can also be classified according to the type of the seizure, neurological findings and electroencephalographic (EEG) analysis [53, 54]. It is well-known that abnormal inhibitory or excitatory currents play a role in the development of epilepsy. Since glutamate is the major excitatory neurotransmitter in the CNS, much work has been focused on its participation in seizures, especially neuronal excitation involving NMDAR [55].

Nateri and coworkers evaluated the function of a constitutively active form of MEK1 (caMEK1) in neurons by using a murine model and showed that ERK activation resulted in spontaneous epileptic seizures. Activation of ERK increased phosphorylation of eukaryotic translation initiation factor 4E (eIF4E) and also the levels of GluN2B subunit, but did not cause alterations in mRNA expression on mice hippocampus and SH-SY5Y neuronal cell line, indicating that events downstream of NMDAR activation are important for epileptogenesis. In addition, the administration of ifenprodil blocked ERK-induced seizures, further suggesting the involvement of NMDAR on seizures and the possibility of pharmacological treatment focused on the NMDAR-ERK pathway [56].

Using a kindling model with tetanic stimulation of the stratum pyramidale of the CA1 region of guinea-pig hippocampus slices, Stelzer and coworkers observed a progressive reduction (up to 25%) of GABAergic inhibition and an increase of excitability due to NMDAR activation, resulting in epileptiform activity. In the presence of D-APV, a competitive antagonist of NMDAR, the tetanic stimuli did not affect the excitatory or inhibitory currents. This situation was fully reverted after removal of the antagonist. Similar results were observed in several other studies using NMDAR antagonists in kindling models of seizures [50, 57, 58].

Neuroinflammation

Neuroinflammation occurs when cells in the CNS, such as microglia and astrocytes, produce higher levels of proinflammatory cytokines [59, 60]. CNS cells also produce anti-inflammatory cytokines (*e.g.* IL-10) to counterbalance the effects of proinflammatory cytokines. Inflammatory cytokines, such as IL-1β and IL-6 produced by neurons, activate microglia and have important roles in CNS function [61]. Although the proinflammatory cytokines are important for normal CNS function, they are also involved in the development and progression of chronic neurodegenerative diseases, such as schizophrenia, Alzheimer's and Parkinson's diseases, and furthermore, many studies suggest a strong correlation between NMDA receptor activity and neuroinflammation in these pathological conditions [62 - 66].

In vivo measurement of activated microglia using positron emission tomography and magnetic resonance imaging showed that the neuroinflammation is an early event in some cases of Alzheimer dementia [59, 67, 68]. The cerebrospinal fluid (CSF) of Alzheimer patients has increased levels of proinflammatory markers such as TNF-α, IL-1β, IL-6 and IL-12, which might augment the toxic effects of Aβ and tau protein [67, 69]. In addition, the inflammatory response is a defense mechanism that starts after Aβ deposition, but in later stages becomes threatening itself, though the release of reactive oxygen species, nitric oxide and excitatory amino acids, such as glutamate that could potentially damage the cells [67, 70].

Neuroinflammation can cause hyperactivation of NMDARs which can positively feedback and lead to additional neuroinflammation and neuronal damage [61]. Using a single dose of lipopolysaccharide (LPS) on mice, as an inflammatory model of Alzheimer, Maher and coworkers showed that LPS caused the overexpression of GluN2B subunit [68]. The authors also observed an increase in NO levels, which could be a result of glutamate interaction with NMDA/GluN2B receptors, since neuronal nitric oxide synthase (nNOS) is in contact with NMDAR through their PDZ domain [24, 31, 71]. Glutamate stimulates NMDAR resulting in Ca2+ influx that activates the nNOS. The increase of NO causes the release of more glutamate from the pre-synaptic neurons and increases levels of Ca2+ in the cell. The treatment of the LPS group with the NOS inhibitor L-NAME blocked the cognitive deficits and showed higher affinity toward nNOS than to inducible nitric oxide synthase (iNOS) which indicated that nNOS is the NOS isoform mainly involved [68].

Chang and coworkers used an excitotoxicity model, based on chronic NMDA administration to rats and analyzed markers of neuroinflammation to study the cross-talk between excitotoxicity and neuroinflammation. The authors observed an increase in levels of interleukin-1β (IL-1β), tumor necrosis factor alpha (TNFα), mRNA and protein levels of glial fibrillary acidic protein (GFAP), a marker of activated astrocytes in neuroinflammation, and iNOS, a marker of activated microglia [62]. These results indicate that persistent activation of NMDAR might also be linked with neuroinflammation in rat frontal cortex.

Liraz-Zaltsman and coworkers used a mouse model of global LPS-induced neuroinflammation to study the protective effects of NMDAR co-agonist D-cycloserine on cognitive and behavioral impairments. The authors showed a long-lasting microgliosis, memory deficits, impaired LTP, and reduced levels of the GluN1 subunits of NMDAR in the LPS-treated group. After administration of D-cycloserine, cognitive function was recovered as well as LTP and GluN1 levels. Thus, neuroinflammation might be linked to NMDAR hypofunction and both can cause persistent cognitive deficits in an animal model, suggesting that NMDAR agonists could be used as treatment for CNS diseases where inflammation is a key component [65].

Acute Damage to the CNS

Increased L-glutamate concentrations following brain trauma are directly correlated with NMDAR hyperactivation and neuronal death in a process called excitotoxicity. The neuronal death, in this case, is the result of excessive Ca^{+2} concentration following receptor activation. Ca^{+2} activates many pathways that damage the neurons, such as activation of caspases pathway, reactive species formation and lipidic peroxidation. NMDAR antagonists have shown promising neuroprotective effects, however, they failed in clinical trials due to serious adverse effects as a result of the blockade of normal receptor function. Thus, many studies have focused on selective NMDAR antagonists, especially for GluN2B subunits, which have been described as more tolerable [11, 17, 72, 73].

In vascular accidents of ischemic or hemorrhagic origin, a large amount of glutamate is released, leading to the constant activation of glutamate receptors. In acute ischemia, the uncontrolled release of glutamate, and the subsequent hyperactivation of the excitatory transmission system is one of the main causes of death and incapacitation of the individual in the long term. The blockage of NMDA receptor can inhibit the calcium influx and reduce neuron injury. Thus, NMDAR antagonists have shown neuroprotective effects against permanent damage of ischemia, even when administered several hours after the ischemic event [74, 75].

Some compounds, such as the ifenprodil and eliprodil prototypes, are effective *in vivo* and *in vitro* to reduce brain damage caused by focal and global ischemia, as well as by acute traumatic injury [76]. Memantine, an ion channel blocker, promoted post-ischemic neurological recovery, brain remodeling, and plasticity in mice. Also, BQ-869, a potent antagonist of NMDA receptors reduced Ca^{2+} influx induced by NMDA, decreased infarction size in focal cerebral ischemia and reduced stroke mortality in rats [77]. Competitive antagonists, such as CGS-19755 and Selfotel were also reported in the literature, but all failed in phase III clinical trials [78].

STRUCTURE AND FUNCTION OF NMDA RECEPTOR AND ITS ANTAGONISTS

The NMDA receptor is a heterotetramer composed by a repertoire of three subunits, GluN1, GluN2 and GluN3. GluN1 subunit exists in eight isoforms (1a-4a and 1b-4b), result from the alternative splicing of a single gene. There are four types of GluN2 subunits (A-D) encoded by four different genes and two GluN3 subunits (A and B) that resulted from two separated genes [11, 79].

The NMDAR subunits show different function and spatiotemporal distribution on the CNS. GluN1 subunit is essential for receptor function and is expressed throughout the CNS after birth. It bears the co-agonist binding site and motifs of interactions with intracellular proteins such as calmodulin and CaMKII, that regulates receptor trafficking and signaling [19]. The four subtypes of GluN2 regulate most of the receptor functional properties and are differentially expressed in the CNS. GluN2B and GluN2D subunits are abundant early in the development. After birth, GluN2A expression sharply rises and becomes predominant in most brain areas. Conversely, GluN2D becomes restricted to the diencephalon and mesencephalon in adult brains. GluN2B expression rises soon after birth and reaches a plateau becoming restricted to the forebrain. GluN2C expression starts after birth, peaks around postnatal day 10 and becomes progressively restricted to the cerebellum and olfactory bulb. GluN3 subunits also differ in expression pattern, GluN3A increases with a peak in early postnatal and GluN3B slowly increases and concentrates in motor neurons [11, 19].

Generally, functional NMDAR are heterodimers of two GluN1 subunits and two GluN2 subunits, and the receptor shows different electrophysiological, kinetic and pharmacological properties according to the subunit composition [14, 20, 80, 81]. Receptors composed of GluN1 and a mixture of GluN2 and GluN3 subunits (*e.g.* GluN1/GluN2B/GluN3A) were described in early stages of development, while GluN1/GluN2A/GluN2B and GluN1/GluN2A/GluN2C were observed in adult brains [11, 19, 20].

Several structures of the NMDAR are available in Protein Data Bank (PDB) (www.rcsb.org) [82]. A quick search in the PDB using the keyword "NMDA" revealed 81 structures in August 2016, of which 57 were obtained by x-ray diffraction, 8 by nuclear magnetic resonance (NMR) and 13 by electron microscopy. Many NMDAR structures have one or more ligands bound to different sites of the receptors. The full-length receptor reveals a balloon-like topology, with a clear boundary between the balloon (N-terminal and ligand-binding domains) and the basket (transmembrane domain) (Fig. **1**). This topology is similar to other iGluR, however, NMDA receptors show a compact packing of NTD and LBD, in contrast to the looser assembly of NTD and LBD of AMPAR and KAR.

The NTD of NMDAR shows a bilobed structure composed of R1 and R2 domains that associate asymmetrically with each other *via* R1-R1 and R2 (GluN2B)-R1 (GluN1) interactions (Fig. **2**) [20, 79]. The NTD is critical for NMDAR function since it regulates channel opening probability and deactivation kinetics [83]. A short peptide links NTD and LBD, resulting in a tight packing of these domains. The LBD is also a bilobed structure formed by D1 and D2 domains that interact with the NTD and TMB, respectively. The receptor ion channel is composed of three transmembrane helices, M1, M3 and M4 and a reentrant helix M2. The M1 and M4 helices form a ring around the M3 ion channel core, while M2 helix defines the selectivity filter [81].

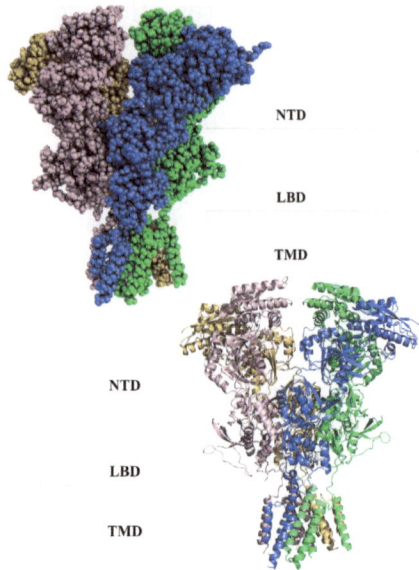

Fig. (1). NMDA receptor structure of *Rattus norvegicus* showing three distinct domains: N-terminal (NTD), ligand binding domain (LBD) and the transmembrane domain (TMD) (Resolution = 3.96 A, PDB code: 4PE5) [20]. The receptor is shown as van der Walls surface (left) and ribbon representation (right). GluN1 subunits are show in light green and pink, while GluN2 are shown in blue and wheat.

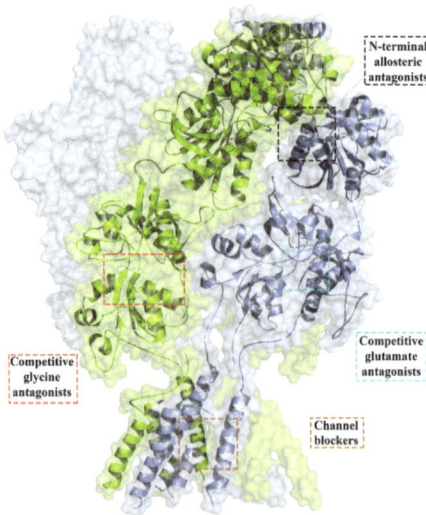

Fig. (2). Full structure of NMDA receptor of *Rattus norvegicus* revealing receptor's architecture and the binding sites of different types of antagonists as: N-terminal allosteric antagonists (black box), competitive glutamate antagonists (blue box), competitive glycine antagonists (red box) and channel blockers (orange box) (GluN1A/GluN2B; Resolution = 3.96 A; PDB: 4PE5) [20]. GluN2B subunit is shown in purple and GluN1 in green.

NMDAR subunits assemble as dimer-of-dimers in all domains with a GluN1-GluN2-GluN1-GluN2 orientation, with two-fold symmetry between NTD and LBD and a pseudo-fourth fold symmetry on the TMD. There is a GluN1-GluN2 swap between NTD and LBD, with the former showing GluN1(1)-GluN2(2) and GluN1(1')-GluN2(2'), while the latter shows GluN1(1)-GluN2(2') and GluN1(1')-GluN2(2) interactions [79, 81].

NMDAR bears binding sites for small molecules that act as positive or negative modulators (Fig. **2**). Three types of modulators were described according to the binding domain: competitive agonist/antagonists, allosteric modulators and channel blockers [12, 31, 84]. Furthermore, protons, zinc ions, and polyamines are potent modulators of NMDAR function. The first compounds studied to modulate NMDAR function were antagonists of the glutamate binding site, such as D-APV (**7**) and CPP (**8**) (Scheme 2). Glycine competitive antagonist (R)-HA966 (**9**) (Scheme 2) and DCKA (**10**) (Scheme 2) were developed after the discovery that glycine was a NMDAR co-agonist. Ion channel blockers as phencyclidine (**11**), ketamine (**12**) and memantine (**13**) (Scheme 2) were discovered between 1980 and 1990. Ifenprodil (**14**) (Scheme 2), a GluN2B-selective antagonist, was the first allosteric modulator targeting the NTD discovered and was followed by several ifenprodil-like compounds, such as traxoprodil (**15**) (Scheme 2) [12, 85].

NMDAR modulators targeting the LBD can be agonists or antagonists of the glutamate (Fig. **3**) or glycine binding sites (Fig. **4**). Many groups have focused their studies on the development of antagonists. Competitive glutamate antagonists usually show the following selectivity order for the NMDAR subunits: GluN2A>GluN2B>GluN2C>GluN2D, even though the glutamate binding site is highly conserved [85]. Some glutamate competitive antagonists were described with different subunit selectivity by adding bulky hydrophobic groups to interact with other pockets of the binding site, such as **7** and analogs, but these compounds also lack selectivity and block normal excitatory synaptic activity and, consequently, have severe adverse effects, such as hallucinations and agitation. Furthermore, for the therapeutic effects, these compounds would need higher doses to compete with high concentrations of glutamate during excitotoxicity. Taken together, these properties and effects of the competitive antagonists hampered the development of broad-spectrum glutamate antagonists [28, 45]. Nonetheless, many compounds are still used in research to gain insights on the different roles of GluN2 subunits.

Compounds targeting the glycine binding site (**9** and **10**) (Fig. **2**) were identified before the receptor structure was elucidated and these compounds showed little to no subunit selectivity leading to problems similar to glutamate competitive antagonists (Fig. **4**) [85, 86].

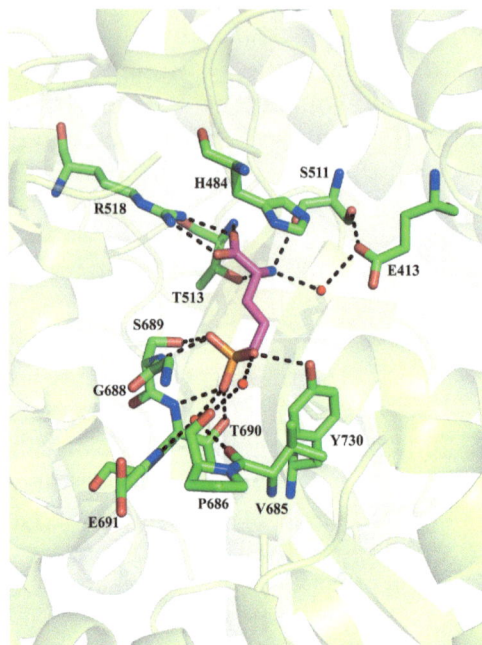

Fig. (3). Binding mode of D-AP5 (purple) in the ligand binding site of GluN2A subunit of *R. norvegicus.* Water molecules are shown as red spheres (Resolution = 1,9 Å; PDB: 4NF5) [189]. Jespersen A, Tajima N, Fernandez-Cuervo G, Garnier-Amblard EC, Furukawa H. Structural insights into competitive antagonism in NMDA receptors. Neuron. 2014; 81(2): 366-78. [http://dx.doi.org/10.1016/j.neuron.2013.11.033] [PMID: 24462099].

Channel blockers, such as ketamine (**12**) and memantine (**13**) (Scheme 2) are used in clinics for anesthesia and AD treatment, respectively. Ketamine is also used as a recreational drug and has antidepressant activity [87, 88]. Zanos and coworkers (2016) described that the antidepressants effects might be attributed to the ketamine metabolite (2S,6S;2R,6R)-hydroxynorketamine (HNK). HNK acts on AMPAR without any activity on NMDAR and anesthetic effects. Furthermore, HNK did not cause significant change in locomotor function or disorientation in animals [87]. Phencyclidine (**11**) is a high-affinity channel blocker that was removed from the market due to dissociative hallucinogenic adverse effects [11, 31, 85]. The most tolerable channel blockers usually show fast off kinetics and preferentially block the ion channel during receptor hyperactivation and glutamate excess (*i.e.* excitotoxicity), and it spares physiological functions in low glutamate concentrations [31, 86]. Memantine was patented by Eli Lilly & Co in the 1960s and was approved by the European Union and the US FDA in 2003 for AD treatment. Memantine mechanism of action consists of blocking ion current without affecting glutamate binding, with a fast off-rate and voltage-dependent blockade which allows the compound to diffuse from the receptor more readily than other high-affinity channel blockers. These properties are associated with memantine clinical efficacy and better tolerability than high-affinity blockers, with low overall adverse effects and low potential for drug-drug interactions [88 - 90].

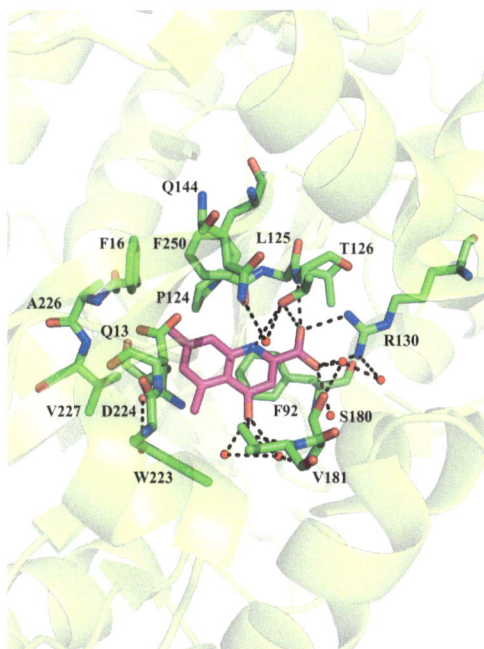

Fig. (4). Binding mode of DCKA (purple) in the ligand binding site of GluN1 subunit of *R. norvegicus*. Water molecules are shown as red spheres (Resolution = 2.0 Å; PDB: 4NF4) [189]. Jespersen A, Tajima N, Fernandez-Cuervo G, Garnier-Amblard EC, Furukawa H. Structural insights into competitive antagonism in NMDA receptors. Neuron. 2014; 81(2): 366-78. [http://dx.doi.org/10.1016/j.neuron.2013.11.033] [PMID: 24462099].

The NTD of NMDAR is the best-characterized region of the receptor. It regulates ion channel function and contributes to the diversity in NMDAR pharmacology [14, 80, 91]. The NTD is the major locus of allosteric binding sites for small ligands and the focus of many studies since the discovery of ifenprodil (**14**) [92]. To date, three binding sites were described in the NTD that interact with endogenous zinc, polyamines and with synthetic compounds such as ifenprodil and its analogs [17].

Zinc ions bind to the inter-domain cleft of the GluN2A or GluN2B subunits, with higher affinity for the former, which can be used to discriminate between receptor subtypes [84]. In addition, zinc makes direct polar interactions with His127 and Glu284 and binds in close proximity to other residues, such as Glu47 and Asp265, which probably help zinc binding with water molecules to stabilize a closed conformation of the NTD bilobed structure (Fig. **5**) [93]. Endogenous polyamines spermine, spermidine and putrescine are highly specific positive allosteric modulators of GluN2B NMDAR that act as a molecular glue between GluN1 and GluN2B subunits by shielding a cluster of negatively-charged residues from both subunits and preventing the NTD clamshell closure [91].

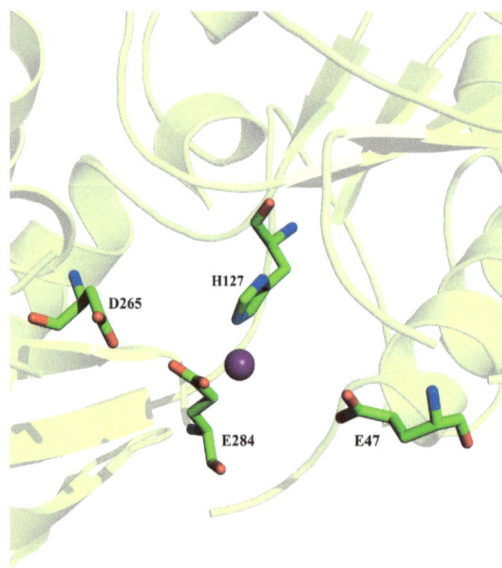

Fig. (5). Crystal structure of zinc (purple sphere) bound in the N-terminal domain of GluN2B subunit of *R. norvegicus* (Resolution = 3.21 Å; PDB: 3JPY) [93].

Synthetic allosteric modulators were first introduced with ifenprodil (Scheme 2), a GluN2B-selective antagonist, which showed neuroprotection against NMDA-mediated excitotoxicity in cultured neurons and *in vivo* in cerebral ischemia [74]. Several series of ifenprodil-like molecules and other classes of related compounds such as traxoprodil (**15**) [94], eliprodil (**16**) [95], radiprodil (**17**) [84], besonprodil (**18**) [96] and Ro-25-6981 (**19**) [97] (Scheme 3) have been developed and are available as pharmacological tools to study receptor structure and function.

These antagonists share a pharmacophore that consists of two aromatic rings separated by a highly variable linker (usually 9-11 Å in length) and a hydrogen bond donor as substituent in one of the rings [39, 57]. Site-directed mutagenesis and virtual screening approaches predicted that ifenprodil-like compounds bind to the NTD of the GluN2B subunits. However, Karakas and coworkers (2011) obtained crystallography structures of ifenprodil and Ro-256981 bound to the interface between GluN1 and GluN2B subunits of the hybrid NMDAR NTD from

Rattus norvegicus and *Xenopus laevis*. The complexes showed that the compounds interact primarily by hydrophobic interactions with residues from GluN1 and GluN2B and make three direct polar interactions with Ser132 of GluN1 and Gln110 and Asp236 of GluN2B (Fig. **6**) [79].

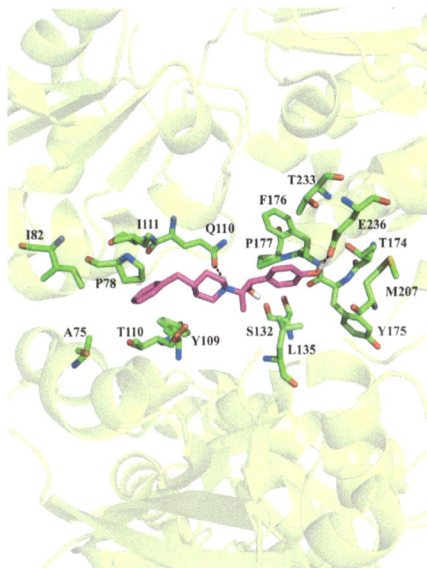

Fig. (6). Binding mode of Ifenprodil (purple) at the cleft between GluN1 (X. laevis) and GluN2B (*R. norvegicus*) N-terminal domain of NMDA receptor (Resolution = 2.6 Å; PDB code: 3QEL) [79].

Since the discovery of ifenprodil, many groups synthesized similar compounds with a variety of potencies and pharmacokinetics profiles. Structurally diverse GluN2B-selective antagonists were also described, such as **20** [98], **21** [99], **22** [100] and **23** [101], showing little resemblance to ifenprodil and raising the possibility of novel binding modes or sites (Scheme 4).

In 2016, two groups proposed a mechanism of action of ifenprodil, showing that ligand binding is followed by structural rearrangement of NTD, with GluN2D-NTD clamshell closure and shortening of GluN1-NTD dimer distances. Ligand binding also strengthens inter-LBD interactions and with the NTD, stabilizing the receptor in an agonist-bound, but desensitized-state [14, 81]. Due to the highly selective action, ifenprodil-like compounds are widely used for pharmacological studies of the receptor and as lead compounds in drug discovery strategies, since they are associated with enhanced tolerability than broad-spectrum NMDAR antagonists [17, 85, 102 - 104]. However, clinical trials with ifenprodil analogs have not been successful yet, possibly due to inhibition of other targets such as Ca^{2+} channels, α1-adrenergic receptors, and 5-HT$_{1a}$ and 5-HT$_2$ receptors [105, 106].

The studies on NMDAR noncompetitive antagonists are primarily focused on GluN2B-selective antagonists. The development of compounds targeting other subunits and/or binding sites has been lagging, with fewer molecules described in the last years [12]. Other allosteric antagonists with different scaffolds than ifenprodil were also described showing selectivity towards GluN2C-D subunits, probably binding to different sites. Quinazolin-4-one derivative (QNZ46, **24**) showed 50-fold selectivity for GluN2C/D over GluN2A/B NMDAR. Studies with chimeric receptors showed a noncompetitive mechanism of action and that the activity was enhanced with L-glutamate binding (Scheme 5) [107].

Bettini and coworkers (2010) described a series of sulfonamide derivatives as allosteric antagonists with selectivity for GluN2A subunit. Among these, TCN-201 (**25**) displayed over 500-fold selectivity for GluN2A subunits showing a unique mechanism of action (Scheme 5). TCN-201 did not compete with glycine (or D-serine) and L-glutamate, indicating a noncompetitive mechanism of action. In addition, some residues on the LBD interface between GluN1-GluN2A were important for TCN-201 activity, such as Val783 in GluN2A, which is substituted by Phe in GluN2B and Leu in GluN2C and D [108, 109], indicating that specific interactions could be explored to develop selective GluN2A antagonists.

Quinazolin-4-ones with pyrazoline scaffold were also described as GluN2C/D antagonists. DQP1105 (4-(5-(4-bromophenyl)-3-(6-methyl-2-oxo-4-phenyl-1,2-dihydroquinolin-3-yl)-4,5-dihydro-1*H*-pyrazol-1-yl)-4-oxobutanoic acid) (**26**) showed a voltage-independent, noncompetitive mechanism of action with 50-fold selectivity for GluN2C and GluN2D receptor over GluN2A and GluN2B (Scheme 5). However, the exact binding site and mechanism of action have not been elucidated yet [110].

DEVELOPMENT OF NEW CNS DRUGS

Previously, the drug discovery process was based on empirical approaches, testing natural products or some synthetic compounds on cells and whole animals. This approach resulted in successful drugs, but only because the compounds tested have already shown drug-like physicochemical properties and good pharmaco-kinetic profile [111].

With the technological advances in chemistry and biology, drug discovery became a rational approach, relying on molecular structure observations and how it correlates with *in vitro* and *in vivo* effects. The development of molecular biology, structural biology, and DNA sequencing led to the identification of novel macromolecular targets associated with diseases [112, 113]. Nuclear magnetic resonance (NMR) and X-ray crystallography enable the knowledge about the three-dimensional structure of many molecular targets [113, 114]. In addition, techniques such as high throughput screening (HTS) increased the number of compounds evaluated [115, 116].

The development of CNS drugs is a highly competitive market with billions of dollars invested in delivering a unique drug to the market. The aging of the population and higher incidence of CNS disorders make the development of new CNS drugs an urgent need, Despite it, the CNS drugs market suffers from the problems common to the discovery of other drugs, such as slow innovation high attrition rates and high costs [117, 118]. In addition, CNS drug discovery is inherently hard because of the drug's permeability through the blood-brain barrier (BBB).

The BBB is composed of brain capillary endothelium and is an extremely selective barrier that blocks most of the small molecules to reach the CNS. This reduced permeability is the result of the tight junctions and lack of fenestration, which prevent paracellular diffusion. Molecules that cross the BBB do it by transcellular diffusion and are subjected to metabolic enzymes, especially cytochrome P450 enzymes (CYP). Thus, oral drugs targeting the CNS must be stable enough to survive intestinal and liver metabolism, but also to cross the BBB. In addition, the drug may be pumped back to the blood by p-glycoproteins efflux pumps [119 - 122].

Thus, for CNS drugs orally administered, the physicochemical properties need to be restricted to avoid permeability problems [119, 120, 123]. Aqueous solubility is one of the most important properties that require optimization in drug development since only the amount of drug in solution can be absorbed and carried through the blood to its target tissue [124, 125]. In some cases, prodrug approaches were developed for increasing the bioavailability of poorly soluble CNS drugs [126]

Lipophilicity (measured as the logarithm of n-octanol/water coefficient, log P) is also important for compounds permeability into the CNS. Compounds with moderate lipophilicity penetrate lipid bilayers more easily. On the other hand, extremely lipophilic compounds tend to bind proteins and lipid membranes [127, 128]. Besides, lipophilic molecules generally show poor solubility. However other factors such as the presence of polar groups and inter and intramolecular interactions may have a strong influence on solubility. Hansch and coworkers observed that log P correlates well with CNS penetration, with optimal value for log P = 2.1 [129]. Michael Hann from GlaxoSmithKline used the term "molecular obesity" to address the tendency to build potency into molecules by increasing lipophilicity, which ultimately decreases solubility [130, 131]. Thus, it is advised to avoid increasing CNS activity by adding more lipophilic groups.

The polar surface area (PSA) is the molecular area occupied by the polar atoms on a molecule (especially O and N) [132]. PSA can be used to evaluate hydrogen bond capacity and now it can be easily calculated and implemented on virtual screening approaches [132]. In general, CNS drugs present lower PSA than other drugs, generally with values between 60-70 $Å^2$, but an upper limit of 90 $Å^2$ can also be used to predict good CNS permeability [124, 127].

The molecular weight (MW) affects CNS penetration, with large molecules being considered less permeable than small polar molecules. For CNS oral drugs, the usual MW cut-off is below 450 Da to preserve penetration and intestinal absorption. In addition, the mean of MW of marketed CNS drugs is below 400 Da [127].

There are many physicochemical guidelines for CNS drug development, involving primarily log P and PSA. However, these should be used with careful, since interesting compounds might be outliers. Nonetheless, molecules that must cross the BBB, in general, tend to be more polar, present low MW and be stable enough to support intestinal, liver and BBB enzymes.

Molecular Modeling as an Important Tool for the Design of New NMDA Receptor Antagonists

The determination of physicochemical properties of the compounds related to pharmacokinetics, and interactions with a molecular target is a highly complex process that costs millions of dollars in drug development [133]. Molecular modeling tools enable to eliminate problematic compounds in the early stages of the process, therefore reducing the cost of lead optimization in the clinical phases. Thus, a combination of activity, selectivity, structure-based design and pharmacokinetic profile is essential for rational drug design [134, 135].

The pharmaceutical industry and some academic groups have adopted the high throughput screening (HTS) to screen large collections of molecules to identify lead compounds and it can be the starting point to develop a new drug [136, 137]. Despite being the method of choice, HTS involves high investments in equipment and screening technology. In addition, HTS suffers from high attrition due to compound interference on assays and false positives, and the various mechanisms of action of hit compounds [117, 138, 139]. Therefore, cost-efficient methods to screen large compounds collections is an active field in drug discovery. In this scenario, computational methodologies have become very important, from target validation to lead optimization [137, 140 - 143].

In silico ADME Prediction for CNS Drug Design

Oral drugs need to cross barriers to reach its target *in vivo* and have the pharmacological effect. These barriers include cell membranes and the BBB for CNS drugs. In addition, oral drugs need to be stable, to remain in the body for enough time to have an effect and must be safe [144, 145]. Pharmacokinetics predictions are now common in early stages of drug discovery, however, until the 1990s, pharmacokinetic studies were addressed only in late phases of development, resulting in high attrition rates and high costs [124, 139]. In modern-day drug discovery, pharmacokinetic failures are much less common. *In silico* predictions of ADME parameters are valuable in drug discovery since it allows early analysis and optimization of the most promising compounds [139, 146, 147].

In silico ADME often relies on good experimental quality (*i.e.* in vitro and in vivo) data to build models. The model can be built using QSAR and QSPR to search for correlations between molecular properties and ADME endpoints. Other methods such as self-organizing maps (SOMs), partial least squares (PLS), decision trees, neural networks are used and the reliability of the model depends on the choice of the most appropriate mathematical representation [144 - 146, 148].

A variety of descriptors have been developed for QSAR and prediction capacity varies with the descriptors used such as MW, lipophilicity, hydrogen bond donors and acceptors, and number of heavy atoms. Lipinski developed the Lipinski's "rule of five", describing that simple descriptors such as MW, log P, hydrogen bond acceptors and donors could be used to predict if a molecule may be oral bioavailable based on drug-like properties (*i.e.* comparing with marketed oral drugs) [124, 149]. Since then, many groups have published guidelines on how molecular descriptors affect pharmacokinetics and these guidelines were applied on commercial and freely available software [125, 150, 151]. These rules help industry and academic groups to compare compounds and prioritize those with better quality and relatively less risk to fail on screening. However, these guidelines should not be followed blindly, since pharmacokinetics is a multifactorial process (Table **2**) [127, 146, 152, 153].

Table 2. General guidelines for CNS drug development.

Recommended physicochemical properties
Log P = 2-3
PSA = 60-90 A^2
MM $\leq$ 450 Da

(Table 2) contd.....

Recommended physicochemical properties
No undesirable functional group
Do not increase potency by increasing lipophilicity

In silico absorption assessments are usually based on simple physicochemical properties such as log P (or log D for ionizable molecules) and PSA to account for hydrogen-bond [154]. Some examples are the Rule of Five [124], Veber's rule [150] and Egan's rule [155]. These guidelines are available in commercial and freely available sources such as GastroPlus™, QikProp, VolSurf [156], admetSAR (http://lmmd.ecust.edu.cn:8000/) [157] and FAF-DRUGS3 (http://fafdrugs3.mti.univ-paris-diderot.fr/) [158].

Many BBB predictions are based on guidelines of physicochemical properties, namely log P, PSA and molecular size, which are readily available and fast to calculate using software such Chemaxon's Marvin suite (http://www.chemaxon.com) and ACD/ChemSketch. Other approaches focus on QSAR models using log BB data to label compounds as CNS-active or non-active, which differ on the training set, the molecular descriptors, classification method and mathematical treatment [123, 159].

In addition to statistical modeling, an emerging field is the ADMET predictions based on structural information of macromolecules relevant to pharmacokinetics such as CYPs. Due to the limitations of QSAR and related models applicability, structure-based ADME predictions are an interesting alternative [146]. This approach involves the use of experimental 3D structures (or homology models) to predict which compounds may interact with a macromolecular target, usually using docking methods. Both experimental data and crystal structures of relevant targets, such as CYPs, hERG channel, sulfotransferases and p-glycoprotein are freely available. In addition, commercial services allow fast screening and evaluation of potential off-target interactions and metabolism [146].

In silico Toxicity Prediction for CNS Drug Design

Toxicity screening is an important part of drug discovery campaigns and early assessment of toxicity can be valuable to prioritize compounds for tests [160]. *In silico* toxicology aims to identify potential risks from chemical structure. The approaches used in toxicological studies can be as simple as substructure identification or more complex such as QSAR models for carcinogenicity [161, 162]. In addition, the main advantages of *in silico* modeling are its low-cost, short time of execution and low investment in equipment (*i.e.* nowadays, any computer can run these calculations).

Guidelines based on compounds from pharmaceutical industry collections have been developed for toxicity prediction. Such rules apply cutoff values for the calculated properties that usually indicate higher probability of toxicity and failure in clinical stages. At Pfizer, an analysis of 245 drug candidates in experimental models of toleration *in vivo* showed that less polar and more lipophilic compounds showed increased toxicity risk. Based on Pfizer 3/75 guideline, molecules with log P > 3 and PSA < 75 Å^2 show higher risks than the ones with log P < 3 and PSA > 75 Å^2 [163]. Another study at GSK identified that compounds with log P > 4 and MW > 400 Da present higher risks of toxicity (GSK 4/400 guideline) [164]. A common aspect of these studies is the definition of lipophilicity (measured by log P) as important for toxic events, which can be related to promiscuous off-target binding. Therefore, it is recommended to keep lipophilicity as low as possible and a practical range would be log P between 2 and 3 (Table **2**).

The chemical structure can give insights of possible toxicity. For example, portions of a molecule (substructure) might be associated with free radical production and covalent binding (*i.e.* cathecols and quinones), while others, such as anilines and phenols, can undergo metabolic activation of reactive species that may damage/disrupt important cellular structures (*i.e.,* the membrane, mitochondria and DNA) and functions [165]. These substructures are known as toxicophores and are a major burden in drug development since they can lead to failure in clinical tests and idiosyncratic adverse reactions [166 - 169]. Many works have addressed toxicophore identifications and databases were developed in order to inform medicinal chemists and structural biologists about which risks a compound may have while comparing it with similar molecules on the database.

FAF-Drugs3 is a server developed to filter and prioritize molecules for screening. The toxicophore alerts in FAF-Drugs3 use a list of 154 SMARTS code for chemical groups associated with toxicity issues, classified according to chemical function. The molecules are then flagged as intermediate, rejected or accepted [158]. Web servers like FAF-Drugs3 are freely available and enable the examination of thousands of virtual molecules.

QSAR models predict the toxicity of new compounds using a training set. These applications can be used to predict specific endpoints, such as mutagenicity or simulate *in vitro* tests like the Ames test. As other QSAR methods, they have an applicability domain (*i.e.* the information on the training set can be used to make predictions for other compounds), but for compounds outside this space, toxicity predictions are not reliable [157, 170]. In addition, models of relatively simple process are easier to develop, since they might involve well-defined mechanisms. On the other hand, complex endpoints such as carcinogenicity and teratogenicity usually present lower accuracy because of the multiplicity of factors that can lead to them and the complexity of these processes. Thus, it is important to know the applicability domain and which endpoint to model for a prediction to be useful [145 - 147].

VIRTUAL SCREENING FOR THE DISCOVERY OF NEW GLUN2B SELECTIVE NMDA RECEPTOR ANTAGONISTS

Virtual screening (VS) is the screening of large compounds collections using computers. One major advantage of VS is that one does not need to synthesize or buy the compounds to generate a hypothesis. In addition, *in silico* methods allow selecting rationally the most promising compounds for *in vitro* and *in vivo* screening, which significantly reduces costs and attrition rates due to interference and false-positive. Many groups reported the discovery of lead compounds using VS strategies validating its valuable role in drug discovery [171]. Virtual screening can be divided into two categories: ligand-based (LBVS) and structure-based (SBVS) [131, 140, 172, 173]. These categories differ in the fact that LBVS does not consider the target structure [84, 131, 174, 175].

Ligand-based virtual screening uses structure-activity relationships (SAR) data to derive mathematical and pharmacophore models, and similarity search in order to correlate structural properties with biological activity. The models are derived from test sets with known activity, and they are used to predict the biological activity of new compounds. In addition, LBVS may be used to search compounds by similarity considering molecular structure, physicochemical properties and pharmacophore [131].

Structure-based virtual screening uses experimentally determined 3D structures of macromolecules, such as proteins (*e.g.* enzymes and receptors), to identify and rank small molecules that may interact with the target [176, 177]. Docking methods are extensively used in SBVS to select compounds based on their binding affinity for the binding site [171, 178]. Usually, the target is kept rigid while entropic and solvation effects are ignored. These approximations are used to speed calculations, especially when screening large compounds collections [179]. The docking method can be divided into two parts; pose prediction (orientation and conformational search) and scoring. Different algorithms are used by the programs for conformational search, divided in systematic search, which explores all degrees of freedom; and stochastic methods, which randomly generate low energy conformers. Although the pose prediction can be an intensive work depending on the ligand and/or proteins degrees of freedom, scoring is a major challenge in docking studies [177, 180, 181].

Scoring functions are used to rank compounds, distinguishing between inactive and active compounds, in order to select the most promising for further tests. These functions estimate the binding energy using approximated expressions of the physical phenomena that drive binding (*i.e.* enthalpy, entropy, solvation and hydrophobic effects), but are usually not reliable to estimate the free energy of binding [182 - 184]. Therefore, it may be difficult to distinguish the correct binding pose of the compounds.

Despite the limitations, docking approaches enable testing hypothesis before experimental assays and SBVS complements HTS in order to find hit compounds [91, 100]. In addition, SBVS allows the screening of millions of virtual molecules in a few hours using high-performance computing and usually shows higher hit-rates than HTS [175, 185]. Despite new information about NMDAR structure, small molecule binding sites and the recent identification of the NMDAR activation mechanism and its inhibition by ifenprodil, there is a lack of published works on VS methods for NMDAR modulators. Most of the works are focused on allosteric GluN2B-selective antagonists either ifenprodil-like or other structural diverse compounds.

Mony and coworkers [186] identified a new series of GluN2B-selective antagonists using pharmacophore-based virtual screening (Scheme 6). Although derivative **27** (IC_{50} = 2.7 µM) showed 10-fold reduced activity compared to ifenprodil, this compound represents a novel scaffold that selectively inhibits GluN2B-containing NMDAR. SAR exploration revealed that adding a second hydroxyl group or changing the double bond configuration resulted in slightly more potent derivatives **28** (IC_{50} = 2.7 µM) and **29** (IC_{50} = 2.7 µM).

Gitto and coworkers [187] described a series of 3-substituted-1H-indoles selective GluN2B-antagonists and used docking studies to explain the biological results. According to the authors, derivatives **30** (IC_{50} = 25 nM), **31** (IC_{50} = 17 nM) and **32** (IC_{50} = 22 nM) (Scheme 7) displayed the highest binding affinity on ifenprodil inhibition assay. In addition, SAR analysis revealed that 6-OH substitution on the indole ring led to compound **31** which is 45-fold more potent than the unsubstituted compound. Docking simulations of these ligands revealed a similar binding mode compared to ifenprodil. However, derivative **31** did not show activity *in vivo*, which could be related to poor BBB penetration. In a follow-up study, Gitto and coworkers [188] designed derivative **33**, bearing a 6-methanosulfonamide group on the indole ring, which showed nanomolar potency on GluN2B receptors (IC_{50} = 8.9 nM), but reduced *in vivo* activity, probably as a result of high PSA (82.27 $Å^2$) (Scheme 7).

Using a combination of site-directed mutagenesis, molecular dynamics and docking simulations, Burger and coworkers [96] investigated the determinants for the binding of ifenprodil-like antagonists to the receptor, such as **34-37** (Scheme 8). According to the authors, the binding pocket for all these compounds was on the GluN1/GluN2B interface, and they presented similar binding modes but different interactions with the subunits. In addition, ligands with a similar binding mode showed different downstream functional effects, suggesting that this class of antagonists could be divided according to changes in receptor motion distant from the binding site.

Stroebel and coworkers [102] used docking to study the binding mode of GluN2B-selective NTD antagonists. The authors described a novel binding mode for compounds structurally diverse from ifenprodil. They only partially overlap with the ifenprodil-like compounds in the binding site, resulting in a different orientation and atomic contacts. In addition, the authors suggested that exploring this structural diversity could facilitate new SAR and drug development prospects for NMDAR modulators.

CONCLUDING REMARKS

The involvement of glutamate and NMDA receptor in several neurodegenerative disorders and CNS acute damage has stimulated the search for safer and effective drugs targeting this receptor. Of particular interest are the GluN2B subunit selective antagonists, which seem to present less adverse effects. However, none of these antagonists discovered so far was approved as a drug. Drug design has been a challenge for medicinal chemistry and the use of molecular modeling combining ligand and structure-based virtual screening with the *in silico* ADME and toxicity prediction may increase the success rates in designing new CNS drugs.

CONFLICT OF INTEREST

The authors (editor) declares no conflict of interest, financial or otherwise.

ACKNOWLEDGMENTS

Marcos Vinicius Santana had a postgraduate scholarship from the Coordination for the Improvement of Higher Education Personnel (CAPES).

REFERENCES

[1]   Fonnum F. Glutamate: a neurotransmitter in mammalian brain. J Neurochem 1984; 42(1): 1-11.
        [http://dx.doi.org/10.1111/j.1471-4159.1984.tb09689.x] [PMID: 6139418]

[2]   Meldrum BS. Glutamate as a neurotransmitter in the brain: review of physiology and pathology. J Nutr 2000; 130(4S Suppl): 1007S-15S.

[3]   Niciu MJ, Ionescu DF, Richards EM, Zarate CA Jr. Glutamate and its receptors in the pathophysiology and treatment of major depressive disorder. J Neural Transm (Vienna) 2014; 121(8): 907-24.

[http://dx.doi.org/10.1007/s00702-013-1130-x] [PMID: 24318540]

[4]   Niswender CM, Conn PJ. Metabotropic glutamate receptors: physiology, pharmacology, and disease. Annu Rev Pharmacol Toxicol 2010; 50(1): 295-322.
[http://dx.doi.org/10.1146/annurev.pharmtox.011008.145533] [PMID: 20055706]

[5]   Ferraguti F, Shigemoto R. Metabotropic glutamate receptors. Cell Tissue Res 2006; 326(2): 483-504.
[http://dx.doi.org/10.1007/s00441-006-0266-5] [PMID: 16847639]

[6]   Kandel ER, Schwartz JH, Jessell TM. Principles of neural science. 5th ed. New York: McGraw-hill 2013; pp. 217-22.

[7]   Wollmuth LP, Sobolevsky AI. Structure and gating of the glutamate receptor ion channel. Trends Neurosci 2004; 27(6): 321-8.
[http://dx.doi.org/10.1016/j.tins.2004.04.005] [PMID: 15165736]

[8]   Lerma J, Marques JM. Kainate receptors in health and disease. Neuron 2013; 80(2): 292-311.
[http://dx.doi.org/10.1016/j.neuron.2013.09.045] [PMID: 24139035]

[9]   Fritsch B, Reis J, Gasior M, Kaminski RM, Rogawski MA. Role of GluK1 kainate receptors in seizures, epileptic discharges, and epileptogenesis. J Neurosci 2014; 34(17): 5765-75.
[http://dx.doi.org/10.1523/JNEUROSCI.5307-13.2014] [PMID: 24760837]

[10]   Kumar J, Mayer ML. Functional insights from glutamate receptor ion channel structures. Annu Rev Physiol 2013; 75(1): 313-37.
[http://dx.doi.org/10.1146/annurev-physiol-030212-183711] [PMID: 22974439]

[11]   Paoletti P, Bellone C, Zhou Q. NMDA receptor subunit diversity: impact on receptor properties, synaptic plasticity and disease. Nat Rev Neurosci 2013; 14(6): 383-400.
[http://dx.doi.org/10.1038/nrn3504] [PMID: 23686171]

[12]   Ogden KK, Traynelis SF. New advances in NMDA receptor pharmacology. Trends Pharmacol Sci 2011; 32(12): 726-33.
[http://dx.doi.org/10.1016/j.tips.2011.08.003] [PMID: 21996280]

[13]   Karakas E, Regan MC, Furukawa H. Emerging structural insights into the function of ionotropic glutamate receptors. Trends Biochem Sci 2015; 40(6): 328-37.
[http://dx.doi.org/10.1016/j.tibs.2015.04.002] [PMID: 25941168]

[14]   Tajima N, Karakas E, Grant T, *et al.* Activation of NMDA receptors and the mechanism of inhibition by ifenprodil. Nature 2016; 534(7605): 63-8.
[http://dx.doi.org/10.1038/nature17679] [PMID: 27135925]

[15]   Gladding CM, Fitzjohn SM, Molnár E. Metabotropic glutamate receptor-mediated long-term depression: molecular mechanisms. Pharmacol Rev 2009; 61(4): 395-412.
[http://dx.doi.org/10.1124/pr.109.001735] [PMID: 19926678]

[16]   Weilinger NL, Maslieieva V, Bialecki J, Sridharan SS, Tang PL, Thompson RJ. Ionotropic receptors and ion channels in ischemic neuronal death and dysfunction. Acta Pharmacol Sin 2013; 34(1): 39-48.
[http://dx.doi.org/10.1038/aps.2012.95] [PMID: 22864302]

[17]   Paoletti P, Neyton J. NMDA receptor subunits: function and pharmacology. Curr Opin Pharmacol 2007; 7(1): 39-47.
[http://dx.doi.org/10.1016/j.coph.2006.08.011] [PMID: 17088105]

[18]   Papouin T, Ladépêche L, Ruel J, *et al.* Synaptic and extrasynaptic NMDA receptors are gated by different endogenous coagonists. Cell 2012; 150(3): 633-46.
[http://dx.doi.org/10.1016/j.cell.2012.06.029] [PMID: 22863013]

[19]   Sanz-Clemente A, Nicoll RA, Roche KW. Diversity in NMDA receptor composition: many regulators, many consequences. Neuroscientist 2013; 19(1): 62-75.
[http://dx.doi.org/10.1177/1073858411435129] [PMID: 22343826]

[20]   Karakas E, Furukawa H. Crystal structure of a heterotetrameric NMDA receptor ion channel. Science

2014; 344(6187): 992-7.
[http://dx.doi.org/10.1126/science.1251915] [PMID: 24876489]

[21]   Mothet J-P, Parent AT, Wolosker H, *et al.* D-serine is an endogenous ligand for the glycine site of the
       N-methyl-D-aspartate receptor. Proc Natl Acad Sci USA 2000; 97(9): 4926-31.
       [http://dx.doi.org/10.1073/pnas.97.9.4926] [PMID: 10781100]

[22]   Nevian T, Sakmann B. Single spine Ca2+ signals evoked by coincident EPSPs and backpropagating
       action potentials in spiny stellate cells of layer 4 in the juvenile rat somatosensory barrel cortex. J
       Neurosci 2004; 24(7): 1689-99.
       [http://dx.doi.org/10.1523/JNEUROSCI.3332-03.2004] [PMID: 14973235]

[23]   Wu Y, Li Y, Gao J, Sui N. Differential effect of NMDA receptor antagonist in the nucleus accumbens
       on reconsolidation of morphine -related positive and aversive memory in rats. Eur J Pharmacol 2012;
       674(2-3): 321-6.
       [http://dx.doi.org/10.1016/j.ejphar.2011.11.011] [PMID: 22119382]

[24]   Hardingham GE, Bading H. Synaptic *versus* extrasynaptic NMDA receptor signalling: implications for
       neurodegenerative disorders. Nat Rev Neurosci 2010; 11(10): 682-96.
       [http://dx.doi.org/10.1038/nrn2911] [PMID: 20842175]

[25]   Okamoto S, Pouladi MA, Talantova M, *et al.* Balance between synaptic *versus* extrasynaptic NMDA
       receptor activity influences inclusions and neurotoxicity of mutant huntingtin. Nat Med 2009; 15(12):
       1407-13.
       [http://dx.doi.org/10.1038/nm.2056] [PMID: 19915593]

[26]   Kaufman AM, Milnerwood AJ, Sepers MD, *et al.* Opposing roles of synaptic and extrasynaptic
       NMDA receptor signaling in cocultured striatal and cortical neurons. J Neurosci 2012; 32(12): 3992-
       4003.
       [http://dx.doi.org/10.1523/JNEUROSCI.4129-11.2012] [PMID: 22442066]

[27]   Hynd MR, Scott HL, Dodd PR. Glutamate-mediated excitotoxicity and neurodegeneration in
       Alzheimer's disease. Neurochem Int 2004; 45(5): 583-95.
       [http://dx.doi.org/10.1016/j.neuint.2004.03.007] [PMID: 15234100]

[28]   Talantova M, Sanz-Blasco S, Zhang X, *et al.* Aβ induces astrocytic glutamate release, extrasynaptic
       NMDA receptor activation, and synaptic loss. Proc Natl Acad Sci USA 2013; 110(27): E2518-27.
       [http://dx.doi.org/10.1073/pnas.1306832110] [PMID: 23776240]

[29]   Hallett PJ, Standaert DG. Rationale for and use of NMDA receptor antagonists in Parkinson's disease.
       Pharmacol Ther 2004; 102(2): 155-74.
       [http://dx.doi.org/10.1016/j.pharmthera.2004.04.001] [PMID: 15163596]

[30]   Milnerwood AJ, Gladding CM, Pouladi MA, *et al.* Early increase in extrasynaptic NMDA receptor
       signaling and expression contributes to phenotype onset in Huntington's disease mice. Neuron 2010;
       65(2): 178-90.
       [http://dx.doi.org/10.1016/j.neuron.2010.01.008] [PMID: 20152125]

[31]   Chen HS, Lipton SA. The chemical biology of clinically tolerated NMDA receptor antagonists. J
       Neurochem 2006; 97(6): 1611-26.
       [http://dx.doi.org/10.1111/j.1471-4159.2006.03991.x] [PMID: 16805772]

[32]   Parsons MP, Raymond LA. Extrasynaptic NMDA receptor involvement in central nervous system
       disorders. Neuron 2014; 82(2): 279-93.
       [http://dx.doi.org/10.1016/j.neuron.2014.03.030] [PMID: 24742457]

[33]   Zhou X, Hollern D, Liao J, Andrechek E, Wang H. NMDA receptor-mediated excitotoxicity depends
       on the coactivation of synaptic and extrasynaptic receptors. Cell Death Dis 2013; 4(3): e560.
       [http://dx.doi.org/10.1038/cddis.2013.82] [PMID: 23538441]

[34]   Ivanov A, Pellegrino C, Rama S, *et al.* Opposing role of synaptic and extrasynaptic NMDA receptors
       in regulation of the extracellular signal-regulated kinases (ERK) activity in cultured rat hippocampal

neurons. J Physiol 2006; 572(Pt 3): 789-98.
[http://dx.doi.org/10.1113/jphysiol.2006.105510] [PMID: 16513670]

[35]   Snyder EM, Nong Y, Almeida CG, *et al.* Regulation of NMDA receptor trafficking by amyloid-beta. Nat Neurosci 2005; 8(8): 1051-8.
[http://dx.doi.org/10.1038/nn1503] [PMID: 16025111]

[36]   Texidó L, Martín-Satué M, Alberdi E, Solsona C, Matute C. Amyloid β peptide oligomers directly activate NMDA receptors. Cell Calcium 2011; 49(3): 184-90.
[http://dx.doi.org/10.1016/j.ceca.2011.02.001] [PMID: 21349580]

[37]   Rönicke R, Mikhaylova M, Rönicke S, *et al.* Early neuronal dysfunction by amyloid β oligomers depends on activation of NR2B-containing NMDA receptors. Neurobiol Aging 2011; 32(12): 2219-28.
[http://dx.doi.org/10.1016/j.neurobiolaging.2010.01.011] [PMID: 20133015]

[38]   Tackenberg C, Grinschgl S, Trutzel A, *et al.* NMDA receptor subunit composition determines beta-amyloid-induced neurodegeneration and synaptic loss. Cell Death Dis 2013; 4: e608.
[http://dx.doi.org/10.1038/cddis.2013.129] [PMID: 23618906]

[39]   Tu S, Okamoto S, Lipton SA, Xu H. Oligomeric Aβ-induced synaptic dysfunction in Alzheimer's disease. Mol Neurodegener 2014; 9(1): 48.
[http://dx.doi.org/10.1186/1750-1326-9-48] [PMID: 25394486]

[40]   Cruz JC, Tsai LH. Cdk5 deregulation in the pathogenesis of Alzheimer's disease. Trends Mol Med 2004; 10(9): 452-8.
[http://dx.doi.org/10.1016/j.molmed.2004.07.001] [PMID: 15350898]

[41]   Brunton L, Chabner B, Knollmann B. Goodman & Gilman's the pharmacological basis of therapeutics 2011.

[42]   Laruelle M. Schizophrenia: from dopaminergic to glutamatergic interventions. Curr Opin Pharmacol 2014; 14(1): 97-102.
[http://dx.doi.org/10.1016/j.coph.2014.01.001] [PMID: 24524997]

[43]   Cohen SM, Tsien RW, Goff DC, Halassa MM. The impact of NMDA receptor hypofunction on GABAergic neurons in the pathophysiology of schizophrenia. Schizophr Res 2015; 167(1-3): 98-107.
[http://dx.doi.org/10.1016/j.schres.2014.12.026] [PMID: 25583246]

[44]   Driesen NR, McCarthy G, Bhagwagar Z, *et al.* Relationship of resting brain hyperconnectivity and schizophrenia-like symptoms produced by the NMDA receptor antagonist ketamine in humans. Mol Psychiatry 2013; 18(11): 1199-204.
[http://dx.doi.org/10.1038/mp.2012.194] [PMID: 23337947]

[45]   Hu W, MacDonald ML, Elswick DE, Sweet RA. The glutamate hypothesis of schizophrenia: evidence from human brain tissue studies. Ann N Y Acad Sci 2015; 1338(1): 38-57.
[http://dx.doi.org/10.1111/nyas.12547] [PMID: 25315318]

[46]   Poels EM, Kegeles LS, Kantrowitz JT. Imaging glutamate in schizophrenia: review of findings and implications for drug discovery. Mol Psychiatry 2015; 152(0): 325-32.
[PMID: 24166406]

[47]   Billingslea EN, Tatard-Leitman VM, Anguiano J, *et al.* Parvalbumin cell ablation of NMDA-R1 causes increased resting network excitability with associated social and self-care deficits. Neuropsychopharmacology 2014; 39(7): 1603-13.
[http://dx.doi.org/10.1038/npp.2014.7] [PMID: 24525709]

[48]   Bygrave AM, Masiulis S, Nicholson E, *et al.* Knockout of NMDA-receptors from parvalbumin interneurons sensitizes to schizophrenia-related deficits induced by MK-801. Transl Psychiatry 2016; 6(4): e778.
[http://dx.doi.org/10.1038/tp.2016.44] [PMID: 27070406]

[49]   Lahti AC, Koffel B, LaPorte D, Tamminga CA. Subanesthetic doses of ketamine stimulate psychosis in schizophrenia. Neuropsychopharmacology 1995; 13(1): 9-19.

[http://dx.doi.org/10.1016/0893-133X(94)00131-I] [PMID: 8526975]

[50] Lahti AC, Weiler MA, Tamara Michaelidis BA, Parwani A, Tamminga CA. Effects of ketamine in normal and schizophrenic volunteers. Neuropsychopharmacology 2001; 25(4): 455-67.
[http://dx.doi.org/10.1016/S0893-133X(01)00243-3] [PMID: 11557159]

[51] Belforte JE, Zsiros V, Sklar ER, *et al.* Postnatal NMDA receptor ablation in corticolimbic interneurons confers schizophrenia-like phenotypes. Nat Neurosci 2010; 13(1): 76-83.
[http://dx.doi.org/10.1038/nn.2447] [PMID: 19915563]

[52] Zhang Y, Behrens MM, Lisman JE. Prolonged exposure to NMDAR antagonist suppresses inhibitory synaptic transmission in prefrontal cortex. J Neurophysiol 2008; 100(2): 959-65.
[http://dx.doi.org/10.1152/jn.00079.2008] [PMID: 18525022]

[53] Chang BS, Lowenstein DH. Epilepsy. N Engl J Med 2003; 349(13): 1257-66.
[http://dx.doi.org/10.1056/NEJMra022308] [PMID: 14507951]

[54] Fisher RS, Acevedo C, Arzimanoglou A, *et al.* ILAE official report: a practical clinical definition of epilepsy. Epilepsia 2014; 55(4): 475-82.
[http://dx.doi.org/10.1111/epi.12550] [PMID: 24730690]

[55] Morimoto K, Fahnestock M, Racine RJ. Kindling and status epilepticus models of epilepsy: rewiring the brain. Prog Neurobiol 2004; 73(1): 1-60.
[http://dx.doi.org/10.1016/j.pneurobio.2004.03.009] [PMID: 15193778]

[56] Nateri AS, Raivich G, Gebhardt C, *et al.* ERK activation causes epilepsy by stimulating NMDA receptor activity. EMBO J 2007; 26(23): 4891-901.
[http://dx.doi.org/10.1038/sj.emboj.7601911] [PMID: 17972914]

[57] Stelzer A, Slater NT, ten Bruggencate G. Activation of NMDA receptors blocks GABAergic inhibition in an *in vitro* model of epilepsy. Nature 1987; 326(6114): 698-701.
[http://dx.doi.org/10.1038/326698a0] [PMID: 2882427]

[58] Bertram E. The relevance of kindling for human epilepsy. Epilepsia 2007; 48(s2) (Suppl. 2): 65-74.
[http://dx.doi.org/10.1111/j.1528-1167.2007.01068.x] [PMID: 17571354]

[59] Min SS, Quan HY, Ma J, Han JS, Jeon BH, Seol GH. Chronic brain inflammation impairs two forms of long-term potentiation in the rat hippocampal CA1 area. Neurosci Lett 2009; 456(1): 20-4.
[http://dx.doi.org/10.1016/j.neulet.2009.03.079] [PMID: 19429126]

[60] Najjar S, Pearlman DM, Alper K, Najjar A, Devinsky O. Neuroinflammation and psychiatric illness. J Neuroinflammation 2013; 10(1): 43.
[http://dx.doi.org/10.1186/1742-2094-10-43] [PMID: 23547920]

[61] Wieck A, Andersen SL, Brenhouse HC. Evidence for a neuroinflammatory mechanism in delayed effects of early life adversity in rats: relationship to cortical NMDA receptor expression. Brain Behav Immun 2013; 28: 218-26.
[http://dx.doi.org/10.1016/j.bbi.2012.11.012] [PMID: 23207107]

[62] Chang YC, Kim HW, Rapoport SI, Rao JS. Chronic NMDA administration increases neuroinflammatory markers in rat frontal cortex: cross-talk between excitotoxicity and neuroinflammation. Neurochem Res 2008; 33(11): 2318-23.
[http://dx.doi.org/10.1007/s11064-008-9731-8] [PMID: 18500552]

[63] Rosi S, Ramirez-Amaya V, Hauss-Wegrzyniak B, Wenk GL. Chronic brain inflammation leads to a decline in hippocampal NMDA-R1 receptors. J Neuroinflammation 2004; 1(1): 12.
[http://dx.doi.org/10.1186/1742-2094-1-12] [PMID: 15285803]

[64] Steullet P, Cabungcal JH, Monin A, *et al.* Redox dysregulation, neuroinflammation, and NMDA receptor hypofunction: A "central hub" in schizophrenia pathophysiology? Schizophr Res 2016; 176(1): 41-51.
[http://dx.doi.org/10.1016/j.schres.2014.06.021] [PMID: 25000913]

[65] Liraz-Zaltsman S, Yaka R, Shabashov D, Shohami E, Biegon A. Neuroinflammation-Induced Memory Deficits Are Amenable to Treatment with D-Cycloserine. J Mol Neurosci 2016; 60(1): 46-62.
[http://dx.doi.org/10.1007/s12031-016-0786-8] [PMID: 27421842]

[66] Sil S, Ghosh T, Ghosh R. NMDA receptor is involved in neuroinflammation in intracerebroventricular colchicine-injected rats. J Immunotoxicol 2016; 13(4): 474-89.
[http://dx.doi.org/10.3109/1547691X.2015.1130760] [PMID: 26788903]

[67] Fernandez-Perez EJ, Peters C, Aguayo LG. Membrane Damage Induced by Amyloid Beta and a Potential Link with Neuroinflammation. Curr Pharm Des 2016; 22(10): 1295-304.
[http://dx.doi.org/10.2174/1381612822101603041111702] [PMID: 26972288]

[68] Maher A, El-Sayed NS, Breitinger H-G, Gad MZ. Overexpression of NMDAR2B in an inflammatory model of Alzheimer's disease: modulation by NOS inhibitors. Brain Res Bull 2014; 109: 109-16.
[http://dx.doi.org/10.1016/j.brainresbull.2014.10.007] [PMID: 25454121]

[69] Hauss-Wegrzyniak B, Dobrzanski P, Stoehr JD, Wenk GL. Chronic neuroinflammation in rats reproduces components of the neurobiology of Alzheimer's disease. Brain Res 1998; 780(2): 294-303.
[http://dx.doi.org/10.1016/S0006-8993(97)01215-8] [PMID: 9507169]

[70] Sutton ET, Thomas T, Bryant MW, Landon CS, Newton CA, Rhodin JA. Amyloid-beta peptide induced inflammatory reaction is mediated by the cytokines tumor necrosis factor and interleukin-1. J Submicrosc Cytol Pathol 1999; 31(3): 313-23.
[PMID: 10626000]

[71] Fujikawa DG. The role of excitotoxic programmed necrosis in acute brain injury. Comput Struct Biotechnol J 2015; 13: 212-21.
[http://dx.doi.org/10.1016/j.csbj.2015.03.004] [PMID: 25893083]

[72] Wroge CM, Hogins J, Eisenman L, Mennerick S. Synaptic NMDA receptors mediate hypoxic excitotoxic death. J Neurosci 2012; 32(19): 6732-42.
[http://dx.doi.org/10.1523/JNEUROSCI.6371-11.2012] [PMID: 22573696]

[73] Lau A, Tymianski M. Glutamate receptors, neurotoxicity and neurodegeneration. Pflugers Arch 2010; 460(2): 525-42.
[http://dx.doi.org/10.1007/s00424-010-0809-1] [PMID: 20229265]

[74] Gotti B, Duverger D, Bertin J, et al. Ifenprodil and SL 82.0715 as cerebral anti-ischemic agents. I. Evidence for efficacy in models of focal cerebral ischemia. J Pharmacol Exp Ther 1988; 247(3): 1211-21.
[PMID: 2849668]

[75] Nikam SS, Meltzer LT. NR2B selective NMDA receptor antagonists. Curr Pharm Des 2002; 8(10): 845-55.
[http://dx.doi.org/10.2174/1381612024607072] [PMID: 11945135]

[76] Başkaya MK, Rao AM, Donaldson D, Prasad MR, Dempsey RJ. Protective effects of ifenprodil on ischemic injury size, blood-brain barrier breakdown, and edema formation in focal cerebral ischemia. Neurosurgery 1997; 40(2): 364-70.
[http://dx.doi.org/10.1097/00006123-199702000-00026] [PMID: 9007871]

[77] Wang YC, Sanchez-Mendoza EH, Doeppner TR, Hermann DM. Post-acute delivery of memantine promotes post-ischemic neurological recovery, peri-infarct tissue remodeling, and contralesional brain plasticity. J Cereb Blood Flow Metab 2017; 37(3): 980-93.
[http://dx.doi.org/10.1177/0271678X16648971] [PMID: 27170698]

[78] Baudy RB, Fletcher H III, Yardley JP, et al. Design, synthesis, SAR, and biological evaluation of highly potent benzimidazole-spaced phosphono-α-amino acid competitive NMDA antagonists of the AP-6 type. J Med Chem 2001; 44(10): 1516-29.
[http://dx.doi.org/10.1021/jm000385w] [PMID: 11334562]

[79] Karakas E, Simorowski N, Furukawa H. Subunit arrangement and phenylethanolamine binding in

GluN1/GluN2B NMDA receptors. Nature 2011; 475(7355): 249-53.
[http://dx.doi.org/10.1038/nature10180] [PMID: 21677647]

[80]   Furukawa H. Structure and function of glutamate receptor amino terminal domains. J Physiol 2012; 590(1): 63-72.
[http://dx.doi.org/10.1113/jphysiol.2011.213850] [PMID: 22106178]

[81]   Zhu S, Stein RA, Yoshioka C, et al. Mechanism of NMDA Receptor Inhibition and Activation. Cell 2016; 165(3): 704-14.
[http://dx.doi.org/10.1016/j.cell.2016.03.028] [PMID: 27062927]

[82]   Berman HM, Westbrook J, Feng Z, et al. The Protein Data Bank. Nucleic Acids Res 2000; 28(1): 235-42.
[http://dx.doi.org/10.1093/nar/28.1.235] [PMID: 10592235]

[83]   Yuan H, Hansen KB, Vance KM, Ogden KK, Traynelis SF. Control of NMDA receptor function by the NR2 subunit amino-terminal domain. J Neurosci 2009; 29(39): 12045-58.
[http://dx.doi.org/10.1523/JNEUROSCI.1365-09.2009] [PMID: 19793963]

[84]   Mony L, Kew JN, Gunthorpe MJ, Paoletti P. Allosteric modulators of NR2B-containing NMDA receptors: molecular mechanisms and therapeutic potential. Br J Pharmacol 2009; 157(8): 1301-17.
[http://dx.doi.org/10.1111/j.1476-5381.2009.00304.x] [PMID: 19594762]

[85]   Monaghan DT, Irvine MW, Costa BM, Fang G, Jane DE. Pharmacological modulation of NMDA receptor activity and the advent of negative and positive allosteric modulators. Neurochem Int 2012; 61(4): 581-92.
[http://dx.doi.org/10.1016/j.neuint.2012.01.004] [PMID: 22269804]

[86]   Lipton SA. Failures and successes of NMDA receptor antagonists: molecular basis for the use of open-channel blockers like memantine in the treatment of acute and chronic neurologic insults. NeuroRx 2004; 1(1): 101-10.
[http://dx.doi.org/10.1602/neurorx.1.1.101] [PMID: 15717010]

[87]   Zanos P, Moaddel R, Morris PJ, et al. NMDAR inhibition-independent antidepressant actions of ketamine metabolites. Nature 2016; 533(7604): 481-6.
[http://dx.doi.org/10.1038/nature17998] [PMID: 27144355]

[88]   Ferris SH. Evaluation of memantine for the treatment of Alzheimer's disease. Expert Opin Pharmacother 2003; 4(12): 2305-13.
[http://dx.doi.org/10.1517/14656566.4.12.2305] [PMID: 14640929]

[89]   Lipton SA. Paradigm shift in neuroprotection by NMDA receptor blockade: memantine and beyond. Nat Rev Drug Discov 2006; 5(2): 160-70.
[http://dx.doi.org/10.1038/nrd1958] [PMID: 16424917]

[90]   Kalia LV, Kalia SK, Salter MW. NMDA receptors in clinical neurology: excitatory times ahead. Lancet Neurol 2008; 7(8): 742-55.
[http://dx.doi.org/10.1016/S1474-4422(08)70165-0] [PMID: 18635022]

[91]   Mony L, Zhu S, Carvalho S, Paoletti P. Molecular basis of positive allosteric modulation of GluN2B NMDA receptors by polyamines. EMBO J 2011; 30(15): 3134-46.
[http://dx.doi.org/10.1038/emboj.2011.203] [PMID: 21685875]

[92]   Williams K. Ifenprodil discriminates subtypes of the N-methyl-D-aspartate receptor: selectivity and mechanisms at recombinant heteromeric receptors. Mol Pharmacol 1993; 44(4): 851-9.
[PMID: 7901753]

[93]   Karakas E, Simorowski N, Furukawa H. Structure of the zinc-bound amino-terminal domain of the NMDA receptor NR2B subunit. EMBO J 2009; 28(24): 3910-20.
[http://dx.doi.org/10.1038/emboj.2009.338] [PMID: 19910922]

[94]   Mott DD, Doherty JJ, Zhang S, et al. Phenylethanolamines inhibit NMDA receptors by enhancing proton inhibition. Nat Neurosci 1998; 1(8): 659-67.

[http://dx.doi.org/10.1038/3661] [PMID: 10196581]

[95]   Avenet P, Léonardon J, Besnard F, *et al.* Antagonist properties of the stereoisomers of ifenprodil at NR1A/NR2A and NR1A/NR2B subtypes of the NMDA receptor expressed in Xenopus oocytes. Eur J Pharmacol 1996; 296(2): 209-13.
[http://dx.doi.org/10.1016/0014-2999(95)00700-8] [PMID: 8838458]

[96]   Burger PB, Yuan H, Karakas E, *et al.* Mapping the binding of GluN2B-selective N-methyl-D-aspartate receptor negative allosteric modulators. Mol Pharmacol 2012; 82(2): 344-59.
[http://dx.doi.org/10.1124/mol.112.078568] [PMID: 22596351]

[97]   Fischer G, Mutel V, Trube G, *et al.* Ro 25-6981, a highly potent and selective blocker of N-methyl-D-aspartate receptors containing the NR2B subunit. Characterization *in vitro*. J Pharmacol Exp Ther 1997; 283(3): 1285-92.
[PMID: 9400004]

[98]   Claiborne CF, McCauley JA, Libby BE, *et al.* Orally efficacious NR2B-selective NMDA receptor antagonists. Bioorg Med Chem Lett 2003; 13(4): 697-700.
[http://dx.doi.org/10.1016/S0960-894X(02)01061-2] [PMID: 12639561]

[99]   Davies DJ, Crowe M, Lucas N, *et al.* A novel series of benzimidazole NR2B-selective NMDA receptor antagonists. Bioorg Med Chem Lett 2012; 22(7): 2620-3.
[http://dx.doi.org/10.1016/j.bmcl.2012.01.108] [PMID: 22366657]

[100] Tewes B, Frehland B, Schepmann D, *et al.* Enantiomerically Pure 2-Methyltetrahydro-3-benzazep-n-1-ols Selectively Blocking GluN2B Subunit Containing N-Methyl-D-aspartate Receptors. J Med Chem 2015; 58(15): 6293-305.
[http://dx.doi.org/10.1021/acs.jmedchem.5b00897] [PMID: 26186074]

[101] Kemp JA, Tasker T. Methods for treating disorders using NMDA NR2B-subtype selective antagonist. US Pat App 12/919,804 2009; 31.

[102] Stroebel D, Buhl DL, Knafels JD, *et al.* A Novel Binding Mode Reveals Two Distinct Classes of NMDA Receptor GluN2B-selective Antagonists. Mol Pharmacol 2016; 89(5): 541-51.
[http://dx.doi.org/10.1124/mol.115.103036] [PMID: 26912815]

[103] Chazot PL, Lawrence S, Thompson CL. Studies on the subtype selectivity of CP-101,606: evidence for two classes of NR2B-selective NMDA receptor antagonists. Neuropharmacology 2002; 42(3): 319-24.
[http://dx.doi.org/10.1016/S0028-3908(01)00191-5] [PMID: 11897110]

[104] Borza I, Domány G. NR2B selective NMDA antagonists: the evolution of the ifenprodil-type pharmacophore. Curr Top Med Chem 2006; 6(7): 687-95.
[http://dx.doi.org/10.2174/156802606776894456] [PMID: 16719809]

[105] Chenard BL, Shalaby IA, Koe BK, *et al.* Separation of alpha 1 adrenergic and N-methyl-D-aspartate antagonist activity in a series of ifenprodil compounds. J Med Chem 1991; 34(10): 3085-90.
[http://dx.doi.org/10.1021/jm00114a018] [PMID: 1681106]

[106] Biton B, Granger P, Carreau A, Depoortere H, Scatton B, Avenet P. The NMDA receptor antagonist eliprodil (SL 82.0715) blocks voltage-operated Ca2+ channels in rat cultured cortical neurons. Eur J Pharmacol 1994; 257(3): 297-301.
[http://dx.doi.org/10.1016/0014-2999(94)90142-2] [PMID: 8088348]

[107] Mosley CA, Acker TM, Hansen KB, *et al.* Quinazolin-4-one derivatives: A novel class of noncompetitive NR2C/D subunit-selective N-methyl-D-aspartate receptor antagonists. J Med Chem 2010; 53(15): 5476-90.
[http://dx.doi.org/10.1021/jm100027p] [PMID: 20684595]

[108] Bettini E, Sava A, Griffante C, *et al.* Identification and characterization of novel NMDA receptor antagonists selective for NR2A- over NR2B-containing receptors. J Pharmacol Exp Ther 2010; 335(3): 636-44.

[http://dx.doi.org/10.1124/jpet.110.172544] [PMID: 20810618]

[109] Hansen KB, Ogden KK, Traynelis SF. Subunit-selective allosteric inhibition of glycine binding to NMDA receptors. J Neurosci 2012; 32(18): 6197-208.
[http://dx.doi.org/10.1523/JNEUROSCI.5757-11.2012] [PMID: 22553026]

[110] Benner A, Bonifazi A, Shirataki C, et al. GluN2B-selective N-methyl-D-aspartate (NMDA) receptor antagonists derived from 3-benzazepines: synthesis and pharmacological evaluation of benzo[7]annulen-7-amines. ChemMedChem 2014; 9(4): 741-51.
[http://dx.doi.org/10.1002/cmdc.201300547] [PMID: 24677663]

[111] Drews J. Drug Discovery: A Historical Perspective. Science 2000; 287(5460): 1960-4.

[112] Butcher EC, Berg EL, Kunkel EJ. Systems biology in drug discovery. Nat Biotechnol 2004; 22(10): 1253-9.
[http://dx.doi.org/10.1038/nbt1017] [PMID: 15470465]

[113] Congreve M, Murray CW, Blundell TL. Structural biology and drug discovery. Drug Discov Today 2005; 10(13): 895-907.
[http://dx.doi.org/10.1016/S1359-6446(05)03484-7] [PMID: 15993809]

[114] Shuker SB, Hajduk PJ, Meadows RP. Discovering High-Affinity Ligands for Proteins: SAR by NMR. Science (80-) 1996; 274(5292): 1531-4.

[115] Silverman L, Campbell R, Broach JR. New assay technologies for high-throughput screening. Curr Opin Chem Biol 1998; 2(3): 397-403.
[http://dx.doi.org/10.1016/S1367-5931(98)80015-X] [PMID: 9691081]

[116] Jorgensen WL. Challenges for Academic Drug Discovery. Angew Chem Int Ed 2012; 51(47): 11680-4.
[http://dx.doi.org/10.1002/anie.201204625]

[117] Bennani YL. Drug discovery in the next decade: innovation needed ASAP. Drug Discov Today 2011; 16(17-18): 779-92.
[http://dx.doi.org/10.1016/j.drudis.2011.06.004] [PMID: 21704185]

[118] Scannell JW, Blanckley A, Boldon H, Warrington B. Diagnosing the decline in pharmaceutical R&D efficiency. Nat Rev Drug Discov 2012; 11(3): 191-200.
[http://dx.doi.org/10.1038/nrd3681] [PMID: 22378269]

[119] Pardridge WM. The blood-brain barrier: bottleneck in brain drug development. NeuroRx 2005; 2(1): 3-14.
[http://dx.doi.org/10.1602/neurorx.2.1.3] [PMID: 15717053]

[120] Abbott NJ, Patabendige AA, Dolman DE, Yusof SR, Begley DJ. Structure and function of the blood-brain barrier. Neurobiol Dis 2010; 37(1): 13-25.
[http://dx.doi.org/10.1016/j.nbd.2009.07.030] [PMID: 19664713]

[121] Abbott NJ. Blood-brain barrier structure and function and the challenges for CNS drug delivery. J Inherit Metab Dis 2013; 36(3): 437-49.
[http://dx.doi.org/10.1007/s10545-013-9608-0] [PMID: 23609350]

[122] Miller DS. Regulation of ABC transporters blood-brain barrier: the good, the bad, and the ugly. Adv Cancer Res 2015; 125: 43-70.
[http://dx.doi.org/10.1016/bs.acr.2014.10.002] [PMID: 25640266]

[123] Shen J, Cheng F, Xu Y, Li W, Tang Y. Estimation of ADME properties with substructure pattern recognition. J Chem Inf Model 2010; 50(6): 1034-41.
[http://dx.doi.org/10.1021/ci100104j] [PMID: 20578727]

[124] Lipinski CA, Lombardo F, Dominy BW, Feeney PJ. Experimental and computational approaches to estimate solubility and permeability in drug discovery and development settings. Adv Drug Deliv Rev 2001; 46(1-3): 3-26.

[http://dx.doi.org/10.1016/S0169-409X(00)00129-0] [PMID: 11259830]

[125] Oprea TI, Allu TK, Fara DC, Rad RF, Ostopovici L, Bologa CG. Lead-like, drug-like or "Pub-like": how different are they? J Comput Aided Mol Des 2007; 21(1-3): 113-9.
[http://dx.doi.org/10.1007/s10822-007-9105-3] [PMID: 17333482]

[126] Müller CE. Prodrug approaches for enhancing the bioavailability of drugs with low solubility. Chem Biodivers 2009; 6(11): 2071-83.
[http://dx.doi.org/10.1002/cbdv.200900114] [PMID: 19937841]

[127] Pajouhesh H, Lenz GR. Medicinal chemical properties of successful central nervous system drugs. NeuroRx 2005; 2(4): 541-53.
[http://dx.doi.org/10.1602/neurorx.2.4.541] [PMID: 16489364]

[128] Mortenson PN, Murray CW. Assessing the lipophilicity of fragments and early hits. J Comput Aided Mol Des 2011; 25(7): 663-7.
[http://dx.doi.org/10.1007/s10822-011-9435-z] [PMID: 21614595]

[129] Hansch C, Björkroth JP, Leo A. Hydrophobicity and central nervous system agents: on the principle of minimal hydrophobicity in drug design. J Pharm Sci 1987; 76(9): 663-87.
[http://dx.doi.org/10.1002/jps.2600760902] [PMID: 11002801]

[130] Hann MM. Molecular obesity, potency and other addictions in drug discovery. Multifaceted Roles Crystallogr Mod Drug Discov 2015; 2(5): 183-96.

[131] Lavecchia A, Di Giovanni C. Virtual screening strategies in drug discovery: a critical review. Curr Med Chem 2013; 20(23): 2839-60.
[http://dx.doi.org/10.2174/09298673113209990001] [PMID: 23651302]

[132] Ertl P, Rohde B, Selzer P. Fast calculation of molecular polar surface area as a sum of fragment-based contributions and its application to the prediction of drug transport properties. J Med Chem 2000; 43(20): 3714-7.
[http://dx.doi.org/10.1021/jm000942e] [PMID: 11020286]

[133] Paul SM, Mytelka DS, Dunwiddie CT, *et al.* How to improve R&D productivity: the pharmaceutical industry's grand challenge. Nat Rev Drug Discov 2010; 9(3): 203-14.
[PMID: 20168317]

[134] Ooms F. Molecular modeling and computer aided drug design. Examples of their applications in medicinal chemistry. Curr Med Chem 2000; 7(2): 141-58.
[http://dx.doi.org/10.2174/0929867003375317] [PMID: 10637360]

[135] Trouiller P, Olliaro P, Torreele E, Orbinski J, Laing R, Ford N. Drug development for neglected diseases: a deficient market and a public-health policy failure. Lancet 2002; 359(9324): 2188-94.
[http://dx.doi.org/10.1016/S0140-6736(02)09096-7] [PMID: 12090998]

[136] Bleicher KH, Böhm HJ, Müller K, Alanine AI. Hit and lead generation: beyond high-throughput screening. Nat Rev Drug Discov 2003; 2(5): 369-78.
[http://dx.doi.org/10.1038/nrd1086] [PMID: 12750740]

[137] Bajorath J. Integration of virtual and high-throughput screening. Nat Rev Drug Discov 2002; 1(11): 882-94.
[http://dx.doi.org/10.1038/nrd941] [PMID: 12415248]

[138] Shoichet BK. Screening in a spirit haunted world. Drug Discov Today 2006; 11(13-14): 607-15.
[http://dx.doi.org/10.1016/j.drudis.2006.05.014] [PMID: 16793529]

[139] Khanna I. Drug discovery in pharmaceutical industry: productivity challenges and trends. Drug Discov Today 2012; 17(19-20): 1088-102.
[http://dx.doi.org/10.1016/j.drudis.2012.05.007] [PMID: 22627006]

[140] Oprea TI, Matter H. Integrating virtual screening in lead discovery. Curr Opin Chem Biol 2004; 8(4): 349-58.

[http://dx.doi.org/10.1016/j.cbpa.2004.06.008] [PMID: 15288243]

[141] Glaab E. Building a virtual ligand screening pipeline using free software: a survey. Brief Bioinform 2016; 17(2): 352-66.
[http://dx.doi.org/10.1093/bib/bbv037] [PMID: 26094053]

[142] Scior T, Bender A, Tresadern G, *et al.* Recognizing pitfalls in virtual screening: a critical review. J Chem Inf Model 2012; 52(4): 867-81.
[http://dx.doi.org/10.1021/ci200528d] [PMID: 22435959]

[143] Ferreira RS, Glaucius O, Andricopulo AD. Integrating virtual and high-throughput screening: opportunities and challenges in drug research and development. Quim Nova 2011; 34(10): 1770-8.
[http://dx.doi.org/10.1590/S0100-40422011001000010]

[144] van de Waterbeemd H, Gifford E. ADMET in silico modelling: towards prediction paradise? Nat Rev Drug Discov 2003; 2(3): 192-204.
[http://dx.doi.org/10.1038/nrd1032] [PMID: 12612645]

[145] Cheng F, Li W, Liu G, Tang Y. In silico ADMET prediction: recent advances, current challenges and future trends. Curr Top Med Chem 2013; 13(11): 1273-89.
[http://dx.doi.org/10.2174/15680266113139990033] [PMID: 23675935]

[146] Moroy G, Martiny VY, Vayer P, Villoutreix BO, Miteva MA. Toward in silico structure-based ADMET prediction in drug discovery. Drug Discov Today 2012; 17(1-2): 44-55.
[http://dx.doi.org/10.1016/j.drudis.2011.10.023] [PMID: 22056716]

[147] Beck B, Geppert T. Industrial applications of in silico ADMET. J Mol Model 2014; 20(7): 2322.
[http://dx.doi.org/10.1007/s00894-014-2322-5] [PMID: 24972798]

[148] O'Brien SE, de Groot MJ. Greater than the sum of its parts: combining models for useful ADMET prediction. J Med Chem 2005; 48(4): 1287-91.
[http://dx.doi.org/10.1021/jm049254b] [PMID: 15715500]

[149] Lipinski CA. Drug-like properties and the causes of poor solubility and poor permeability. J Pharmacol Toxicol Methods 2000; 44(1): 235-49.
[http://dx.doi.org/10.1016/S1056-8719(00)00107-6] [PMID: 11274893]

[150] Veber DF, Johnson SR, Cheng HY, Smith BR, Ward KW, Kopple KD. Molecular properties that influence the oral bioavailability of drug candidates. J Med Chem 2002; 45(12): 2615-23.
[http://dx.doi.org/10.1021/jm020017n] [PMID: 12036371]

[151] Walters WP, Ajay , Murcko MA. Recognizing molecules with drug-like properties. Curr Opin Chem Biol 1999; 3(4): 384-7.
[http://dx.doi.org/10.1016/S1367-5931(99)80058-1] [PMID: 10419858]

[152] Lajiness MS, Vieth M, Erickson J. Molecular properties that influence oral drug-like behavior. Curr Opin Drug Discov Devel 2004; 7(4): 470-7.
[PMID: 15338956]

[153] Gleeson MP, Hersey A, Hannongbua S. In-silico ADME models: a general assessment of their utility in drug discovery applications. Curr Top Med Chem 2011; 11(4): 358-81.
[http://dx.doi.org/10.2174/156802611794480927] [PMID: 21320065]

[154] Lin JH, Lu AY. Role of pharmacokinetics and metabolism in drug discovery and development. Pharmacol Rev 1997; 49(4): 403-49.
[PMID: 9443165]

[155] Egan WJ, Merz KM Jr, Baldwin JJ. Prediction of drug absorption using multivariate statistics. J Med Chem 2000; 43(21): 3867-77.
[http://dx.doi.org/10.1021/jm000292e] [PMID: 11052792]

[156] Cruciani G, Crivori P, Carrupt P, Testa B. Molecular fields in quantitative structure–permeation relationships: the VolSurf approach. J Mol Struct 2000.

[157] Cheng F, Li W, Zhou Y, *et al.* admetSAR: a comprehensive source and free tool for assessment of chemical ADMET properties. J Chem Inf Model 2012; 52(11): 3099-105.
[http://dx.doi.org/10.1021/ci300367a] [PMID: 23092397]

[158] Lagorce D, Sperandio O, Baell JB, Miteva MA, Villoutreix BO. FAF-Drugs3: a web server for compound property calculation and chemical library design. Nucleic Acids Res 2015; 43(W1): W200-7.
[http://dx.doi.org/10.1093/nar/gkv353] [PMID: 25883137]

[159] Goodwin JT, Clark DE. In silico predictions of blood-brain barrier penetration: considerations to "keep in mind". J Pharmacol Exp Ther 2005; 315(2): 477-83.
[http://dx.doi.org/10.1124/jpet.104.075705] [PMID: 15919767]

[160] Li AP. Screening for human ADME/Tox drug properties in drug discovery. Drug Discov Today 2001; 6(7): 357-66.
[http://dx.doi.org/10.1016/S1359-6446(01)01712-3] [PMID: 11267922]

[161] Roncaglioni A, Toropov AA, Toropova AP, Benfenati E. In silico methods to predict drug toxicity. Curr Opin Pharmacol 2013; 13(5): 802-6.
[http://dx.doi.org/10.1016/j.coph.2013.06.001] [PMID: 23797035]

[162] Ekins S. Progress in computational toxicology. J Pharmacol Toxicol Methods 2014; 69(2): 115-40.
[http://dx.doi.org/10.1016/j.vascn.2013.12.003] [PMID: 24361690]

[163] Hughes JD, Blagg J, Price DA, *et al.* Physiochemical drug properties associated with *in vivo* toxicological outcomes. Bioorg Med Chem Lett 2008; 18(17): 4872-5.
[http://dx.doi.org/10.1016/j.bmcl.2008.07.071] [PMID: 18691886]

[164] Gleeson MP. Generation of a set of simple, interpretable ADMET rules of thumb. J Med Chem 2008; 51(4): 817-34.
[http://dx.doi.org/10.1021/jm701122q] [PMID: 18232648]

[165] Baell J, Walters MA. Chemistry: Chemical con artists foil drug discovery. Nature 2014; 513(7519): 481-3.
[http://dx.doi.org/10.1038/513481a] [PMID: 25254460]

[166] Kazius J, McGuire R, Bursi R. Derivation and validation of toxicophores for mutagenicity prediction. J Med Chem 2005; 48(1): 312-20.
[http://dx.doi.org/10.1021/jm040835a] [PMID: 15634026]

[167] Baell JB, Holloway GA. New substructure filters for removal of pan assay interference compounds (PAINS) from screening libraries and for their exclusion in bioassays. J Med Chem 2010; 53(7): 2719-40.
[http://dx.doi.org/10.1021/jm901137j] [PMID: 20131845]

[168] Whitty A. Growing PAINS in academic drug discovery. Future Med Chem 2011; 3(7): 797-801.
[http://dx.doi.org/10.4155/fmc.11.44] [PMID: 21644825]

[169] Dahlin JL, Nissink JW, Strasser JM, *et al.* PAINS in the assay: chemical mechanisms of assay interference and promiscuous enzymatic inhibition observed during a sulfhydryl-scavenging HTS. J Med Chem 2015; 58(5): 2091-113.
[http://dx.doi.org/10.1021/jm5019093] [PMID: 25634295]

[170] Maunz A, Gütlein M, Rautenberg M, Vorgrimmler D, Gebele D, Helma C. lazar: a modular predictive toxicology framework. Front Pharmacol 2013; 4(April): 38.
[PMID: 23761761]

[171] Grinter SZ, Zou X. Challenges, applications, and recent advances of protein-ligand docking in structure-based drug design. Molecules 2014; 19(7): 10150-76.
[http://dx.doi.org/10.3390/molecules190710150] [PMID: 25019558]

[172] Triballeau N, Acher F, Brabet I, Pin J. Virtual screening workflow development guided by the

"receiver operating characteristic" curve approach. Application to high-throughput docking on metabotropic. J Med 2005.

[173]   Ferreira LG, Dos Santos RN, Oliva G, Andricopulo AD. Molecular docking and structure-based drug design strategies. Molecules 2015; 20(7): 13384-421.
[http://dx.doi.org/10.3390/molecules200713384] [PMID: 26205061]

[174]   Koes DR, Camacho CJ. ZINCPharmer: pharmacophore search of the ZINC database. Nucleic Acids Res 2012; 40(Web Server issue): W409-14.
[PMID: 22553363]

[175]   Braga RC, Alves VM, Silva AC, *et al.* Virtual screening strategies in medicinal chemistry: the state of the art and current challenges. Curr Top Med Chem 2014; 14(16): 1899-912.
[http://dx.doi.org/10.2174/1568026614666140929120749] [PMID: 25262801]

[176]   Kalyaanamoorthy S, Chen YP. Structure-based drug design to augment hit discovery. Drug Discov Today 2011; 16(17-18): 831-9.
[http://dx.doi.org/10.1016/j.drudis.2011.07.006] [PMID: 21810482]

[177]   Meng X-Y, Zhang H-X, Mezei M, Cui M. Molecular docking: a powerful approach for structure-based drug discovery. Curr Comput Aided Drug Des 2011; 7(2): 146-57.
[http://dx.doi.org/10.2174/157340911795677602] [PMID: 21534921]

[178]   Yuriev E, Ramsland PA. Latest developments in molecular docking: 2010-2011 in review. J Mol Recognit 2013; 26(5): 215-39.
[http://dx.doi.org/10.1002/jmr.2266] [PMID: 23526775]

[179]   Kitchen DB, Decornez H, Furr JR, Bajorath J. Docking and scoring in virtual screening for drug discovery: methods and applications. Nat Rev Drug Discov 2004; 3(11): 935-49.
[http://dx.doi.org/10.1038/nrd1549] [PMID: 15520816]

[180]   Verdonk ML, Giangreco I, Hall RJ, Korb O, Mortenson PN, Murray CW. Docking performance of fragments and druglike compounds. J Med Chem 2011; 54(15): 5422-31.
[http://dx.doi.org/10.1021/jm200558u] [PMID: 21692478]

[181]   Warren GGL, Andrews CW, Capelli AA-M, *et al.* A critical assessment of docking programs and scoring functions. J Med Chem 2006; 49(20): 5912-31.

[182]   Goodsell DS, Morris GM, Olson AJ. Automated docking of flexible ligands: applications of AutoDock. J Mol Recognit 1996; 9(1): 1-5.
[http://dx.doi.org/10.1002/(SICI)1099-1352(199601)9:1<1::AID-JMR241>3.0.CO;2-6]   [PMID: 8723313]

[183]   Trott O, Olson AJ. AutoDock Vina: improving the speed and accuracy of docking with a new scoring function, efficient optimization, and multithreading. J Comput Chem 2010; 31(2): 455-61.
[PMID: 19499576]

[184]   Verdonk ML, Cole JC, Hartshorn MJ, Murray CW, Taylor RD. Improved protein-ligand docking using GOLD. Proteins 2003; 52(4): 609-23.
[http://dx.doi.org/10.1002/prot.10465] [PMID: 12910460]

[185]   Cheng T, Li Q, Zhou Z, Wang Y, Bryant SH. Structure-based virtual screening for drug discovery: a problem-centric review. AAPS J 2012; 14(1): 133-41.
[http://dx.doi.org/10.1208/s12248-012-9322-0] [PMID: 22281989]

[186]   Mony L, Triballeau N, Paoletti P, Acher FC, Bertrand H-O. Identification of a novel NR2B-selective NMDA receptor antagonist using a virtual screening approach. Bioorg Med Chem Lett 2010; 20(18): 5552-8.
[http://dx.doi.org/10.1016/j.bmcl.2010.07.043] [PMID: 20692832]

[187]   Gitto R, De Luca L, Ferro S, *et al.* Synthesis and biological characterization of 3-substituted--H-indoles as ligands of GluN2B-containing N-methyl-D-aspartate receptors. J Med Chem 2011; 54(24): 8702-6.

[http://dx.doi.org/10.1021/jm2008002] [PMID: 22050212]

[188] Gitto R, De Luca L, Ferro S, *et al.* Synthesis and biological characterization of 3-substituted 1H-indoles as ligands of GluN2B-containing N-methyl-D-aspartate receptors. Part 2. J Med Chem 2012; 55(23): 10532-9.
[http://dx.doi.org/10.1021/jm301508d] [PMID: 23140383]

[189] Jespersen A, Tajima N, Fernandez-Cuervo G, Garnier-Amblard G, Furukawa H. Structural insights into competitive antagonism in NMDA receptors. Neuron 2014; 81(2): 366-78.
[http://dx.doi.org/10.1016/j.neuron.2013.11.033] [PMID: 24462099]

CHAPTER 7

Therapeutic Potential of Plant Alkaloids as Antidepressant

Haroon Khan[*]

Department of Pharmacy, Abdul Wali Khan University, Mardan 23200, Pakistan

Abstract: Depression represents an assorted mood disorder, which has been recognized as a global issue. According to statistics, the life style of millions of individuals is distorted by mental or behavioral disorder. Prompt medicinal intervention is extremely required in major depression as millions of patients are reported for suicidal attempt each year. In clinics, several synthetic agents are in practice for the treatment of various types of depression, but their uses have been facing several limitations like slow onsite of action, efficacy in wide-range patients, side effects/toxicity *etc*. Unlike synthetic therapeutics, phytopharmaceuticals are extensively used world-wide due to their diverse applications, efficacious, and minimum side effects which lead to patient compliance. The alkaloids are indeed the most widely distributed and studied plant secondary metabolites that have established therapeutic history. This book chapter is primarily based on preclinical studies of various isolated alkaloids, their underlying possible mechanism(s) where discussed, structural edges and future directions towards clinical trials for the discovery of new effective and acceptable antidepressants.

Keywords: Alkaloids, Antidepressant effects, Better pharmacokinetic profile, Candidates for clinical trials, Diverse efficacy, Future therapies, Lead compounds, Medicinal Plants, Natural antidepressant, Patient compliance, Preclinical studies, SAR studies, Underlying mechanisms, Various classes, Various families.

INTRODUCTION

Depression is a state of poor mood disorder with general immobility which could ultimately influence an individual's thinking pattern, line of action, overall emotions and sense of well-being [1, 2]. In depression, it has been observed that individuals are mostly disturbed and restless having scattered feeling, anxiety, perception of guiltiness, lack of confidence in actions while considering themselves as a burden on community and worthless. Such individuals may lose concentration or interest in activities, and overall change in eating habits is most

---

[*] **Corresponding author Haroon Khan:** Department of Pharmacy, Abdul Wali Khan University, Mardan 23200, Pakistan; Tel: +92-3329123171; E-mail: hkdr2006@gmail.com

Atta-ur-Rahman (Ed.)

frequently encountered during different phases of depression; such as complete loss, decrease or increase of appetite [3]. Along with mood changes, physical and social abnormalities may be observed in different stages of depression, chiefly in severe depression. For instance, sleeping disorders such as insomnia or over sleeping and abnormal eating behavior may lead to weight loss or vice versa, loss of energy and libido, social interaction and relationships may also be affected due to disturbance of hormonal and many endocrine functions [4]. The psychiatrists describe major depression as a facet of some psychiatric syndromes, however sometimes, it could be a normal human response to different events that happen in surrounding for instance bereavement, a symptom of some bodily ailments or an unwanted effect of a therapeutic agent.

In 1988, Gold *et al.* found that patients with major depression have symptoms that cause alteration in the brain affecting the neurotransmitters levels, specifically norepinephrine, serotonin, and dopamine [5]. It is also observed that some features of depressive disorder are identical to the anxiety disorders, including severe phobias, generalized anxiety disorder, social anxiety disorder, post-traumatic stress disorder, and obsessive-compulsive disorder [6]. The disorder is also often associated with suicide and there occur between 10 and 20 million suicide attempts every year [7, 8]. According to World Health Organization statistics, about 450 million people suffer from a mental or behavioral disorder. This is responsible for12.3% of the global burden of disease and predicted to rise up to 15% by 2020 [9].

Despite recent technological development in treatment strategies and application of novel treatment approaches, the patients did not give optimal response to the treatments and continue to experience depressive relapses. The current antidepressant drugs are facing many challenges as mentioned by Prof. Robinson (2003) in his review published on the same issue [10]. It has been observed that even the first line antidepressants such as the selective serotonin reuptake inhibitors (SSRIs) and serotonin-norepinephrine reuptake inhibitors have limitation like poor response and remission rates, slow onset of action, poor tolerability, persistent adverse effects, and the potential for clinically significant pharmacokinetic drug interactions. In order to overcome above pharmacokinetic limitations which ultimately affect patient compliance, new effective and safe antidepressants are most warrant.

ALKALOIDS AS THERAPEUTIC AGENTS

Alkaloids represent a large group of naturally occurring secondary metabolites that possess mostly basic nitrogen atoms in their main skeleton. Similarly, some of the compounds of the group are neutral in character [11] and even few are weakly acidic in nature [12] therefore the most specific definition of alkaloids is extremely difficult. Alkaloids are being isolated from a variety of organisms, contrary to initial belief that lemmatize it to plant including microorganisms and animals, during different metabolic reactions/process as secondary metabolites. However, plants were the first and foremost most dominant source even today [13].

The general and common procedure of alkaloid isolation from extracts is involved acid-base treatment/extraction. Foremost, alkaloids are renowned for their toxic nature [14]. Mostly, alkaloid demonstrated pharmacological effects and therefore, are using as therapeutic agents [15], as recreational drugs, or in entheogenic rituals. In this chapter, we have discussed in detail the antidepressant like activity of various plants derived alkaloids, their antidepressant mechanism, clinical trials where avalible and future clinical prospects or as lead compounds.

ALKALOIDS—AS ANTIDEPRESSANT AGENTS

The preclinical studies on the antidepressant like effect of different classes of alakoids derived from plants along with their mechanism(s) (Fig. **1**) which have been reported in the literature. A team of the researcher isolated Akuammidine, rhaziminine, and tetrahydrosecamine from *Rhazya stricta*. The different concentration of leaf extract of the plant produced varying effects. However, the pretreatment of the lyophilized extract of *R. stricta* illustrated marked antidepressant-like effect in various *in vitro* models [16]. Interestingly, treatment with different doses caused marked variation in overall activity. The experimental results showed that the Akuammidine, rhaziminine, and tetrahydrosecamine caused a significant inhibition on the MAOA over expression and thus produced antidepressant like effect [16].

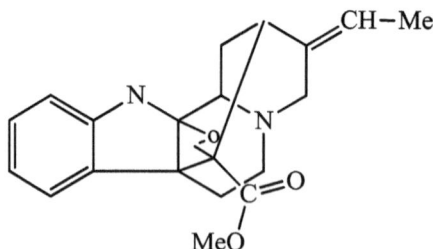

The pramipexole (2-amino-4,5,6,7-tetrahydro-6-propyl-amino-benzt-iazole-dihydrochloride), is a non-ergoline alkaloid showed significant reduction in immobility time in animal studies in forced swim test [17]. It elicited predominant action on dopamine D3 receptor and the overall effects was like conventional antidepressant drugs. Letter on, this dopaminergic agent showed efficacy in a double blind clinical trial in comparison placebo in bipolar and unipolar depressive patients at low doses (0.5-0.75 mg/day). It had a positive effect on the severity as well as duration of depressive attacks at the sixth week of the pramipexole administration as compared to the placebo. Moreover, it was found safe; without any serious unwanted effect during the treatment period. On the basis of these clinical studies, the recommended dose of the compound is 0.125-0.9 mg. Similar to other dopaminergic agonists, pramipexole also interfere with the synthesis of dopamine and thus reduces turnover in the brain, mediated through up regulation of dopamine auto-receptors. As a result, it leads to down regulation of dopamine catabolic oxidation and neurotoxic oxygen radicals. Furthermore, the proposed genetic expression for this compound may be neurotrophic effects *via* up-regulation of anti-apoptotic protein bcl-2 [18, 19].

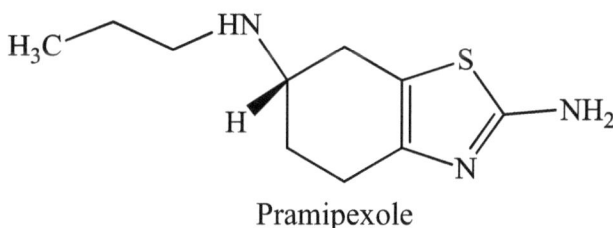

Pramipexole

Farzin and Mansouri N (2006) studied β-carboline alkaloids in animal model of forced swim test. In this study, the β-carboline alkaloids such harmane, norharmane and harmine elicited marked reduction in the immobility time when studied in the mouse forced swim test at various doses and thus produced significant an antidepressant-like effect [20]. The harmane at a dose of 5-15 mg/kg, norharmane 2.5-10 mg/kg and harmine (5-15 mg/kg, provoked antidepressant-like effects as an emerging antidepressant compounds. The mechanistic studies have shown that it produced its effect by down regulation of MAO-A expression *via* significant modulation of serotonin receptor 2A and benzodiazepine receptors located at cell-surface [21, 22].

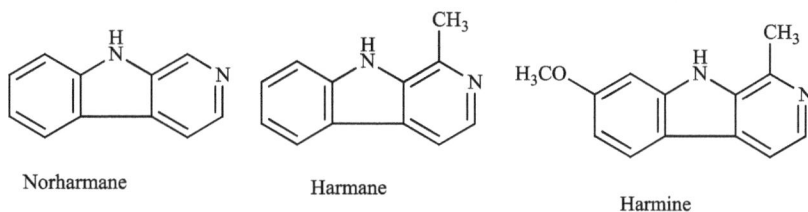

Norharmane   Harmane

Harmine

The phytochemical studies of Xu and colleugues (2006) isolated protopine from a Chinese plant, *Dactylicapnos scandens* Hutch. The isolation was followed by extract's marked antidepressant like effect. It produced significant dose-dependent reduction in the immobility time when tested in the tail suspension test [23]. The mechanistic studies suggested that protopine exhibited selective inhibition of both serotonin transporter and noradrenaline transporter in an *in vitro* assays. However, up to the concentration of 10 µg/ml, it had no effect on GABA transporter 1 and dopamine transporter [23] therefore, further detail studies are required to ascertain its underlying mechanism.

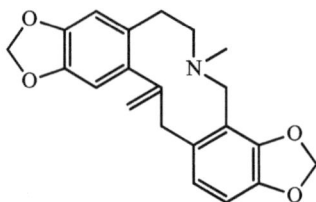

Protopine

Wattanathorn and team members (2008) find out that piperine was a major alkaloid, isolated from *Piper nigrum*. The compound showed pronounced antidepressant activity in mice exposed to both chronic and acute stress. During the course of experiment, Piperine produced a significant change in both immobility and swimming times [24]. The compound has been reported for upstream regulation of the serotonin level in the cerebral cortex and limbic areas. This lead to the anti-depression like activity of piperine. However, further investigations about precise underlying mechanism are still required [24]. Pal et al (2011) showed that piperine has even a positive effect in epilepsy-induced depression. The proposed mechanism for depression in epilepsy may be changes in the expression and or interfering with monoaminergic and GABAergic pathways. Furthermore, the antidepressant like effect of piperine in post-status-epilepticus when studied in rodents, may be due to its inhibition on MAO expression, accompanied by its neuroprotective action [25]. The study of Huang et al (2013) further strengthens the belief of MAO involvement in its effect. Here, a *trans*-resveratrol was combined with piperine, which partly potentiated the activation of monoaminergic system in the brain [26].

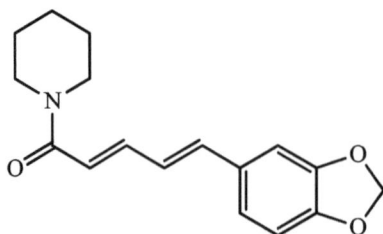

Piperine

Leatispicine, an amide-alkaloid has been isolated by a group of Chines researchers from *Piper laetispicum*. When tested in the forced swim test (5-40 mg/kg), it produced a significant dose-dependent reduction in mobility time when studied at various test doses and thus characterized as an antidepressant agent [27]. The compound also exhibited antinociceptive effect and thus, could be a useful agonist in depression-induced pain. Leatispicine mechanisms of action were presumed to be acting on the central nervous system and modulating monoaminergic neurotransmitters [27, 28].

The structural features of these amide alkaloids showed similarity, except some double bond allocation in the carbon chain. Both the leatispiamide-A, and Laetispicine exhibited conjugated double bond located at 2–3 and 4–5. Of two, Leatispiamide A and Laetispicine have isolated double bond from benzene ring; correspondingly, it has a conjugated double bond with benzene ring double bond in carbon chain. Thus, it is, concluded that the antidepressant like activities of these are attributed to isolated double bond from benzene ring and conjugated double bond located at 2–3 and 4–5 [28].

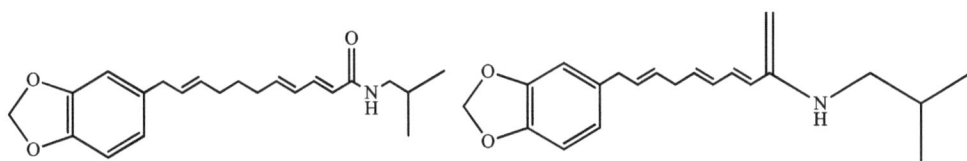

Leatispicine   Leatispiamide A

Similarly, Malaysian researchers led by Idayu (2011) isolated mitagynine from *Mitagyna spicosa*. The intraperitionial administation of mitagynine produced significant reduction in the immobility time of experimental animals when tested in both forced swim test and tail suspension test without any significant effect on locomotor activity [29]. The possible mechanism for antidepressant-like effect of mitagynine was the down regulation of corticosterone and the interaction with neuroendocrine HPA axis systems [29], because the injection of corticosterone caused depression like effect in experimental animal in a dose dependent manner by interfering with the HPA axis normal functioning [30].

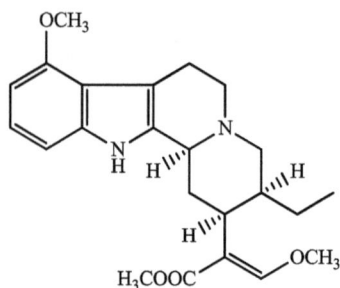

Mitagynine

The scientists of Chinese National Academy of Sciences has isolated and characterized mauritine A from *Ziziphus apetala*. This compound showed strong inhibition on the activity of 11-β-hydroxysteroid dehydrogenase *in vitro* [31]. The inhibition of the enzyme explain the possible mechanism for antidepressant like effect is Mauritine A [31]; however, further detailed studies are required to make any decision regarding its mechanism.

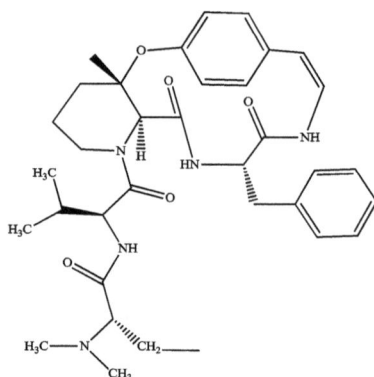

mauritine A

Similarly, various diterpene alkaloids have been isolated from *Aconitum baicalens* which were characterized as napelline, songorine, hypaconitine, and mesaconitine. These isolated alkaloids when studied in animal models of depression, they exhibited significant antidepressant-like effect at test dose of 0.025 mg/kg [32]. The songorine was most active at test dose. The overall behavior and motor activity was not altered by the treatment of this alkaloid in open field test. The proposed mechanism for diterpene alkaloids of *A. baicalense* was up regulation of serotonergic system activity in an animal model of depression being a widespread mediator in mood expression and elevation [32].

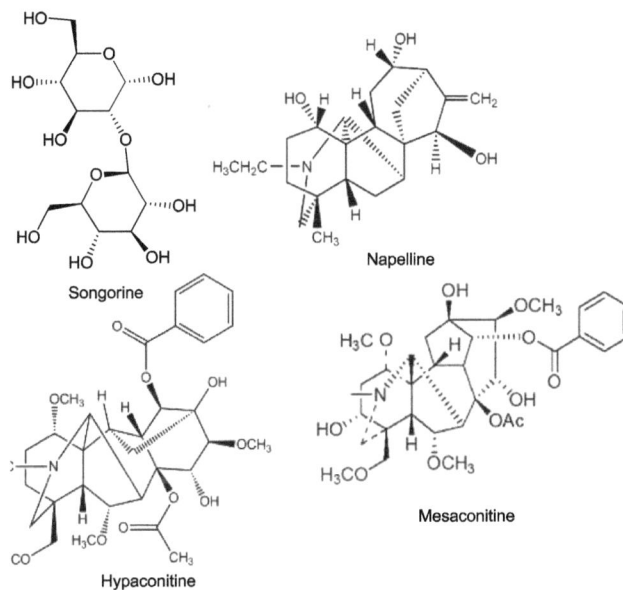

Songorine

Napelline

Hypaconitine

Mesaconitine

The phytochemical invesigation of a Brazilian research group on *Psychotria myriantha* Mull isolated a compound which was charterized as strictosidinic acid. It demonstrated marked antidepressant-like effect on a 5-HT system in rat hippocampus. The strictosidinic acid treatment *via* intra-hippocampal injection caused downstream regulation of 5-HT levels, while when injected through intraperitoneal route, it reduced DOPAC expression [33]. Moreover, it is estimated in rat brain that strictosidinic acid mediate its antidepressant effect due to inhibition of monoamine oxidase activity [33].

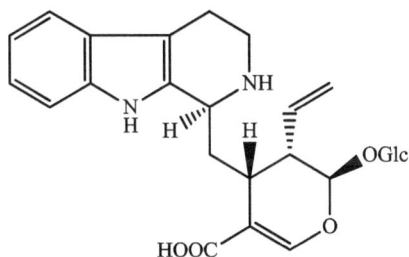

Berberine, an isoquinoline alkaloid is known for its multiple pharmacological actions. In 2012, the experimental findings of Lee *et al.* on berberine in the forced swim test illustrated that its injection causes down regulation of immobility, while it was observed to improve climbing behavior of mice. However, berberine treatment not interfered with swimming time of mice, but marked increased in open-arm exploration in the elevated plus maze test. All together, these finding supportive of the antidepressant-like activity [34]. The antidepressant like effect of berberine was not dose dependent. It caused marked change in immobility time when studied in mice using standard experimental protocols (forced swim and tail-suspension tests) at 5, 10 and 20 mg/kg intraperitoneal treatment [35]. Berberine followed several underlying mechanisms for its antidepressant effect, that includes serotonergic, noradrenergic, and dopaminergic intervention [34, 35].

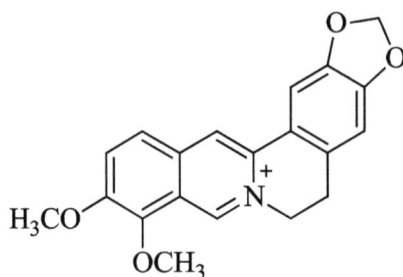

Berberine

Martinez-Vazquez and his co-workers (2012) isolated certain alkaloids from *Annona cherimolia*, including 1,2-dimethoxy-5,6,6a,7-tetrahydr--4H-dibenzoquinoline-3,8,9,10-tetraol, anonaine, liriodenine, and nornuciferine. The results showed that repeated treatment with this plant produced an antidepressant-like action in mice [36]. Furthermore, 1,2-dimethoxy-5,6,6a-7-tetrahydro-4H-dibenzoquinoline-3,8,9,10-tetraol, anonaine, liriodenine, and nornuciferine produced an antidepressant-like action from a generalized increase in monoaminergic turnover [36]. It has been reported that anonaine exhibited good selectivity for ^{3}H-dopamine uptake and shown inhibitory effect.

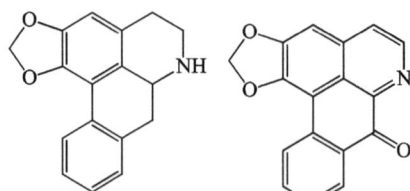

Anonaine   Liriodenine

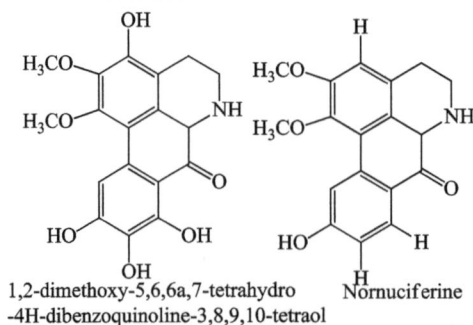

1,2-dimethoxy-5,6,6a,7-tetrahydro   Nornuciferine
-4H-dibenzoquinoline-3,8,9,10-tetraol

In 2014, an Indian research group phytochemically explored *Boerhaavia diffusa* Linn that led to the isolation of an alkaloid named punaravine. It's treatment caused significant antidepressant like activity in unstressed and stressed mice in different models used for the assessment of antidepressant like effect at after 14 days oral administration at 20 and 40 mg/kg [37]. Punaravine mediated its antidepressant like effect through downstream of MAOA expression in animal brain and, in addition, overall reduction was also observed in the plasma corticosterone expression. The alkaloid also decreased nitrite level in animal plasma, reflecting its antioxidant effect [37, 38].

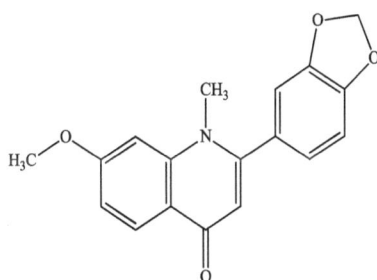

Punaravine

In the same year, a group of researcher led by Loria (2014) from the USA experimentally proved that Mesembrine an alkaloid isolated *Sceletium tortuosum* possessed antidepressant like action in animal studies at 10-80 mg/kg without any effect on the locomotor activity [39]. From mechanistic point of view, it was suggested that the underlying mechanism for antidepressant like effect of mesembrine significant inhibition of 5HT reuptake [39].

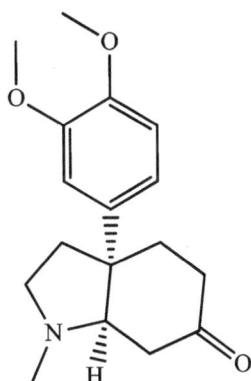

Mesembrine

Jianga and his co-workers (2015) purified evodamine from *Evodia fructus* and studied its effect in chronic unpredictable mild stress animal model. It was found that evodamine overturn the reduction of sucrose preference, a number of crossing, 5-HT, and Na level and also increase immobility time [40]. Furthermore, experiment has shown that evodamine mediated its effect through modulating effects on the monoamine transmitters and BDNF-TrkB signaling in the hippocampus [40].

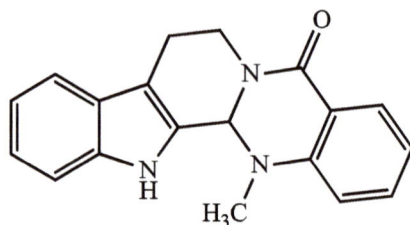

Evodamine

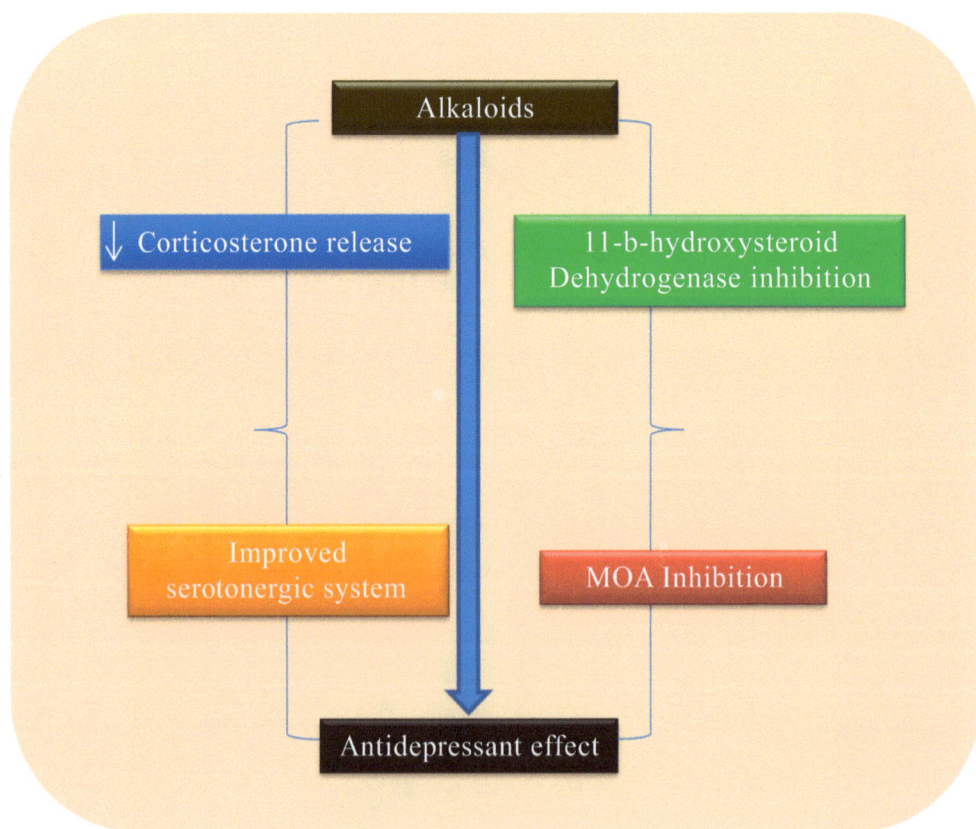

Fig. (1). Proposed mechanisms of antidepressant effect of plant alkaloids.

ETHNOMEDICINE IN DRUG DISCOVERY

The use of plants by humans takes us to 60,000 years back to the Paleolithic age [41] and extensively benefited humanity against diseases. Such usage over time has led to the in-depth knowledge of ethnomedicine and this approach has been widely applied in new drug discovery. (1) it provides basis for various biochemical and pharmacological evaluations, (2) based on pharmacological effects, isolation of bioactive compounds, (3) to identify the lead compounds and trim it in the light of structure activity relationship for obtaining low toxic or highly active compound, (4) use of whole plant or part of it as herbal remedy.

The statistic revealed that more than 120 compounds, derived from less than 100 plant species used as drugs globally and 80% of these have ethnomedicinal basis [42]. This has provided foundation for the establishment of a data-base for worldwide information and collaboration regarding ethnomedicinal uses of plant derived drug named as The Traditional Medicine Collection Tracking System [43].

Different international drug discovery programs/projects randomly used plant extracts in animal studies such as National Cancer Institute of United States, but the overall results were very discouraging. After screening about 35,000 plant species, only few compounds reached to clinical trials where they failed to produce significant results [42]. Therefore, the best approach is to design project based on ethnomedicinal uses of plant. However, the ethnomedicinal studies methods have limitations [44] that need to improve to get best results.

The current findings from various medicinal plants have shown the outstanding antidepressant like activities of isolated compounds. The bioactivity guided isolation is the basis of these result. Further, detailed studies on such compounds could lead to new effective agents (Fig. **2**).

STRUCTURE ACTIVITY RELATIONSHIPS (SAR)

The Structure Activity Relationships (SAR) is the key in pharmacological and toxicological investigations of chemical agents. Herein, different structural features are studied and correlated with the resulting effects. SAR are extensively used in pharmaceutical industry for new drug designing in order to obtain desirable actions.

The general view of the structures of amide alkaloids isolated from *Piper laetispicum* showed similarity in structures, except some double bond locations in the carbon chain. The leatispiamide-A, and Laetispicine both have conjugated double bonds located at 2–3 and 4–5. Of two, Leatispiamide A and Laetispicine have isolated double bond from benzene ring; correspondingly, it has a conjugated double bond with benzene ring in carbon chain. Thus, it is concluded that the antidepressant like activities of these are attributed to isolated double bond from benzene ring and conjugated double bond located at 2–3 and 4–5 [28].

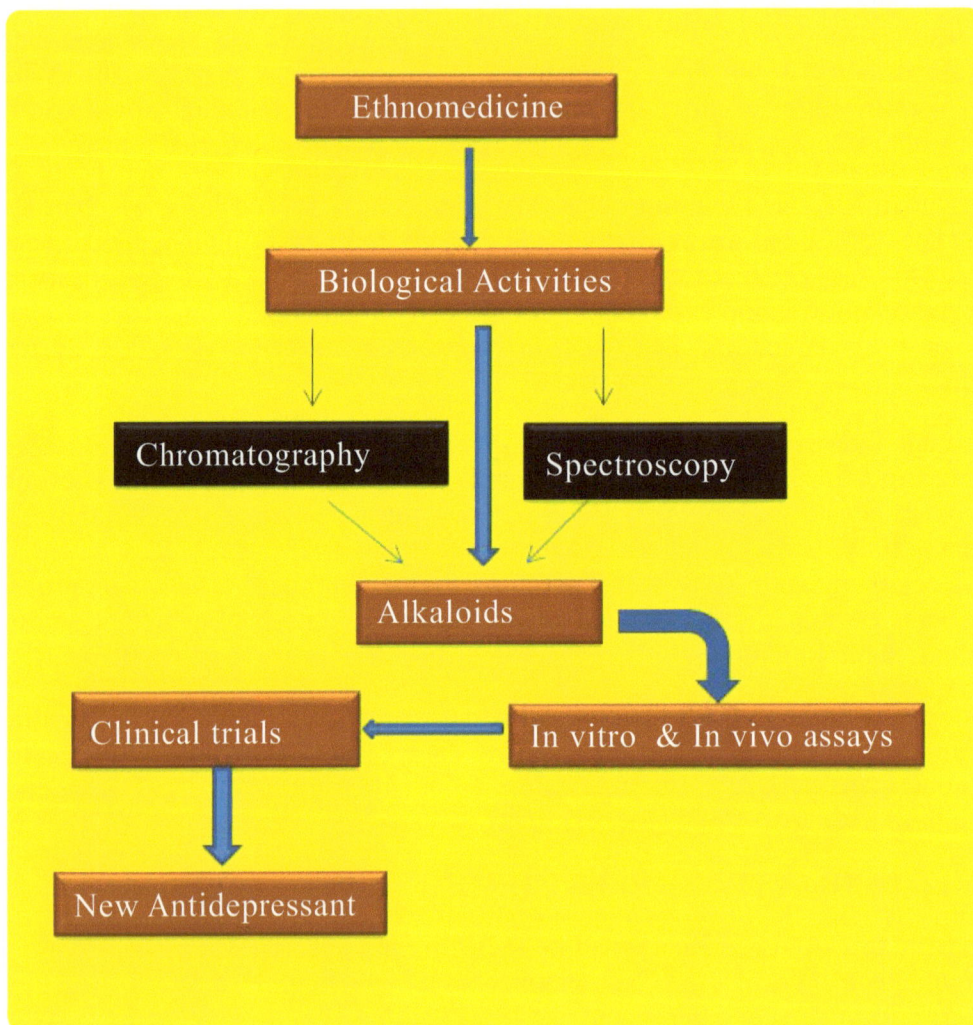

Fig. (2). Showing drug discovery steps from medicinal plant.

The β-carboline alkaloids, norharmane, harmine and harmaline showed strong MAO-A inhibition, however norharmane was most potent. SAR of these compounds suggested that a flat geometry of these compounds might be more feasible to interact with the different parts of the MAOs active sites [45].

Moreover, a detailed SAR of different alkaloids has been described by Passos *et al.* (2014) in their book [46].

FUTURE PROSPECTS/CONCLUSION

Our analysis has been based on the available literature and it has been concluded that alkaloids could play an outstanding role as natural antidepressants. While considering their diversity in nature and free availability, these alkaloids could be an economical source of healing various type of depressions including mild to complicated disorders. Indeed, the available therapeutic agents failed to produce effect in all patients, and 30-40% failure has been reported with even most commonly used antidepressant drugs. Additionally, these agents are very slow in action and mostly need months to produce some meaningful effects and thus, encounter multiple side effects.

In this regard, several alkaloids are already in clinical practice, producing outstanding results in different therapeutic classes since long time. These reported alkaloids though evoked antidepressant like effects in various *in vitro* and *in vitro* studies, thus are potential candidates to be explored in clinical studies. Additionally, modern and latest sophisticated techniques can be used to optimize the different structural and pharmacological parameters of these reported alkaloids such as potency and toxicity. Indeed, such characteristics are key to the success of any drug in clinical trials.

In short, comprehensive scientific evidences based on various *in vitro* and *in vitro* studies around the world reported in this chapter strongly support the idea that the plant-based alkaloids can serve as leads for antidepressant drug discovery, with strong potency and general efficacy. In this view point, it is recommended to subject these alkaloids to further clinical studies on their efficacy, potency, and safety to ensure their clinical status.

CONFLICT OF INEREST

The author (editor) declares no conflict of interest, financial or otherwise.

ACKNOWLEDGMENT

Declared none

REFERENCES

[1]　Rosenbaum D, Hagen K, Deppermann S, *et al.* State-dependent altered connectivity in late-life depression: a functional near-infrared spectroscopy study. Neurobiol Aging 2016; 39: 57-68.
[http://dx.doi.org/10.1016/j.neurobiolaging.2015.11.022] [PMID: 26923402]

[2]   Kalmbach DA, Arnedt JT, Swanson LM, Rapier JL, Ciesla JA. Reciprocal dynamics between self-rated sleep and symptoms of depression and anxiety in young adult women: a 14-day diary study. Sleep Med 2017; 33: 6-12.
        [http://dx.doi.org/10.1016/j.sleep.2016.03.014] [PMID: 28449907]

[3]   Ingram R. Depression.Encyclopedia of Mental Health. 2nd ed. Oxford: Academic Press 2016; pp. 26-33.
        [http://dx.doi.org/10.1016/B978-0-12-397045-9.00084-7]

[4]   Tondo L, Isacsson G, Baldessarini R. Suicidal behaviour in bipolar disorder: risk and prevention. CNS Drugs 2003; 17(7): 491-511.
        [http://dx.doi.org/10.2165/00023210-200317070-00003] [PMID: 12751919]

[5]   Gold PW, Goodwin FK, Chrousos GP. Clinical and biochemical manifestations of depression. Relation to the neurobiology of stress (2). N Engl J Med 1988; 319(7): 413-20.
        [http://dx.doi.org/10.1056/NEJM198808183190706] [PMID: 3041279]

[6]   Muhammad N, Saeed M, Khan H, Haq I. Evaluation of n-hexane extract of *Viola betonicifolia* for its neuropharmacological properties. J Nat Med 2013; 67(1): 1-8.
        [http://dx.doi.org/10.1007/s11418-012-0636-0] [PMID: 22359189]

[7]   Calvó-Perxas L, Vilalta-Franch J, Turró-Garriga O, López-Pousa S, Garre-Olmo J. Gender differences in depression and pain: A two year follow-up study of the Survey of Health, Ageing and Retirement in Europe. J Affect Disord 2016; 193: 157-64.
        [http://dx.doi.org/10.1016/j.jad.2015.12.034] [PMID: 26773909]

[8]   Gadassi R, Mor N. Confusing acceptance and mere politeness: Depression and sensitivity to Duchenne smiles. J Behav Ther Exp Psychiatry 2016; 50: 8-14.
        [http://dx.doi.org/10.1016/j.jbtep.2015.04.007] [PMID: 25958338]

[9]   Ridout KK, Ridout SJ, Price LH, Sen S, Tyrka AR. Depression and telomere length: A meta-analysis. J Affect Disord 2016; 191: 237-47.
        [http://dx.doi.org/10.1016/j.jad.2015.11.052] [PMID: 26688493]

[10]   Robinson MJ. Antidepressant psychopharmacology: Current limitations and future directions. Prim Psychiatry 2003; 10(1): 43-9.

[11]   McNaught. Wilkinson A IUPAC Compendium of Chemical Terminology. 2nd ed., Oxford: Blackwell Scientific Publications 1997.

[12]   Manske RH. The Alkaloids Chemistry and Physiology. New York: Academic Press 1965.

[13]   Khan H. Alkaloids: potential therapeuty modality in the management of asthma. Journal of Ayurvedic and Herbal Medicine 2015; 1(1): 3-3.

[14]   Lu J-J, Bao J-L, Chen X-P, Huang M, Wang Y-T. Alkaloids isolated from natural herbs as the anticancer agents. Evidence-based Complementary and Alternative Medicine : eCAM 2012; 2012: 485042.
        [http://dx.doi.org/10.1155/2012/485042]

[15]   Amirkia V, Heinrich M. Alkaloids as drug leads – A predictive structural and biodiversity-based analysis. Phytochem Lett 2014; 10(0): xlviii-liii.
        [http://dx.doi.org/10.1016/j.phytol.2014.06.015]

[16]   Ali BH, Bashir AK, Tanira MO, *et al.* Effect of extract of Rhazya stricta, a traditional medicinal plant, on rat brain tribulin. Pharmacol Biochem Behav 1998; 59(3): 671-5.
        [http://dx.doi.org/10.1016/S0091-3057(97)00464-4] [PMID: 9512070]

[17]   Maj J, Rogóż Z, Skuza G, Kołodziejczyk K. Antidepressant effects of pramipexole, a novel dopamine receptor agonist. Journal of Neural Transmission 1997; 104(4): 525-33.
        [http://dx.doi.org/10.1007/BF01277669]

[18]   Zarate CA Jr, Payne JL, Singh J, *et al.* Pramipexole for bipolar II depression: a placebo-controlled

proof of concept study. Biol Psychiatry 2004; 56(1): 54-60.
[http://dx.doi.org/10.1016/j.biopsych.2004.03.013] [PMID: 15219473]

[19]   Akdeniz F, Aldemir E, Vahip S. [The role of low-dose pramipexole in the treatment of treatment-resistant bipolar depression: a case report]. Turk Psikiyatr Derg 2009; 20(1): 94-8.
[PMID: 19306131]

[20]   Farzin D, Mansouri N. Antidepressant-like effect of harmane and other β-carbolines in the mouse forced swim test. Eur Neuropsychopharmacol 2006; 16(5): 324-8.
[http://dx.doi.org/10.1016/j.euroneuro.2005.08.005] [PMID: 16183262]

[21]   Kim DH, Jang YY, Han ES, Lee CS. Protective effect of harmaline and harmalol against dopamine- and 6-hydroxydopamine-induced oxidative damage of brain mitochondria and synaptosomes, and viability loss of PC72 cells. Eur J Neurosci 2001; 13(10): 1861-72.
[http://dx.doi.org/10.1046/j.0953-816x.2001.01563.x] [PMID: 11403679]

[22]   Glennon RA, Dukat M, Grella B, *et al.* Binding of β-carbolines and related agents at serotonin (5-HT(2) and 5-HT(1A)), dopamine (D(2)) and benzodiazepine receptors. Drug Alcohol Depend 2000; 60(2): 121-32.
[http://dx.doi.org/10.1016/S0376-8716(99)00148-9] [PMID: 10940539]

[23]   Xu L-F, Chu W-J, Qing X-Y, *et al.* Protopine inhibits serotonin transporter and noradrenaline transporter and has the antidepressant-like effect in mice models. Neuropharmacology 2006; 50(8): 934-40.
[http://dx.doi.org/10.1016/j.neuropharm.2006.01.003] [PMID: 16530230]

[24]   Wattanathorn J, Chonpathompikunlert P, Muchimapura S, Priprem A, Tankamnerdthai O. Piperine, the potential functional food for mood and cognitive disorders. Food Chem Toxicol 2008; 46(9): 3106-10.
[http://dx.doi.org/10.1016/j.fct.2008.06.014] [PMID: 18639606]

[25]   Pal A, Nayak S, Sahu PK, Swain T. Piperine protects epilepsy associated depression: a study on role of monoamines. Eur Rev Med Pharmacol Sci 2011; 15(11): 1288-95.
[PMID: 22195361]

[26]   Huang W, Chen Z, Wang Q, *et al.* Piperine potentiates the antidepressant-like effect of trans-resveratrol: involvement of monoaminergic system. Metabolic Brain Disease 2013; 28(4): 585-95.
[http://dx.doi.org/10.1007/s11011-013-9426-y]

[27]   Yao CY, Wang J, Dong D, Qian FG, Xie J, Pan SL. Laetispicine, an amide alkaloid from Piper laetispicum, presents antidepressant and antinociceptive effects in mice. Phytomedicine 2009; 16(9): 823-9.
[http://dx.doi.org/10.1016/j.phymed.2009.02.008] [PMID: 19447013]

[28]   Xie H, Yan MC, Jin D, *et al.* Studies on antidepressant and antinociceptive effects of ethyl acetate extract from Piper laetispicum and structure-activity relationship of its amide alkaloids. Fitoterapia 2011; 82(7): 1086-92.
[http://dx.doi.org/10.1016/j.fitote.2011.07.006] [PMID: 21787850]

[29]   Idayu NF, Hidayat MT, Moklas MA, *et al.* Antidepressant-like effect of mitragynine isolated from Mitragyna speciosa Korth in mice model of depression. Phytomedicine 2011; 18(5): 402-7.
[http://dx.doi.org/10.1016/j.phymed.2010.08.011] [PMID: 20869223]

[30]   Johnson SA, Fournier NM, Kalynchuk LE. Effect of different doses of corticosterone on depression-like behavior and HPA axis responses to a novel stressor. Behav Brain Res 2006; 168(2): 280-8.
[http://dx.doi.org/10.1016/j.bbr.2005.11.019] [PMID: 16386319]

[31]   Han J, Ji C-J, He W-J, *et al.* Cyclopeptide Alkaloids from Ziziphus apetala. J Nat Prod 2011; 74(12): 2571-5.
[http://dx.doi.org/10.1021/np200755t] [PMID: 22148241]

[32]   Nesterova YV, Povetieva TN, Suslov NI, Semenov AA, Pushkarskiy SV. Antidepressant activity of

diterpene alkaloids of *Aconitum baicalense* Turcz. Bull Exp Biol Med 2011; 151(4): 425-8.
[http://dx.doi.org/10.1007/s10517-011-1347-3] [PMID: 22448357]

[33]   Farias FM, Passos CS, Arbo MD, *et al.* Strictosidinic acid, isolated from Psychotria myriantha Mull. Arg. (Rubiaceae), decreases serotonin levels in rat hippocampus. Fitoterapia 2012; 83(6): 1138-43.
[http://dx.doi.org/10.1016/j.fitote.2012.04.013] [PMID: 22546150]

[34]   Lee B, Sur B, Yeom M, Shim I, Lee H, Hahm D-H. Effect of berberine on depression- and anxiety-like behaviors and activation of the noradrenergic system induced by development of morphine dependence in rats. Korean J Physiol Pharmacol 2012; 16(6): 379-86.
[http://dx.doi.org/10.4196/kjpp.2012.16.6.379] [PMID: 23269899]

[35]   Kulkarni SK, Dhir A. On the mechanism of antidepressant-like action of berberine chloride. Eur J Pharmacol 2008; 589(1-3): 163-72.
[http://dx.doi.org/10.1016/j.ejphar.2008.05.043] [PMID: 18585703]

[36]   Martínez-Vázquez M, Estrada-Reyes R, Araujo Escalona AG, *et al.* Antidepressant-like effects of an alkaloid extract of the aerial parts of Annona cherimolia in mice. J Ethnopharmacol 2012; 139(1): 164-70.
[http://dx.doi.org/10.1016/j.jep.2011.10.033] [PMID: 22101086]

[37]   Dhingra D, Valecha R. Punarnavine, an alkaloid isolated from ethanolic extract of Boerhaavia diffusa Linn. reverses depression-like behaviour in mice subjected to chronic unpredictable mild stress. Indian J Exp Biol 2014; 52(8): 799-807.
[PMID: 25141543]

[38]   Perviz S, Khan H, Pervaiz A. Plant alkaloids as an emerging therapeutic alternative for the treatment of depression. Front Pharmacol 2016; 7: 28.
[http://dx.doi.org/10.3389/fphar.2016.00028] [PMID: 26913004]

[39]   Loria MJ, Ali Z, Abe N, Sufka KJ, Khan IA. Effects of Sceletium tortuosum in rats. J Ethnopharmacol 2014; 155(1): 731-5.
[http://dx.doi.org/10.1016/j.jep.2014.06.007] [PMID: 24930358]

[40]   Jiang M-L, Zhang Z-X, Li Y-Z, Wang X-H, Yan W, Gong G-Q. Antidepressant-like effect of evodiamine on chronic unpredictable mild stress rats. Neurosci Lett 2015; 588: 154-8.
[http://dx.doi.org/10.1016/j.neulet.2014.12.038] [PMID: 25545553]

[41]   Patwardhan B. Traditional Medicine-Inspired Evidence-Based Approaches to Drug Discovery A2 - Mukherjee, Pulok K Evidence-Based Validation of Herbal Medicine. Boston: Elsevier 2015; pp. 259-72.

[42]   Fabricant DS, Farnsworth NR. The value of plants used in traditional medicine for drug discovery. Environ Health Perspect 2001; 109 (Suppl. 1): 69-75.
[http://dx.doi.org/10.1289/ehp.01109s169] [PMID: 11250806]

[43]   Harris ES, Erickson SD, Tolopko AN, *et al.* Traditional Medicine Collection Tracking System (TM-CTS): a database for ethnobotanically driven drug-discovery programs. J Ethnopharmacol 2011; 135(2): 590-3.
[http://dx.doi.org/10.1016/j.jep.2011.03.029] [PMID: 21420479]

[44]   Albuquerque UP, de Medeiros PM, Ramos MA, Júnior WS, Nascimento AL, Avilez WM, *et al.* Are ethnopharmacological surveys useful for the discovery and development of drugs from medicinal plants? Rev Bras Farmacogn 2014; 24(2): 110-5.
[http://dx.doi.org/10.1016/j.bjp.2014.04.003]

[45]   Samoylenko V, Rahman MM, Tekwani BL, *et al.* Banisteriopsis caapi, a unique combination of MAO inhibitory and antioxidative constituents for the activities relevant to neurodegenerative disorders and Parkinson's disease. J Ethnopharmacol 2010; 127(2): 357-67.
[http://dx.doi.org/10.1016/j.jep.2009.10.030] [PMID: 19879939]

[46]   Passos C, Simoes-Pires C, Henriques A, Cuendet M, Carrupt P-A, Christen P. Alkaloids as Inhibitors
        of Monoamine Oxidases and Their Role in the Central Nervous System. Studies in Natural Products
        Chemistry 2014; p. 123.
        [http://dx.doi.org/10.1016/B978-0-444-63430-6.00004-7]

SUBJECT INDEX

www.ingramcontent.com/pod-product-compliance
Lightning Source LLC
Chambersburg PA
CBHW050816220326
41598CB00006B/227